THIRD EDITION

Structural Engineer's Pocket Book

Eurocodes

T0172700

THIRD EDITION

Structural Engineer's Pocket Book

Eurocodes

FIONA COBB

CRC Press
Taylor & Francis Group
Boca Raton London New York

CRC Press is an imprint of the
Taylor & Francis Group, an **informa** business

A SPON BOOK

Revised printing October 2017

CRC Press
Taylor & Francis Group
6000 Broken Sound Parkway NW, Suite 300
Boca Raton, FL 33487-2742

First issued in hardback 2017

© 2015 by Taylor & Francis Group, LLC
CRC Press is an imprint of Taylor & Francis Group, an informa business

No claim to original U.S. Government works

ISBN-13: 978-0-0809-7121-6 (pbk)
ISBN-13: 978-1-1384-7018-7 (hbk)

Library of Congress Cataloging-in-Publication Data

Cobb, Fiona.
    Structural engineer's pocket book : Eurocodes / Fiona Cobb. -- Third edition.
       pages cm
    Includes bibliographical references and index.
    ISBN 978-0-08-097121-6 (paperback)
    1. Building--Handbooks, manuals, etc. 2. Building--Europe--Handbooks, manuals, etc. 3. Structural engineering--Handbooks, manuals, etc. I. Title.

TH151.C623 2014
624.1--dc23   2014028245

Visit the Taylor & Francis Web site at
http://www.taylorandfrancis.com

and the CRC Press Web site at
http://www.crcpress.com

Contents

Preface to Third Edition (Eurocodes)

This, heavily revised, third edition of the Structural Engineer's Pocket Book (SEPB) is not 'yet another guide to the Eurocodes'. It is intended as a day-to-day reference and hopes to satisfy a number of different groups of engineers in the transition from British Standards to Eurocodes. The first group includes those who are 'fluent' in Eurocodes and are looking for a source of Euro-compatible facts and figures for day-to-day use. Next are the students or graduates who have been taught Eurocodes at university, who are looking for a practical data book to complement their theoretical knowledge. Last, but not least, are the practising engineers who have been content using British Standards who will want quick access to the 'new' codes, with the differences, potential pitfalls and advantages clearly highlighted.

It has been difficult to strike a balance to cater for all of these groups, particularly when many authors choose not to include references to British Standard notation. However, based on the rationale that I used for the first edition, I have simply included what I find helpful, for example, using both British Standard and Eurocode notation at the head of the steel tables as an aide memoire, a quick reference table of differences between the different codes and inclusion of older codes of practice for assessment of historic structures.

It should also be said that in the United Kingdom the simultaneous introduction of a completely new and radical set of structural codes of practice, to replace all existing codes, has never been attempted before. It is an immensely ambitious task. Despite the withdrawal of British Standards in 2010, Eurocodes are not in widespread use and much of the innovative code content derived from research has not been widely tested in the field. Key figures involved in the drafting of the Eurocodes acknowledge that it will take time to find the 'wrinkles'. All engineers should be encouraged to 'read the codes of practice like the devil reads the bible' as whichever code of practice is used, it is no substitute for sound engineering judgement.

The Eurocodes use considerable notation and symbols. The front and back covers include fold-out leaves which summarise the most commonly used symbols for easy reference alongside the main text.

Once again, I would be interested to receive any comments, corrections or suggestions on the content of the book by email at sepocketbook@gmail.com

Fiona Cobb

Preface to Second Edition

When the *Structural Engineer's Pocket Book* was first conceived, I had no idea how popular and widely used it would become. Thanks to all those who took the time to write to me with suggestions. I have tried to include as many as I can, but as the popularity of the book is founded on a delicate balance between size, content and cover price, I have been unable to include everything asked of me. Many readers will notice that references to Eurocodes are very limited. The main reason being that the book is not intended as a text book and is primarily for use in scheme design (whose sizes do not vary significantly from those determined using British Standards). However, Eurocode data will be included in future editions once the codes (and supporting documents) are complete, the codes have completed industry testing and are more widely used.

As well as generally updating the British Standards revised since 2002, the main additions to the second edition are a new chapter on sustainability, addition of BS8500, revised 2007 Corus steel section tables (including 20 new limited release UB and UC sections) and a summary of Eurocode principles and load factors.

Once again, I should say that I would be interested to receive any comments, corrections or suggestions on the content of the book by email at sepb@inmyopinion.co.uk

Fiona Cobb

Preface to First Edition

As a student or graduate engineer, it is difficult to source basic design data. Having been unable to find a compact book containing this information, I decided to compile my own after seeing a pocket book for architects. I realised that a *Structural Engineer's Pocket Book* might be useful for other engineers and construction industry professionals. My aim has been to gather useful facts and figures for use in preliminary design in the office, on site or in the IStructE Part 3 exam, based on UK conventions.

The book is not intended as a textbook; there are no worked examples and the information is not prescriptive. Design methods from British Standards have been included and summarised, but obviously these are not the only way of proving structural adequacy. Preliminary sizing and shortcuts are intended to give the engineer a 'feel' for the structure before beginning design calculations. All of the data should be used in context, using engineering judgement and current good practice. Where no reference is given, the information has been compiled from several different sources.

Despite my best efforts, there may be some errors and omissions. I would be interested to receive any comments, corrections or suggestions on the content of the book by email at sepb@inmyopinion.co.uk. Obviously, it has been difficult to decide what information can be included and still keep the book a compact size. Therefore, any proposals for additional material should be accompanied by a proposal for an omission of roughly the same size—the reader should then appreciate the many dilemmas that I have had during the preparation of the book! If there is an opportunity for a second edition, I will attempt to accommodate any suggestions which are sent to me and I hope that you find the *Structural Engineer's Pocket Book* useful.

Fiona Cobb

Acknowledgements

Thanks to the following people and organisations:

Price & Myers for giving me varied and interesting work, without which this book would not have been possible! Paul Batty David Derby, Sarah Fawcus, Step Haiselden, Simon Jewell, Chris Morrisey, Mark Peldmanis, Sam Price, Helen Remordina, Harry Stocks and Paul Toplis for their comments and help reviewing chapters. Colin Ferguson, Derek Fordyce, Phil Gee, Alex Hollingsworth, Paul Johnson, Deri Jones, Robert Myers, Dave Rayment and Andy Toohey for their help, ideas, support, advice and/or inspiration at various points in the preparation of the book. Renata Corbani, Rebecca Rue and Sarah Hunt at Elsevier. The technical and marketing representatives of the organisations mentioned in the book. Last but not least, thanks to Jim Cobb, Elaine Cobb, Iain Chapman for his support and the loan of his computer and Jean Cobb for her help with typing and proof reading.

Additional help on the second edition:

Lanh Te, Prashant Kapoor, Meike Borchers and Dave Cheshire.

Special thanks on the third edition:

Rob Thomas, Alisdair Beal, Trevor Draycott, Peter Bullman, Stephen Aleck, Liz Burton and Tony Moore. For Fraser and Flora, whose arrivals punctuated the second and third editions.

Previous editions contained information on the Historical Use of Building Materials (in this edition on pages 32 and 33) which was reproduced incorrectly. Date lines were not plotted to scale and should not be relied upon.

Text and Illustration Credits

1
General Information

Metric system

The most universal system of measurement is the International System of Units, referred to as SI, which is an absolute system of measurement based upon the fundamental quantities of mass, length and time, independent of where the measurements are made. This means that while mass remains constant, the unit of force (Newton) will vary with location. The acceleration due to gravity on earth is 9.81 m/s^2.

The system uses the following basic units:

Length	**m**	Metre
Time	**s**	Second
Luminous intensity	**cd**	Candela
Quantity/substance	**mol**	mole (6.02×10^{23} particles of substance (Avogadro's number))
Mass	**kg**	Kilogram
Temperature	**K**	Kelvin (0°C = 273 K)
Unit of plane angle	**rad**	Radian

The most commonly used prefixes in engineering are:

giga	**G**	1,000,000,000	1×10^9
mega	**M**	1,000,000	1×10^6
kilo	**k**	1000	1×10^3
centi	**c**	0.01	1×10^{-2}
milli	**m**	0.001	1×10^{-3}
micro	μ	0.000001	1×10^{-6}
nano	**N**	0.000000001	1×10^{-9}

The base units and the prefixes listed above imply a system of supplementary units which forms the convention for noting SI measurements, such as the Pascal for measuring pressure where 1 Pa = 1 N/m^2 and 1 MPa = 1 N/mm^2.

Typical metric units for UK structural engineering

Mass of material	kg
Density of material	kg/m^3
Bulk density	kN/m^3
Weight/force/point load	kN
Bending moment	kNm
Load per unit length	kN/m
Distributed load	kN/m^2
Wind loading	kN/m^2
Earth pressure	kN/m^2
Stress	N/mm^2
Modulus of elasticity	kN/mm^2
Deflection	mm
Span or height	m
Floor area	m^2
Volume of material	m^3
Reinforcement spacing	mm
Reinforcement area	mm^2 or mm^2/m
Section dimensions	mm
Moment of inertia	cm^4 or mm^4
Section modulus	cm^3 or mm^3
Section area	cm^2 or mm^2
Radius of gyration	cm or mm

Imperial units

In the British Imperial System the unit of force (pound) is defined as the weight of a certain mass which remains constant, independent of the gravitational force. This is the opposite of the assumptions used in the metric system where it is the mass of a body which remains constant. The acceleration due to gravity is 32.2 ft/s^2, but this is rarely needed. While on the surface it appears that the UK building industry is using metric units, the majority of structural elements are produced to traditional Imperial dimensions which are simply quoted in metric.

The standard units are:

Length

1 mile	= 1760 yards
1 furlong	= 220 yards
1 yard (yd)	= 3 feet
1 foot (ft)	= 12 inches
1 inch (in)	= 1/12 foot

Area

1 sq. mile	= 640 acres
1 acre	= 4840 sq. yd
1 sq. yd	= 9 sq. ft
1 sq. ft	= 144 sq. in
1 sq. in	= 1/144 sq. ft

Weight

1 ton	= 2240 pounds
1 hundredweight (cwt)	= 112 pounds
1 stone	= 14 pounds
1 pound (lb)	= 16 ounces
1 ounce	= 1/16 pound

Capacity

1 bushel	= 8 gallons
1 gallon	= 4 quarts
1 quart	= 2 pints
1 pint	= 1/2 quart
1 fl. oz	= 1/20 pint

Volume

1 cubic yard	= 27 cubic feet
1 cubic foot	= 1/27 cubic yards
1 cubic inch	= 1/1728 cubic feet

Nautical measure

1 nautical mile	= 6080 feet
1 cable	= 600 feet
1 fathom	= 6 feet

Conversion factors

Given the dual use of SI and British Imperial Units in the UK construction industry, quick and easy conversion between the two systems is essential. A selection of useful conversion factors are:

Mass	1 kg	= 2.205 lb	1 lb	= 0.4536 kg
	1 tonne	= 0.9842 tons	1 ton	= 1.016 tonnes
Length	1 mm	= 0.03937 in	1 in	= 25.4 mm
	1 m	= 3.281 ft	1 ft	= 0.3048 m
	1 m	= 1.094 yd	1 yd	= 0.9144 m
Area	1 mm^2	= 0.00153 in^2	1 in^2	= 645.2 mm^2
	1 m^2	= 10.764 ft^2	1 ft^2	= 0.0929 m^2
	1 m^2	= 1.196 yd^2	1 yd^2	= 0.8361 m^2
Volume	1 mm^3	= 0.000061 in^3	1 in^3	= 16390 mm^3
	1 m^3	= 35.32 ft^3	1 ft^3	= 0.0283 m^3
	1 m^3	= 1.308 yd^3	1 yd^3	= 0.7646 m^3
Density	1 kg/m^3	= 0.06242 lb/ft^3	1 lb/ft^3	= 16.02 kg/m^3
	1 tonne/m^3	= 0.7524 ton/yd^3	1 ton/yd^3	= 1.329 tonne/m^3
Force	1 N	= 0.2248 lbf	1 lbf	= 4.448 N
	1 kN	= 0.1004 tonf	1 tonf	= 9.964 kN
Stress and pressure	1 N/mm^2	= 145 lbf/in^2	1 lbf/in^2	= 0.0068 N/mm^2
	1 N/mm^2	= 0.0647 tonf/in^2	1 tonf/in^2	= 15.44 N/mm^2
	1 N/m^2	= 0.0208 lbf/ft^2	1 lbf/ft^2	= 47.88 N/m^2
	1 kN/m^2	= 0.0093 tonf/ft^2	1 tonf/ft^2	= 107.3 kN/m^2
Line loading	1 kN/m	= 68.53 lbf/ft	1 lbf/ft	= 0.0146 kN/m
	1 kN/m	= 0.03059 tonf/ft	1 tonf/ft	= 32.69 kN/m
Moment	1 Nm	= 0.7376 lbf ft	1 lbf ft	= 1.356 Nm
Modulus of elasticity	1 N/mm^2	= 145 lbf/in^2	1 lbf/in^2	= 6.8×10^{-3} N/mm^2
	1 kN/mm^2	= 145032 lbf/in^2	1 lbf/in^2	= 6.8×10^{-6} kN/mm^2
Section modulus	1 mm^3	= 61.01×10^{-6} in^3	1 in^3	= 16390 mm^3
	1 cm^3	= 61.01×10^{-3} in^3	1 in^3	= 16.39 cm^3
Second moment of area	1 mm^4	= 2.403×10^{-6} in^4	1 in^4	= 416200 mm^4
	1 cm^4	= 2.403×10^{-2} in^4	1 in^4	= 41.62 cm^4
Temperature	x°C	= $[(1.8x + 32)]$°F	y°F	= $[(y-32)/1.8]$°C

Notes:
[a]   1 tonne = 1000 kg = 10 kN.
[b]   1 ha = 10,000 m^2.

Measurement of angles

There are two systems for the measurement of angles commonly used in the United Kingdom.

English system

The English or sexagesimal system which is universal:

1 right angle = 90° (degrees)
1° (degree) = 60′ (minutes)
1′ (minute) = 60″ (seconds)

International system

Commonly used for the measurement of plane angles in mechanics and mathematics, the radian is a constant angular measurement equal to the angle subtended at the centre of any circle, by an arc equal in length to the radius of the circle.

π radians   = 180° (degrees)

1 radian $= \dfrac{180°}{\pi} = \dfrac{180°}{3.1416} = 57°17′44″$

Equivalent angles in degrees, radians and trigonometric ratios

Angle θ in radians	0	$\dfrac{\pi}{6}$	$\dfrac{\pi}{4}$	$\dfrac{\pi}{3}$	$\dfrac{\pi}{2}$
Angle θ in degrees	0°	30°	45°	60°	90°
sin θ	0	$\dfrac{1}{2}$	$\dfrac{1}{\sqrt{2}}$	$\dfrac{\sqrt{3}}{2}$	1
cos θ	1	$\dfrac{\sqrt{3}}{2}$	$\dfrac{1}{\sqrt{2}}$	$\dfrac{1}{2}$	0
tan θ	0	$\dfrac{1}{\sqrt{3}}$	1	$\sqrt{3}$	∞

Construction documentation and procurement

Construction documentation

The members of the design team each produce drawings, specifications and schedules which explain their designs to the contractor. The drawings indicate in visual form how the design is to look and how it is to be put together. The specification describes the design requirements for the materials and workmanship, and additional schedules set out sizes and co-ordination information not already covered in the drawings or specifications. The quantity surveyor uses all of these documents to prepare bills of quantities, which are used to help break down the cost of the work. The drawings, specifications, schedules and bills of quantities form the tender documentation. 'Tender' is when the bills and design information are sent out to contractors for their proposed prices and construction programmes. 'Procurement' simply means the method by which the contractor is to be chosen and employed, and how the building contract is managed.

Certain design responsibilities can be delegated to contractors and subcontractors (generally for items which are not particularly special or complex, e.g. precast concrete stairs or concealed steelwork connections, etc.) using a Contractor Design Portion (CDP) within the specifications. The CDP process reduces the engineer's control over the design, and therefore it is generally quicker and easier to use CDPs only for concealed/straightforward structural elements. CDPs are generally unsuitable for anything new or different (when there is perhaps something morally dubious about trying to pass off the design responsibility anyway).

With the decline of traditional contracts, many quantity surveyors are becoming confused about the differences between CDP and preliminaries requirements – particularly in relation to temporary works. Although temporary works should be allowed for in the design of permanent works, designing and detailing them is included as the contractor's responsibility in the contract preliminaries (normally NBS clause A36/320). Temporary works should not be included as a CDP as it is not the designer's responsibility to delegate. If it is mistakenly included, the designer (and hence the client) takes on additional responsibilities regarding the feasibility and co-ordination of the temporary works with the permanent works.

Traditional procurement

Once the design is complete, tender documentation is prepared and sent out to the selected contractors (three to six depending on how large the project is) who are normally only given a month to absorb all the information and return a price for the work. Typically, a main contractor manages the work on site and has no labour of his own. The main contractor gets prices for the work from subcontractors and adds profit and preliminaries before returning the tenders to the design team. The client has the option to choose any of the tenderers, but the selection in the United Kingdom is normally on the basis of the lowest price. The client will be in contract with the main contractor, who in turn is in contract with the subcontractors. The architect normally acts as the contract administrator for the client. The tender process is sometimes split into overlap part of the design phase with a first stage tender and to achieve a quicker start on site than with a conventional tender process.

Construction management

Towards the end of the design process, the client employs a management contractor to oversee the construction. The management contractor takes the tender documentation, splits the information into packages and chooses trade contractors (a different name for a subcontractor) to tender for the work. The main differences between construction management and traditional procurement are that the design team can choose which trade contractors are asked to price and the trade contractors are directly contracted to the client. While this type of contractual arrangement can work well for straightforward buildings, it is not ideal for refurbishment or very complex jobs where it is not easy to split the job into simple 'trade packages'.

Design and Build

This procurement route is preferred by clients who want cost security and it is generally used for projects which have cost certainty, rather than quality of design, as the key requirement. There are two versions of Design and Build. This first is for the design team to work for the client up to the tender stage before being 'novated' to work for the main contractor. (A variant of this is a fixed sum contract where the design team remains employed by the client, but the cost of the work is fixed.) The second method is when the client tenders the project to a number of consortia on an outline description and specification. A consortium is typically led by a main contractor who has employed a design team. This typically means that the main contractor has much more control over the construction details than with other procurement routes.

Partnering

Partnering is difficult to define, and can take many different forms, but often means that the contractor is paid to be included as a member of the design team, where the client has set a realistic programme and budget for the size and quality of the building required. Partnering generally works best for teams who have worked together before, where the team members are all selected on the basis of recommendations and past performance. Ideally the contractor can bring his experience in co-ordinating and programming construction operations to advise the rest of the team on the choice of materials and construction methods. Normally detailing advice can be more difficult as the main contractors tend to rely on their subcontractors for the fine details. The actual contractual arrangement can be like any of those previously mentioned and sometimes the main contractor will share the risk of cost increases with the client on the basis that they can take a share of any cost savings.

Drawing conventions

Drawing conventions provide a common language so that those working in the construction industry can read the technical content of the drawings. It is important that everyone uses the same drawing conventions, to ensure clear communication. Construction industry drawing conventions are covered by BS EN ISO 7519, which takes over from the withdrawn BS 1192 and BS 308.

A drawing can be put to the best use if the projections/views are carefully chosen to show the most amount of information with maximum clarity. Most views in construction drawings are drawn orthographically (drawings in two dimensions), but isometric (30°) and axonometric (45°) projections should not be forgotten when dealing with complicated details. Typically drawings are split into: location, assembly and component. These might be contained in only one drawing for a small job. Drawing issue sheets should log issue dates, drawing revisions and the reasons for the issue.

Appropriate scales need to be picked for the different types of drawings:

Location/site plans – Used to show site plans, site levels, roads layouts, etc. Typical scales: 1:200, 1:500 and up to 1:2500 if the project demands.

General arrangement (GA) – Typically plans, sections and elevations set out as orthographic projections (i.e., views on a plane surface). The practical minimum for tender or construction drawings is usually 1:50, but 1:20 can also be used for more complicated plans and sections.

Details – Used to show the construction details referenced in the plans to show how individual elements or assemblies fit together. Typical scales: 1:20, 1:10, 1:5, 1:2 or 1:1.

Structural drawings should contain enough dimensional and level information to allow detailing and construction of the structure.

For small jobs or early in the design process, 'wobbly line' hand drawings can be used to illustrate designs to the design team and the contractor. The illustrations in this book show that type of freehand scale drawings which can be done using different line thicknesses and without using a ruler. These types of sketches can be quicker to produce and easier to understand than computer-drawn information, especially in the preliminary stages of design.

Typical sheet sizes

A0	1189 × 841 mm
A1	841 × 594 mm
A2	594 × 420 mm
A3	420 × 297 mm
A4	297 × 210 mm
A5	210 × 148 mm

Line thicknesses

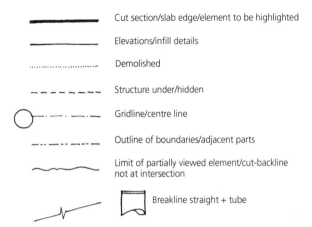

Cut section/slab edge/element to be highlighted

Elevations/infill details

Demolished

Structure under/hidden

Gridline/centre line

Outline of boundaries/adjacent parts

Limit of partially viewed element/cut-backline
not at intersection

Breakline straight + tube

Hatching

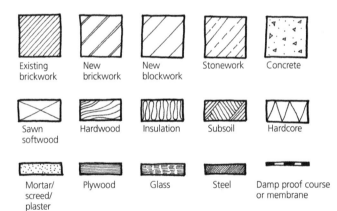

Existing brickwork

New brickwork

New blockwork

Stonework

Concrete

Sawn softwood

Hardwood

Insulation

Subsoil

Hardcore

Mortar/ screed/ plaster

Plywood

Glass

Steel

Damp proof course or membrane

Steps, ramps and slopes

Stairs

Ramp

Landscape slope

Slope/pitch

Arrow indicates 'up'

Common arrangement of work sections

The Common Arrangement of Work Sections for Building Work (CAWS) is intended to provide a standard for the production of specifications and bills of quantities for building projects, so that the work can be divided up more easily for costing and for distribution to subcontractors. The full document is very extensive, with sections to cover all aspects of the building work including the contract, structure, fittings, finishes, landscaping and mechanical and electrical services. The following sections are extracts from CAWS to summarise the sections most commonly used by structural engineers:

A Preliminaries/general conditions	A1	The project generally	A2	The contract
	A3	Employer's requirements	A4	Contractor's general costs
C Existing site/ buildings/services	C1	Demolition	C2	Alteration – composite items
	C3	Alteration – support	C4	Repairing/renovating concrete/masonry
	C5	Repairing/renovating metal/timber		
D Groundwork	D1	Investigation/ stabilisation/dewatering	D2	Excavation/filling
	D3	Piling	D4	Diaphragm walling
	D5	Underpinning		
E In situ concrete/large precast concrete	E1	In situ concrete	E2	Formwork
	E3	Reinforcement	E4	In situ concrete sundries
	E5	Precast concrete large units	E6	Composite construction
F Masonry	F1	Brick/block walling	F2	Stone walling
	F3	Masonry accessories		
G Structural/carcassing in metal or timber	G1	Structural/carcassing metal	G2	Structural/carcassing timber
	G3	Metal/timber decking		
R Disposal systems	R1	Drainage	R2	Sewerage

Note: There is a very long list of further subheadings which can be used to cover sections in more detail (e.g. F10 is specifically for Brick/block walling). However, the list is too extensive to be included here.

Source: CPIC. 1998.

Summary of ACE conditions of engagement

The Association of Consulting Engineers (ACE) represents the consulting sector of the engineering profession in the UK. The ACE Conditions of Engagement, Agreement B(1) (2004) is used where the engineer is appointed directly to the client and works with an architect who is the lead consultant or the contract administrator. A summary of the Normal Services from Agreement B(1) is given below with references to the work stages defined by the Royal Institute of British Architects (including references to both the 2007 lettered stages and the 2013 numbered stages for clarity).

Feasibility		
Stage 0/1 (previously A)	Appraisal	Identification of client requirements and development constraints by the Lead Consultant, with an initial appraisal to allow the client to decide whether to proceed and to select the probable procurement method.
Stage 0/1 (previously B)	Strategic briefing	Confirmation of key requirements and constraints for or by the client, including any topographical, historical or contamination constraints on the proposals. Consider the effect of public utilities and transport links for construction and post-construction periods on the project. Prepare a site investigation desk study and, if necessary, bring the full site investigation forward from Stage C. Identify the Project Brief, establish design team working relationships and lines of communication and discuss with the client any requirements for site staff or resident engineers. Collaborate on the design with the design team and prepare a stage report if requested by the client or lead consultant.
Preconstruction phase		
Stage 2 (previously C)	Outline proposals	Visit the site and study any reports available regarding the site. Advise the client on the need and extent of site investigations, arrange quotes and proceed when quotes are approved by the client. Advise the client of any topographical or dimensional surveys that are required. Consult with any local or other authorities about matters of principle and consider alternative outline solutions for the proposed scheme. Provide advice, sketches, reports or outline specifications to enable the Lead Consultant to prepare his outline proposals and assist the preparation of a Cost Plan. Prepare a report and, if required, present it to the client.
Stage 3 (previously D)	Detailed proposals	Develop the design of the detailed proposals with the design team for submission of the Planning Application by the Lead Consultant. Prepare drawings, specifications, calculations and descriptions in order to assist the preparation of a Cost Plan. Prepare a report and, if required, present it to the client
Stage 3 (previously E)	Final proposals	Develop and co-ordinate all elements of the project in the overall scheme with the design team, and prepare calculations, drawings, schedules and specifications as required for presentation to the client. Agree on a programme for the design and construction of the Works with the client and the design team.

continued

(continued) Summary of ACE conditions of engagement

Preconstruction phase

Stage 3/4 (previously F)	Production information	Develop the design with the design team and prepare drawings, calculations, schedules and specifications for the Tender Documentation and for Building Regulations Approval. Prepare any further drawings and schedules necessary to enable Contractors to carry out the Works, excluding drawings and designs for temporary works, formwork and shop fabrication details (reinforcement details are not always included as part of the normal services). Produce a Designer's Risk Assessment in line with Health & Safety CDM Regulations. Advise the Lead Consultant on any special tender or contract conditions.
Stage 3/4 (previously G)	Tender documents	Assist the Lead Consultant in identifying and evaluating potential contractors and/or specialists for the construction of the project.
Stage 3/4 (previously H)	Tender action	Assist the selection of contractors for the tender lists, assemble Tender Documentation and issue it to the selected tenderers. On return of tenders, advise on the relative merits of the contractors proposals, programmes and tenders.

Construction phase

Stage 5 (previously J)	Mobilisation	Assist the Client and Lead Consultant in letting the building contract, appointing the contractor and arranging site handover to the contractor. Issue construction information to the contractor and provide further information to the contractor as and when reasonably required. Comment on detailed designs, fabrication drawings, bar bending schedules and specifications submitted by the Contractors, for general dimensions, structural adequacy and conformity with the design. Advise on the need for inspections or tests arising during the construction phase and the appointment and duties of Site Staff.
Stage 5 (previously K)	Construction to practical completion	Assist the Lead Consultant in examining proposals, but not including alternative designs for the Works, submitted by the Contractor. Attend relevant site meetings and make other periodic visits to the site as appropriate to the stage of construction. Advise the Lead Consultant on certificates for payment to Contractors. Check that work is being executed generally adhering to the control documents and with good engineering practice. Inspect the construction on completion and, in conjunction with any Site Staff, record any defects. On completion, deliver one copy of each of the final structural drawings to the planning supervisor or client. Perform work or advise the Client in connection with any claim in connection with the structural works.
Stage 6 (previously L)	After practical completion	Assist the Lead Consultant with any administration of the building contract after practical completion. Make any final inspections in order to help the Lead Consultant settle the final account.
Stage 7	Post occupancy evaluation	A new RIBA Stage not yet covered by the ACE conditions.

Source: ACE. 2004.

2
Statutory Authorities and Permissions

Planning

Planning regulations control individuals' freedom to alter their property in an attempt to protect the environment in UK towns, cities and countryside, in the public interest. Different regulations and systems of control apply in the different UK regions. Planning permission is not always required, and in such cases the planning department will issue a Lawful Development Certificate on request and for a fee.

England and Wales

The main legislation that sets out the planning framework in England and Wales is the Town and Country Planning Act 1990. The government's statements of planning policy may be found in White Papers, Planning Policy Guidance Notes (PPGs), Mineral Policy Guidance Notes, Regional Policy Guidance Notes, departmental circulars and ministerial statements published by the Department for Communities and Local Government (DCLG).

Scotland

The First Minister for Scotland is responsible for the planning framework. The main planning legislation in Scotland is the Town and Country Planning Act (Scotland) 1997 and the Planning (Listed Buildings and Conservation Areas) (Scotland) Act 1997. The legislation is supplemented by the Scottish Government who publish National Planning Policy Guidelines (NPPGs) which set out the Scottish policy on land use and other issues. In addition, a series of Planning Advice Notes give guidance on how best to deal with matters such as local planning, rural housing design and improving small towns and town centres.

Northern Ireland

The Planning (NI) Order 1991 could be said to be the most significant of the many different Acts which make up the primary and subordinate planning legislation in Northern Ireland. As in the other UK regions, the Northern Ireland Executive publishes policy guidelines called Planning Policy Statements (PPSs) which set out the regional policies to be implemented by the local authority.

Building regulations and standards

Building regulations have been around since Roman times and are now used to ensure reasonable standards of construction, health and safety, energy efficiency and access for the disabled. Building control requirements, and their systems of control, are different for the different UK regions.

The legislation is typically set out under a Statutory Instrument, empowered by an Act of Parliament. In addition, the legislation is further explained by the different regions in explanatory booklets, which also describe the minimum standards 'deemed to satisfy' the regulations. The 'deemed to satisfy' solutions do not preclude designers from producing alternative solutions provided that they can be supported by calculations and details to satisfy the local authority who implement the regulations. Building control fees vary across the country but are generally calculated on a scale in relation to the cost of the work.

England and Wales

England and Wales have had building regulations since about 1189 when the first version of a London Building Act was issued. Today the relevant legislation is the Building Act 1984 and the Statutory Instrument Building Regulations 2010. The Approved Documents published by the DCLG are a guide to the minimum requirements of the regulations.

Applications may be made as 'full plans' submissions well before work starts, or for small elements of work as a 'building notice' 48 hours before work starts. Completion certificates demonstrating Building Regulations Approval can be obtained on request. Third parties can become approved inspectors and provide building control services.

Approved documents (as amended)

A	Structure
	A1 Loading
	A2 Ground Movement
	A3 Disproportionate Collapse
B	Fire Safety (volumes 1 and 2)
C	Site Preparation and Resistance to Moisture
D	Toxic Substances
E	Resistance to the Passage of Sound
F	Ventilation
G	Hygiene
H	Drainage and Waste Disposal
J	Combustion Appliances and Fuel Storage Systems
K	Protection from Falling, Collision and Impact
L	Conservation of Fuel and Power
	L1A New Dwellings
	L1B Existing Dwellings
	L2A New Buildings (other than dwellings)
	L2B Existing Buildings (other than dwellings)
M	Access to and Use of Buildings
N	Glazing
P	Electrical Safety – Dwellings
Regulation 7	Materials and workmanship

Scotland

Building standards have been in existence in Scotland since around 1119 with the establishment of the system of Royal Burghs. The three principal documents which currently govern building control are the Building (Scotland) Act 2003 and the Technical Standards 2009 – the explanatory guide to the regulations published by the Scottish Government.

Applications for all building and demolition works must be made to the local authority, who assess the proposals for compliance with the technical standards before issuing a building warrant, which is valid for five years. For simple works a warrant may not be required, but the regulations still apply. Unlike the other regions in the United Kingdom, work may start on a site only after a warrant has been obtained. Buildings may be occupied at the end of the construction period only after the local authority has issued a completion certificate. Building control departments typically will only assess very simple structural proposals and for more complicated work, qualified engineers must 'self-certify' their proposals, overseen by the Scottish Building Standards Agency (SBSA). Technical handbooks are to be updated annually and are free to download from the SBSA's website.

Technical handbooks (domestic or non-domestic)

0	General
1	Structure
2	Fire
3	Environment
4	Safety
5	Noise
6	Energy
Appendix A	Defined Terms
Appendix B	List of Standards and other Publications

Northern Ireland

The main legislation, policy and guidelines in Northern Ireland are the Building Regulations (Northern Ireland) Order 1979 as amended by the Planning and Building Regulations (Northern Ireland) (Amendment) Order 1990; the Building Regulations (NI) 2000 and the technical booklets – which describe the minimum requirements of the regulations published by the Northern Ireland Executive.

Building regulations in Northern Ireland are the responsibility of the Department of Finance and Personnel and are implemented by the district councils. Until recently the regulations operated on strict prescriptive laws, but the system is now very similar to the system in England and Wales. Applicants must demonstrate compliance with the 'deemed to satisfy' requirements. Applications may be made as a 'full plans' submission well before work starts, or as a 'building notice' for domestic houses just before work starts. Builders must issue stage notices for local authority site inspections. Copies of the stage notices should be kept with the certificate of completion by the building owner.

Technical booklets

A	Interpretation and General
B	Materials and Workmanship
C	Preparation of Sites and Resistance to Moisture
D	Structure E Fire Safety
F	Conservation of Fuel and Power
	F1 Dwellings
	F2 Buildings other than dwellings
G	Sound
	G1 Sound (Conversions)
H	Stairs, Ramps, Guarding and Protection from Impact
J	Solid Waste in Buildings
K	Ventilation
L	Combustion Appliances and Fuel Storage Systems
N	Drainage
P	Sanitary Appliances and Unvented Hot Water Storage Systems
R	Access to and Use of Buildings
V	Glazing

Listed Buildings

In the United Kingdom, buildings of 'special architectural or historic interest' can be Listed to ensure that their features are considered before any alterations to the exterior or interior are agreed. Buildings may be Listed because of their association with an important architect, person or event or because they are a good example of design, building type, construction or use of material. Listed Building Consent must be obtained from the local authority before any work is carried out on Listed Building. In addition, there may be special conditions attached to ecclesiastical, or old ecclesiastical, buildings or land by the local diocese or the Home Office.

England and Wales

English Heritage (EH) in England and CADW in Wales work for the government to identify buildings of 'special architectural or historic interest'. All buildings built before 1700 (and most buildings between 1700 and 1840) with a significant number of original features will be Listed. A building normally must be over 30 years old to be eligible for Listing. There are three grades: I, II* and II, and there are approximately 500,000 buildings listed in England, with about 13,000 in Wales. Grades I and II* are eligible for grants from EH for urgent major repairs and residential Listed Buildings may be VAT zero rated for approved alterations.

Scotland

Historic Scotland maintains the lists and schedules for the Scottish Government. All buildings before 1840 of substantially unimpaired character can be Listed. There are over 40,000 Listed buildings divided into three grades: A, B and C. Grade A is used for buildings of national or international importance or little altered examples of a particular period, style or building type, while a Grade C building would be of local importance or be a significantly altered example of a particular period, style or building type.

Northern Ireland

The Environment and Heritage Service (EHS) within the Northern Ireland Executive has carried out a survey of all the building stock in the region and keeps the Northern Ireland Buildings Database. Buildings must be at least 30 years old to be Listed and there are currently about 8500 Listed Buildings. There are three grades of Listing: A, B+ and B (with two further classifications B1 and B2) which have similar qualifications to the other UK regions.

Conservation areas

Local authorities have a duty to designate conservation areas in any area of 'special architectural or historic interest' where the character or appearance of the area is worth preserving or enhancing. There are around 8500 conservation areas in England and Wales, 600 in Scotland and 30 in Northern Ireland. The character of an area does not just come from buildings and so the road and path layouts, greens and trees, paving and building materials and public and private spaces are protected. Conservation area consent is required from the local authority before work starts, to ensure any alterations do not detract from the area's appearance.

Tree preservation orders

Local authorities have specific powers to protect trees by making Tree Protection Orders (TPOs). Special provisions also apply to trees in conservation areas. A TPO makes it an offence to cut down, lop, top, uproot, wilfully damage or destroy the protected tree without the local planning authority's permission. All of the UK regions operate similar guidelines with slightly different notice periods and penalties.

The owner remains responsible for the tree(s), their condition and any damage they may cause, but only the planning authority can give permission to work on them. Arboriculturalists (who can give advice on work which needs to be carried out on trees) and contractors (who are qualified to work on trees) should be registered with the Arboricultural Association. In some cases (including if the tree is dangerous) no permission is required, but notice (about 5 days or 6 weeks in a conservation area) depending on the UK region) must be given to the planning authority. When it is agreed that a tree can be removed, this is normally on the condition that a similar tree is planted as a replacement. Permission is generally not required to cut down or work on trees with a trunk of less than 75 mm diameter (measured at 1.5 m above ground level) or 100 mm diameter if thinning is done to help the growth of other trees. Fines of up to £20,000 can be levied if work is carried out without permission.

Archaeology and ancient monuments

Archaeology in Scotland, England and Wales is protected by the Ancient Monuments and Archaeology Areas Act 1979, while the Historic Monuments and Archaeology Objects (NI) Order 1995 applies in Northern Ireland.

Archaeology in the United Kingdom can represent every period from the camps of hunter gatherers 10,000 years ago to the remains of twentieth century industrial and military activities. Sites include places of worship, settlements, defences, burial grounds, farms, fields and sites of industry. Archaeology in rural areas tends to be very close to the ground surface, but in urban areas, deep layers of deposits built up as buildings were demolished and new buildings were built directly on the debris. These deposits, often called 'medieval fill', are an average of 5 m deep in places like the City of London and York.

Historic or ancient monuments are those structures which are of national importance. Typically monuments are in private ownership but are not occupied buildings. Scheduled monument consent is required for alterations and investigations from the regional heritage bodies: EH, Historic Scotland, CADW in Wales and EHS in Northern Ireland.

Each of the UK regions operates very similar guidelines in relation to archaeology, but through different frameworks and legislation. The regional heritage bodies develop the policies which are implemented by the local authorities. These policies are set out in PPG 16 for England and Wales, NPPG 18 for Scotland and PPS 6 for Northern Ireland. These guidance notes are intended to ensure that:

1. Archaeology is a material consideration for a developer seeking planning permission.
2. Archaeology strategy is included in the urban development plan by the local planning authority.
3. Archaeology is preserved, where possible, in situ.
4. The developer pays for the archaeological investigations, excavations and reporting.
5. The process of assessment, evaluation and mitigation is a requirement of planning permission.
6. The roles of the different types of archaeologists in the processes of assessment, evaluation and mitigation are clearly defined.

Where 'areas of archaeological interest' have been identified by the local authorities, the regional heritage bodies act as curators (EH, Historic Scotland, Cadw in Wales and EHS in Northern Ireland). Any developments within an area of archaeological interest will have archaeological conditions attached to the planning permission to ensure that the following process is put into action:

1. Early consultation between the developers and curators so that the impact of the development on the archaeology (or vice versa) can be discussed and the developer can get an idea of the restrictions which might be applied to the site, the construction process and the development itself.
2. Desk study of the site by an archaeologist.
3. Field evaluation by archaeologists using field walking, trial pits, boreholes and/or geophysical prospecting to support the desk study.
4. Negotiation between the site curators and the developer's design team to agree the extent of archaeological mitigation. The developer must submit plans for approval by the curators.
5. Mitigation – either preservation of archaeology in situ or excavation of areas to be disturbed by development. The archaeologists may have either a watching brief over the excavations carried out by the developer (where they monitor construction work for finds) or on significant sites, carry out their own excavations.
6. Post-excavation work to catalogue and report on the archaeology and either store or display the findings.

Generally the preliminary and field studies are carried out by private consultants and contractors employed by the developers to advise the local authority planning department. In some areas, advice can also be obtained from a regional archaeologist. In Northern Ireland, special licences are required for every excavation which must be undertaken by a qualified archaeologist. In Scotland, England and Wales, the archaeological contractors or consultants can have a 'watching brief'.

Field evaluations can often be carried out using geotechnical trial pits with the excavations being done by the contractor or the archaeologist depending on the importance of the site. If an interesting find is made in a geotechnical trial pit and the archaeologists would like to keep the pit open for inspection by, say, the curators, the developer does not have to comply if there would be inconvenience to the developer or building users, or for health and safety reasons.

Engineers should ensure, for the field excavation and mitigation stages, that the archaeologists record **all** the features in the excavations up to this century's interventions as these records can be very useful to the design team. Positions of old concrete footings could have as much of an impact on proposed foundation positions as archaeological features!

Party Wall etc. Act

The Party Wall etc. Act 1996 came into force in 1997 throughout England and Wales. In 2014 there is no equivalent legislation in Northern Ireland. In Scotland, The Tenements (Scotland) Act 2004 applies, but this seems to relate more to management and maintenance of shared building assets.

Different sections of the Party Wall Act apply, depending on whether you propose to carry out work to an existing wall or structure shared with another property; build a freestanding wall or the wall of a building astride a boundary with a neighbouring property, and/or excavate within 3 m of a neighbouring building or structure. Work can fall within several sections of the Act at one time. A building owner must notify his neighbours and agree the terms of a Party Wall Award before starting any work.

The Act refers to two different types of Party Structure: 'Party Wall' and 'Party Fence Wall'. Party Walls are loosely defined as a wall on, astride or adjacent to a boundary enclosed by building on one side or both sides. Party Fence Walls are walls astride a boundary but not part of a building; it does not include things like timber fences. A Party Structure is a wide term which can sometimes include floors or partitions.

The notice periods and sections 1, 2 and 6 of the Act are most commonly used, and are described below.

Notice periods and conditions

In order to exercise rights over the Party Structures, the Act says that the owner must give notice to adjoining owners; the building owner must not cause unnecessary inconvenience, must provide compensation for any damage and must provide temporary protection for buildings and property where necessary. The owner and the adjoining owner in the Act are defined as anyone with an interest greater than a tenancy from year to year. Therefore, this can include shorthold tenants, long leaseholders and freeholders for any one property.

A building owner, or surveyor acting on his behalf, must send a notice in advance of the start of the work. Different notice periods apply to different sections of the Act, but work can start within the notice period with the written agreement of the adjoining owner. A notice is only valid for one year from the date that it is served and must include the owner's name and address; the building's address (if different); a clear statement that the notice is under the provisions of the Act (stating the relevant sections); full details of the proposed work (including plans where appropriate) and the proposed start date for the work.

The notice can be served by post, in person or fixed to the adjoining property in a 'conspicuous part of the premises'. Once the Notice has been served, the adjoining owner can consent in writing to the work or issue a counter notice setting out any additional work he would like to carry out. The owner must respond to a counter notice within 14 days. If the owner has approached the adjoining owners and discussed the work with them, the terms of a Party Wall Award may have already been agreed in writing before a notice is served.

If a notice is served and the adjoining owner does not respond within 14 days, a dispute is said to have arisen. If the adjoining owner refuses to discuss terms or appoint a surveyor to act on his behalf, the owner can appoint a surveyor to act on behalf of the adjoining owner. If the owners discuss, but cannot agree terms they can jointly appoint a surveyor (or they can each appoint one) to draw up the Party Wall Award. If two surveyors cannot agree, a nominated third surveyor can be called to act impartially. In complex cases, this can often take over a year to resolve and in such cases the notice period can run out, meaning that the process must begin again by serving another notice. In all cases, the surveyors are appointed to consider the rights of the owner over the wall and not to act as advocates in the negotiation of compensation! The building owner covers the costs associated with all of the surveyors and experts who were asked about the work.

When the terms have been agreed, the Party Wall Award should include a description (in drawings and/or writing) of what, when and how work is to be carried out; a record of the condition of the adjoining owner's property before work starts; arrangements to allow access for surveyors to inspect while the works are going on and say who will pay for the cost of the works (if repairs are to be carried out as a shared cost or if the adjoining owner has served a counter notice and is to pay for those works). Either owner has 14 days to appeal to the County Court against an Award if an owner believes that the person who has drafted the Award has acted beyond their powers.

An adjoining owner can ask the owner for a 'bond'. The bond money becomes the property of the adjoining owner (until the work has been completed in accordance with the Award) to ensure that funds are available to pay for the completion of the works in case the owner does not complete the works.

The owner must give 14 days' notice if his representatives are to access the adjoining owner's property to carry out or inspect the works. It is an offence to refuse entry or obstruct someone who is entitled to enter the premises under the Act if the offender knows that the person is entitled to be there. If the adjoining property is empty, the owner's workmen and own surveyor or architect may enter the premises if they are accompanied by a police officer.

Section 1: new building on a boundary line

Notice must be served to build on or astride a boundary line, but there is no right to build astride if your neighbour objects. You can build foundations on the neighbouring land if the wall line is immediately adjacent to the boundary, subject to supervision. The Notice is required at least **1 month** before the proposed start date.

Section 2: work on existing party walls

The most commonly used rights over existing Party Walls include cutting into the wall to insert a DPC or support a new beam bearing; raising, underpinning, demolishing and/or rebuilding the Party Wall and/or providing protection by putting a flashing from the higher over the lower wall. Minor works such as fixing shelving, fitting electrical sockets or replastering are considered to be too trivial to be covered in the Act.

A building owner, or Party Wall Surveyor acting on the owner's behalf, must send a Notice at least **2 months** in advance of the start of the work.

Section 6: excavation near neighbouring buildings

Notice must be served at least **1 month** before an owner intends to excavate or construct a foundation for a new building or structure within 3 m of an adjoining owner's building where that work will go deeper than the adjacent owner's foundations, or within 6 m of an adjoining owner's building where that work will cut a line projecting out at 45° from the bottom of that building's foundations. This can affect neighbours who are not immediately adjacent. The Notice must state whether the owner plans to strengthen or safeguard the foundations of the adjoining owner. Adjoining owners must agree specifically in writing to the use of 'special foundations' – these include reinforced concrete foundations. After work has been completed, the adjoining owner may request particulars of the work, including plans and sections.

Source: DETR. 1997.

CDM

The Construction Design & Management (CDM) Regulations 2007 were developed to assign responsibilities for health and safety to the client, the design team and the principal contractor. The Approved Code of Practice is published by the Health and Safety Executive for guidance on the Regulations.

The client is required to appoint a CDM co-ordinator (CDMC) who has overall responsibility for co-ordinating health and safety aspects of the design and planning stages of a project. The duties of the CDMC can theoretically be carried out by any of the traditional design team professionals. The CDMC must ensure that the designers avoid, minimise or control health and safety risks for the construction and maintenance of the project, as well as ensuring that the contractor is competent to carry out the work and briefing the client on health and safety issues during the works.

The CDMC prepares the pre-construction information for inclusion in the tender documents which should include project-relevant health and safety information gathered from the client and designers. This should highlight any unusual aspects of the project (also highlighted on the drawings) that a competent contractor would not be expected to know. This document is taken on by the successful principal contractor and developed into the construction phase health and safety plan by the addition of the contractor's health and safety policy, risk assessments and method statements as requested by the designers. The health and safety plan is intended to provide a focus for the management and prevention of health and safety risks as the construction proceeds.

The health and safety file is generally compiled at the end of the project by the contractor and the CDMC who collect the design information relevant to the life of the building. The CDMC must ensure that the file is compiled and passed to the client or the people who will use, operate, maintain and/or demolish the project. A good health and safety file will be a relatively compact maintenance manual including information to alert those who will be owners, or operators of the new structure, about the risks which must be managed when the structure and associated plant are maintained, repaired, renovated or demolished. After handover the client is responsible for keeping the file up to date.

Full CDM regulation provisions apply to projects over 30 days or involve 500 person days of construction work, but not to projects with domestic (i.e., owner–occupier) clients.

In 2014 a complete overhaul of the CDM 2007 Regulations is expected. This is intended to:

- bring the regulations into line with European Temporary or Mobile Construction Sites Directive (TMCSD),
- reduce unnecessary 'red tape' in line with current UK Government policy, and
- address the 'two tier' industry which has emerged since the Regulations were first introduced where larger sites have made significant improvements, but where smaller sites have disproportionately higher rates of serious and fatal accidents.

The final details were not available at the time of publication but are likely to include:

- changes to make the Regulations easier to understand;
- CDM co-ordinator role replaced by the principal designer;
- Approved Code of Practice (ACOP) replaced with targeted guidance,
- prescriptive requirements for assessment of competency replaced with more generic tests;
- align notification requirements with the TMSCD, and
- application of the Regulations to domestic clients, but in a proportionate way.

Building Information Modelling (BIM)

With the UK construction industry estimated to account for 7% of GDP (of which 40% is public procurement), the UK Government is keen to ensure that it gets value for money. Historically, the UK construction industry has a reputation for cost and programme over-runs. Initiatives to improve performance can be traced back to Sir Michael Latham's 1994 report 'Constructing the Team' and Sir John Egan's 1998 report 'Rethinking Construction'. Both were critical of construction industry performance and suggested a number of different ways to improve the efficiency.

Past initiatives have focussed on collaborative working, innovative forms of contract and lean/off-site manufacturing techniques to eliminate waste and improve performance. With the widespread developments in IT in recent years (but with very little software standardisation to maximise efficiency), it is hoped that Building Information Modelling (BIM) will solve old and new problems in one stroke.

BIM is defined by the Construction Project Information Committee as *'a digital representation of physical and functional characteristics of a facility creating a shared knowledge resource for information about it forming a reliable basis for decisions during its life-cycle; from earliest conception to demolition'*.

In very simple terms, BIM is a 3D drawing model used to design 'a facility', record lifetime changes/maintenance/operation and then assist in its demolition. The aim is that information exchange will be made more efficient, interdisciplinary working will be easier and clients will have easy access to their 'asset data' if all parties use a shared pool of information. Different levels of integration and collaboration occur in practice, but the interactive virtual model is intended to give more than just measurements. For example, in the model, a wall might have a database entry/link for estimating information, details of its fire rating and its energy performance characteristics. Information can be exported to create schedules and other documents, while hyperlinks will point to specifications or a manufacturer's information sheets not contained within the model itself. In theory alterations to one part of the model will be automatically amended elsewhere to create a coordinated and integrated building model. A BIM 'model manager' role is generally created within the design teams using this system.

From 2016, 'fully collaborative' or 'Level 3' BIM will be a requirement on the UK Government projects, with all project and asset information, documentation and data being electronic. Electronic building data will be required in a neutral 'COBie' format to ensure interoperability. There are a number of contractual and liability issues to be resolved before Level 3 BIM will be widespread. It will probably be a generation before it can be seen whether digital files will be managed as well as traditional archives, and whether BIM delivers its full potential.

3
Design Data

Design data checklist

The following design data checklist is a useful reminder of all of the limiting criteria which should be considered when selecting an appropriate structural form:

- Description/building use
- Client brief and requirements
- Site constraints
- Loadings
- Structural form: load transfer, stability and robustness
- Materials
- Movement joints
- Durability
- Fire resistance
- Performance criteria: deflection, vibration, etc.
- Temporary works and construction issues
- Soil conditions, foundations and ground slab
- Miscellaneous issues

Structural form

It is worth trying to remember the different structural forms when developing a scheme design. A particular structural form might fit the vision for the form of the building. Force or moment diagrams might suggest a building shape. The following diagrams of structural form are intended as useful reminders:

Trusses

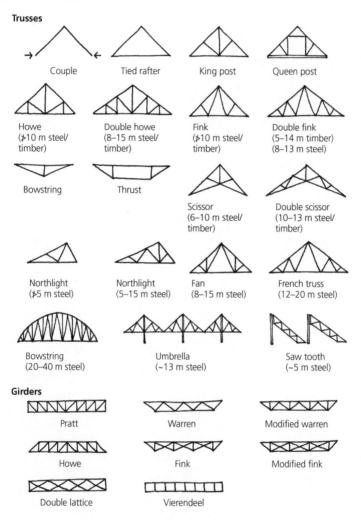

Couple

Tied rafter

King post

Queen post

Howe
(≯10 m steel/
timber)

Double howe
(8–15 m steel/
timber)

Fink
(≯10 m steel/
timber)

Double fink
(5–14 m timber)
(8–13 m steel)

Bowstring

Thrust

Scissor
(6–10 m steel/
timber)

Double scissor
(10–13 m steel/
timber)

Northlight
(≯5 m steel)

Northlight
(5–15 m steel)

Fan
(8–15 m steel)

French truss
(12–20 m steel)

Bowstring
(20–40 m steel)

Umbrella
(~13 m steel)

Saw tooth
(~5 m steel)

Girders

Pratt

Warren

Modified warren

Howe

Fink

Modified fink

Double lattice

Vierendeel

Portal frames

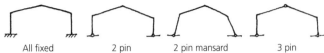

| All fixed | 2 pin | 2 pin mansard | 3 pin |

Arches

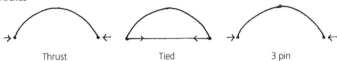

| Thrust | Tied | 3 pin |

Suspension

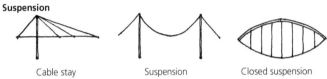

| Cable stay | Suspension | Closed suspension |

Walls

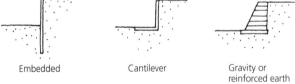

| Solid | Piers | Chevron | Diaphragm |

Timber

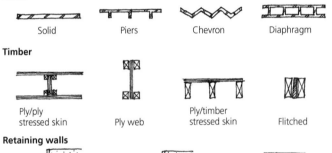

Ply/ply
stressed skin   Ply web   Ply/timber
stressed skin   Flitched

Retaining walls

| Embedded | Cantilever | Gravity or reinforced earth |

Structural movement joints

Joints should be provided to control temperature, moisture, acoustic and ground movements. Movement joints can be difficult to waterproof and detail and therefore should be kept to a minimum. The positions of movement joints should be considered for their effect on the overall stability of the structure.

Primary movement joints

Primary movement joints are required to prevent cracking where buildings (or parts of buildings) are large, where a building spans different ground conditions, changes height considerably or where the shape suggests a point of natural weakness. Without detailed calculations, joints should be detailed to permit 15–25 mm movement. Advice on joint spacing for different building types can be variable and conflicting. The following figures are some approximate guidelines based on the building type:

Concrete	25 m (e.g. for roofs with large thermal differentials)–50 m c/c
Steel industrial buildings	100 m typical–150 m maximum c/c
Steel commercial buildings	50 m typical–100 m maximum c/c
Masonry	40 m–50 m c/c

Secondary movement joints

Secondary movement joints are used to divide structural elements into smaller elements to deal with the local effects of temperature and moisture content. Typical joint spacings are:

Clay bricks	Up to 12 m c/c on plan (6 m from corners) and 9 m vertically or every three storeys if the building is greater than 12 m or four storeys tall (in cement mortar)
Concrete blocks	3 m–7 m c/c (in cement mortar)
Hardstanding	70 m c/c
Steel roof sheeting	20 m c/c down the slope, no limit along the slope

Fire resistance periods for structural elements

Fire resistance of structure is required to maintain structural integrity to allow time for the building to be evacuated. Generally, roofs do not require protection. Architects typically specify fire protection in consultation with the engineer.

		Minimum period of fire resistance minutes					
		Basement[g] storey including floor over		Ground or upper storey			
		Depth of a lowest basement		Height of top floor above ground, in a building or separated part of a building			
Building types		>10 m	<10 m	>5 m	<18 m	<30 m	>30 m
Residential flats and maisonettes		90	60	30[a]	60[b]	90[b]	120[b]
Residential houses		N/A	30[a]	30[a]	60[c]	N/A	N/A
Institutional residential[d]		90	60	30[a]	60	90	120[e]
Office	Not sprinklered	90	60	30[a]	60	90	X
	Sprinklered	60	60	30[a]	30[a]	90	120[e]
Shops and commercial	Not sprinklered	90	60	60	60	90	X
	Sprinklered	60	60	30[a]	60	90	X
Assembly and recreation	Not sprinklered	90	60	60	60	90	X
	Sprinklered	60	60	30[a]	60	60	120[e]
Industrial	Not sprinklered	120	90	60	90	120	X
	Sprinklered	90	60	30[a]	60	90	120[e]
Storage and other non-residential	Not sprinklered	120	90	60	90	120	X
	Sprinklered	90	60	30[a]	60	90	120[e]
Car park for light vehicles	Open sided	N/A	N/A	15[a]	15[a,h]	15[a,h]	60
	All others	90	60	30[a]	60	90	120[e]

Notes:

X Not permitted.

[a] Increased to 60 minutes for compartment walls with other fire compartments or 30 min for elements protecting a means of escape.

[b] Reduced to 30 minutes for a floor in a maisonette not contributing to the support of the building.

[c] To be 30 minutes in the case of three-storey houses and 60 minutes for compartment walls separating buildings.

[d] NHS hospitals should have a minimum of 60 minutes.

[e] Reduced to 90 minutes for non-structural elements.

[f] Should comply with Building Regulations: B3 Section 12.

[g] The uppermost floor over basements should meet provision for ground and upper floors if higher.

[h] Fire engineered steel elements with certain Hp/A ratios are deemed to satisfy. See Table 2.A2, Approved Document B for full details.

Source: Building Regulations Approved Document B. 2007.

Typical building tolerances

Space between walls

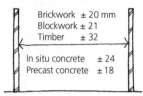

Brickwork ± 20 mm
Blockwork ± 21
Timber   ± 32

In situ concrete   ± 24
Precast concrete   ± 18

Space between columns

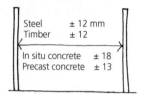

Steel   ± 12 mm
Timber   ± 12

In situ concrete   ± 18
Precast concrete   ± 13

Wall verticality

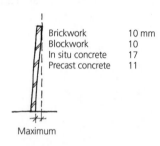

Brickwork   10 mm
Blockwork   10
In situ concrete   17
Precast concrete   11

Maximum

Column verticality

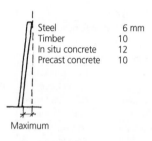

Steel   6 mm
Timber   10
In situ concrete   12
Precast concrete   10

Maximum

Vertical position of beams

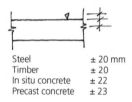

Steel   ± 20 mm
Timber   ± 20
In situ concrete   ± 22
Precast concrete   ± 23

Vertical position of floors

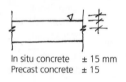

In situ concrete   ± 15 mm
Precast concrete   ± 15

Plan position

Brickwork   ± 10 mm
Steel   ± 10
Timber   ± 10
In situ concrete   ± 12
Precast concrete   ± 10

Flatness of floors

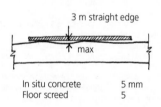

3 m straight edge

max

In situ concrete   5 mm
Floor screed   5

Source: BS 5606:1990.

Historical use of building materials

Masonry and timber

	Georgian including William IV	Victorian	Edwardian	Inter Wars	Post Wars
	1714 — 1800	1837 — 1901	1919	1945	

Masonry

Material	Markers
Bonding timbers	(Georgian–early)
Non-hydraulic lime mortar	(Georgian to Edwardian)
Mathematical tiles	84 / 50s
Hydraulic lime mortar	50s 96 / 80s
Clinker concrete blocks	60s
Cavity walls	00s / 50 51 / 10
Pressed bricks	51
Flettons	70s
Concrete bricks	20
Dense concrete blocks	50s
Sand line bricks	20
Stretcher bond	20s
Mild steel cavity wall ties	60 / 40s
Galvanised steel cavity wall ties	60 / 45 80s
Stainless steel cavity wall ties	65 80s
Lightweight concrete blocks	53 60s

Timber

Material	Markers
Trussed timber girders	33 / 50 97
King + queen post trusses	50 / 30s
Wrought iron flitched beams	10s / 70s
Belfast trusses	60 / 40s
Trussed rafters	50s
Ply stressed skin panels	60s

Concrete and steel

	Georgian including William IV (1714 – 1800 – 1837)	Victorian (1837 – 1901)	Edwardian (1901 – 1919)	Inter Wars (1919 – 1945)	Post Wars (1945 –)
Concrete					
Limecrete/Roman cement	56 ---- 96	80s			
Jack arch floors	96	62			
Portland cement	24	51			
Filler joists		62 – 70s		30	
Clinker concrete		80		30	
RC framed buildings		54 ---- 97			
RC shells + arches				20s	
Hollow pot slabs				25	80
Flat slabs			00s	31	
Lightweight concrete				32	50
Precast concrete floors					50
Composite metal deck slabs					70s
Woodwool permanent shutters					69 90s
Waffle/coffered stabs					60s
Composite steel + concrete floors with shear keys					70s
Cast Iron (CI) + Wrought Iron (WI)					
CI columns	70s – 92			30s	
CI beams	96	65			
WI rods + flats	---------- 10s	80			
WI roof trusses		37			
WI built-up beams		40			
WI rolled sections		50s			
'Cast steel' columns		90s	10s		
Mild steel					
Plates + rods		80			60
Riveted sections		90s			60
Hot-rolled sections (up to 24" deep)		83			
Roof trusses		90s			
Steel-framed buildings		96			
Welds					55
Castellated beams				38	
High strength friction grip bolts (HSFG)					50s
Hollow sections + rolled sections up to 1100 mm deep					60
Stainless steel					
Bolts, straps, lintels, shelf angles, etc.			13		70s

Selection of materials

Material	Advantage	Disadvantage
Aluminium	Good strength to dead weight ratio for long spans Good corrosion resistance Often from recycled sources	Cannot be used where stiffness is critical Stiffness is a third of that of steel About two to three times the price of steel
Concrete	Design is tolerant to small, late alterations Integral fire protection Integral corrosion protection Provides thermal mass if left exposed Client pays as the site work progresses: 'pay as you pour'	Dead load limits scope Greater foundation costs Greater drawing office and detailing costs Only precasting can accelerate site work Difficult to post-strengthen elements Fair faced finish needs very skilled contractors and carefully designed joints
Masonry	Provides thermal mass The structure is also the cladding Can be decorative by using a varied selection of bricks Economical for low rise buildings Inherent sound, fire and thermal properties Easy repair and maintenance	Skilled site labour required Long construction period Less economical for high rise Large openings can be difficult Regular movement joints Uniform appearance can be difficult to achieve
Steelwork	Light construction reduces foundation costs Intolerant to late design changes Fast site programme Members can be strengthened easily Ideal for long spans and transfer structures	Design needs to be fixed early Needs applied insulation, fire protection and corrosion protection Skilled workforce required Early financial commitment required from client to order construction materials Long lead-ins Vibrations can govern design
Timber	Traditional/low-tech option Sustainable material Cheap and quick with simple connections Skilled labour not an absolute requirement Easily handled	Limited to 4–5 storeys maximum construction height Requires fire protection Not good for sound insulation Must be protected against insects and moisture Connections can carry relatively small loads

Note: See sustainability chapter for additional considerations.

Selection of floor construction

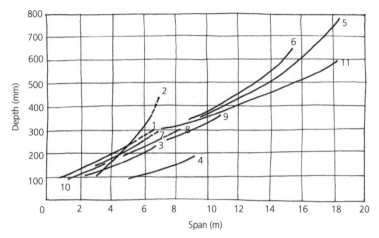

1. Timber joists at 400 c/c
2. Stressed skin ply panel
3. One way reinforced concrete slab
4. Precast prestressed concrete plank
5. Precast double tee beams
6. Coffered concrete slab
7. Beam + block floor
8. Reinforced concrete flat slab
9. Post tensioned flat slab
10. Concrete metal deck slab
11. Composite steel beams

Transportation

Although the transport of components is not usually the final responsibility of the design engineer, it is important to consider the limitations of the available modes of transport early in the design process using Department for Transport (DfT) information. Specific cargo handlers should be consulted for comments on sea and air transport, but a typical shipping container is 2.4 m wide, 2.4–2.9 m high and can be 6, 9, 12 or 13.7 m in length. Transportation of items which are likely to exceed 20 m by 4 m should be very carefully investigated. Private estates may have additional and more onerous limitations on deliveries and transportation. Typical road and rail limitations are listed below as the most common form of UK transport, but the relevant authorities should be contacted to confirm the requirements for specific projects.

Rail transportation

Railtrack can carry freight in shipping containers or on flat bed wagons. The maximum load on a four-axle flat wagon is 66 tonnes. The maximum height of a load is 3.9 m above the rails and wagons are generally between 1.4 and 1.8 m high. All special requirements should be discussed with Railtrack Freight or Network Rail.

Road transport

The four main elements of legislation that cover the statutory controls on length, width, marking, lighting and police notification for large loads are the Motor Vehicles (Construction and Use) Regulations 1986, the Motor Vehicles (Authorization of Special Types) General Order 1979, the Road Vehicles Lighting Regulations 1989 and the Road Traffic Act 1972. A summary of the requirements is set out below.

Height of load

There is no statutory limit governing the overall height of a load; however, where possible it should not exceed 4.95 m from the road surface to maximise use of the motorway and trunk road network (where the average truck flat bed is about 1.7 m). Local highway authorities should be contacted for guidance on proposed routes avoiding head height restrictions on minor roads for heights exceeding 3.0–3.6 m.

Weight of vehicle or load

Gross weight of vehicle, W (kg)	Notification requirements
$44{,}000 < W \leq 80{,}000$ or has any axle weight greater than permitted by the Construction & Use Regulations	2 days' clear notice with indemnity to the Highway and Bridge Authorities
$80{,}000 < W \leq 150{,}000$	2 days' clear notice to the police and 5 days' clear notice with indemnity to the Highway and Bridge Authorities
$W > 150{,}000$	DfT Special Order BE16 (allow 10 weeks for application processing) plus 5 days' clear notice to the police and 5 days' clear notice with indemnity to the Highway and Bridge Authorities

Width of load

Total loaded width[a], B (m)	Notification requirements
$B \leq 2.9$	No requirement to notify police
$2.9 < B \leq 5.0$	2 days' clear notice to police
$5.0 < B \leq 6.1$	DfT permission VR1 (allow 10 days for application processing) and 2 days' clear notice to police
$B > 6.1$	DfT Special Order BE16 (allow 8 weeks for application processing) and 5 days' clear notice to police and 5 days' clear notice with indemnity to Highway and Bridge Authorities

Note:
[a] A load may project over one or both sides by up to 0.305 m, but the overall width is still limited as above.

Loads with a width of over 2.9 m or with loads projecting more than 0.305 m on either side of the vehicle must be marked to comply with the requirements of the Road Vehicles Lighting Regulations 1989.

Length of load

Total loaded length, L (m)	Notification requirements
$L < 18.75$	No requirement to notify police
$18.75 \leq L < 27.4$	Rigid or articulated vehicles.[a] 2 days' clear notice to police
(Rigid vehicle) $L > 27.4$	DfT Special Order BE16 (allow 8 weeks for application processing) and 5 days' clear notice to police and 5 days' clear notice with indemnity to Highway and Bridge Authorities
(All other trailers) $L > 25.9$	All other trailer combinations carrying the load. 2 days' clear notice to police

Note:
[a] The length of the front of an articulated motor vehicle is excluded if the load does not project over the front of the motor vehicle.

Projection of overhanging loads

Overhang position	Overhang length, L (m)	Notification requirements
Rear	$L < 1.0$	No special requirement
	$1.0 < L < 2.0$	Load must be made clearly visible
	$2.0 < L < 3.05$	Standard end marker boards are required
	$L > 3.05$	Standard end marker boards are required plus police notification and an attendant is required
Front	$L < 1.83$	No special requirement
	$2.0 < L < 3.05$	Standard end marker boards are required plus the driver is required to be accompanied by an attendant
	$L > 3.05$	Standard end marker boards are required plus police notification and the driver is required to be accompanied by an attendant

Typical vehicle sizes and weights

Vehicle type		Weight, W (kg)	Length, L	Width, L (m)	Height, H (m)	Turning circle (m)
3.5 tonne van		3500	5.5	2.1	2.6	13.0
7.5 tonne van		7500	6.7	2.5	3.2	14.5
Single decker bus		16,260	11.6	2.5	3.0	20.0
Refuse truck		16,260	8.0	2.4	3.4	17.0
2-axle tipper		16,260	6.4	2.5	2.6	15.0

continued

(continued) Typical vehicle sizes and weights

Vehicle type		Weight, W (kg)	Length, L	Width, L (m)	Height, H (m)	Turning circle (m)
Van (up to 16.3 tonnes)		16,260	8.1	2.5	3.6	17.5
Skiploader		16,260	6.5	2.5	3.7	14.0
Fire engine		16,260	7.0	2.4	3.4	15.0
Bendy bus		17,500	18.0	2.6	3.1	23.0

Temporary/auxiliary works toolkit

Steel trench prop load capacities

Better known as 'Acrow' props, these adjustable props should conform to BS 4704 or BS EN 1065. Verticality of the loads greatly affects the prop capacity and fork heads can be used to eliminate eccentricities. Props exhibiting any of the following defects should not be used:

- A tube with a bend, crease or noticeable lack of straightness.
- A tube with more than superficial corrosion.
- A bent head or base plate.
- An incorrect or damaged pin.
- A pin not properly attached to the prop by the correct chain or wire.

Steel trench 'acrow' prop sizes and reference numbers to BS 4074

Prop size/reference[a]	Height range minimum (m)	Maximum (m)
0	1.07	1.82
1	1.75	3.12
2	1.98	3.35
3	2.59	3.96
4	3.20	4.87

Note:
[a] The props are normally identified by their length.

Steel trench prop load capacities

A prop will carry its maximum safe load when it is plumb and concentrically loaded as shown in the charts in BS 4074. A reduced safe working load should be used for concentric loading with an eccentricity, $e \leq 1.5°$ out of plumb as follows:

Capacity of props with $e \leq 1.5°$ (kN)									
Height (m)	< 2.75	3.00	3.25	3.50	3.75	4.00	4.25	4.50	4.75
Prop size 0, 1, 2 and 3	17	16	13	11	10	–	–	–	–
Prop size 4	–	–	17	14	11	10	9	8	7

Soldiers

Slim soldiers, also known as slimshors, can be used horizontally and vertically and have more load capacity than steel trench props. Lengths of 0.36, 0.54, 0.72, 0.9, 1.8, 2.7 or 3.6 m are available. Longer units can be made by joining smaller sections together. A connection between units with four M12 bolts will have a working moment capacity of about 12 kN m, which can be increased to 20 kN m if stiffeners are used.

Slimshor section properties

Area cm^2	I_{xx} cm^4	I_{yy} cm^4	Z_{xx} cm^3	Z_{yy} cm^3	r_x cm	r_y cm	$M_{max\,x}$ kN m	$M_{max\,y}$ kN m
19.64	1916	658	161	61	9.69	5.70	38	7.5

Slimshor compression capacity

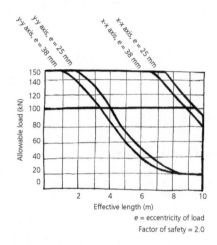

e = eccentricity of load

Factor of safety = 2.0

Slimshor moment capacity

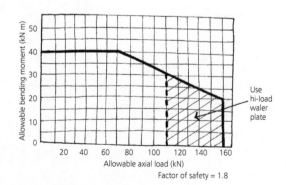

Factor of safety = 1.8

Source: RMD Kwikform. 2002.

Ladder beams

Used to span horizontally in scaffolding or platforms, ladder beams are made in 48.3ϕ 3.2 CHS, 305 mm deep, with rungs at 305 mm centres. All junctions are saddle welded. Ladder beams can be fully integrated with scaffold fittings. Bracing of both the top and bottom chords is required to prevent buckling. Standard lengths are 3.353 m (11'), 4.877 m (16') and 6.400 m (21').

Manufacturers should be contacted for loading information. However, if the tension chord is tied at 1.5 m centres and the compression chord is braced at 1.8 m centres the moment capacity for working loads is about 8.5 kN m. If the compression chord bracing is reduced to 1.5 m centres, the moment capacity will be increased to about 12.5 kN m. The maximum allowable shear is about 12 kN.

Unit beams

Unit beams are normally about 615 mm deep, are about 2.5 times stronger than ladder beams and are arranged in a similar way to a warren girder. Loads should only be applied at the node points. They may be used to span between scaffolding towers or as a framework for temporary buildings. As with ladder beams, bracing of both the top and bottom chords is required for preventing buckling, but diagonal plan bracing should be provided to the compression flange. Units can be joined together with M24 bolts to make longer length beams. Standard lengths are 1.8 m (6'), 2.7 m (9') and 3.6 m (12').

Manufacturers should be contacted for loading information. However, if the tension chord is tied at 3.6 m centres and the compression chord is braced at 2.4 m centres the moment capacity for working loads is about 13.5 kN m. If the compression bracing is reduced to 1.2 m centres, the moment capacity will be increased to about 27.5 kN m. The maximum allowable shear is about 14 kN.

4
Basic and Shortcut Tools for Structural Analysis

Load factors and limit states

There are two design considerations: strength and stiffness. The structure must be strong enough to resist the worst loading conditions without collapse and be stiff enough to resist normal working conditions without excessive deflection or deformation. Typically the requirements for strength and stiffness are split between the following 'limit states':

Ultimate limit state: Strength (including yielding, rupture, buckling and forming a mechanism), stability against overturning and swaying, fracture due to fatigue and brittle fracture.

Serviceability limit state: Deflection, vibration, wind-induced oscillation and durability.

A factor of safety against structural failure of 2.0–10.0 will be chosen depending on the materials and workmanship. There are three main methods of applying the factor of safety to structural design:

Allowable or permissible stress design: Where the ultimate strengths of the materials are divided by a factor of safety to provide design stresses for comparison with unfactored loads. Normally the design stresses stay within the elastic range. This method is not strictly applicable to plastic (e.g. steel) or semi-plastic (e.g. concrete or masonry) materials and there is one factor of safety to apply to all conditions of materials, loading and workmanship. This method has also been found to be unsafe in some conditions when considering the stability of structures in relation to overturning.

Load factor design: Where working loads are multiplied by a factor of safety for comparison with the ultimate strength of the materials. This method does not consider variability of the materials and as it deals with ultimate loads, it cannot be used to consider deflection and serviceability under working loads.

Ultimate loads or limit state design: The applied loads are multiplied by partial factors of safety and the ultimate strengths of the materials are divided by further partial factors of safety to cover variation in the materials and workmanship. This method allows a global factor of safety to be built up using the partial factors at the designer's discretion, by varying the amount of quality control which will be available for the materials and workmanship. The designer can therefore choose whether to analyse the structure with working loads in the elastic range, or in the plastic condition with ultimate loads. Serviceability checks are generally made with unfactored working loads, although Eurocodes have adopted different combinations of actions defending on the duration of load phenomena.

Geometric section properties

Section	A (mm²)	C_y (mm)	C_z (mm)	I_y (cm⁴)	I_z (cm⁴)	J (Approx.) cm⁴
	b^2	$\dfrac{b}{2}$	$\dfrac{b}{2}$	$\dfrac{b^4}{12}$	$\dfrac{b^4}{12}$	$\dfrac{5b^4}{36}$
	bd	$\dfrac{d}{2}$	$\dfrac{b}{2}$	$\dfrac{bd^3}{12}$	$\dfrac{db^3}{12}$	$\dfrac{d^3}{3}\left[\dfrac{b-0.63d}{}\times\left(1-\dfrac{d^4}{12b^4}\right)\right]$ for $d > b$
	$\dfrac{\pi d^2}{4}$	$\dfrac{d}{2}$	$\dfrac{d}{2}$	$\dfrac{\pi d^4}{64}$	$\dfrac{\pi d^4}{64}$	$\dfrac{\pi d^4}{32}$

Shape						
Triangle	$\dfrac{bd}{2}$	$\dfrac{d}{3}$	$\dfrac{b}{2}$	$\dfrac{bd^3}{36}$	$\dfrac{db^3}{48}$	$\dfrac{b^3 d^3}{(15b^2 + 20d^2)}$ for $\dfrac{2}{3} < \dfrac{b}{d} < \sqrt{3}$
Square tube	$b^2 - (b - 2t)^2$	$\dfrac{b}{2}$	$\dfrac{b}{2}$	$\dfrac{b^4 - (b - 2t)^4}{12}$	$\dfrac{b^4 - (b - 2t)^4}{12}$	$(b - t^3)t$
Pipe	$\dfrac{\pi(d^4 - (d - 2t)^2)}{4}$	$\dfrac{d}{2}$	$\dfrac{d}{2}$	$\dfrac{\pi(d^4 - (d - 2t)^4)}{64}$	$\dfrac{\pi(d^4 - (d - 2t)^4)}{64}$	$\dfrac{\pi(d - t)^3 t}{4}$
I-beam (H)	$2bt_1 + t_2(d - 2t_1)$	$\dfrac{d}{2}$	$\dfrac{b}{2}$	$\dfrac{bd^3 - (b - t_2)(d - 2t_1)^3}{12}$	$\dfrac{2t_2 b^3 - (d - 2t_1)t_2^3}{12}$	$\dfrac{2t_1^3 b + t_2^3 d}{3}$

continued

(continued) Geometric section properties

Section	A (mm²)	C_y (mm)	C_z (mm)	I_y (cm⁴)	I_z (cm⁴)	J (Approx.) cm⁴
	$bd - 2bt_1 - (d - 2t_1)t_2$	$\dfrac{d}{2}$	$\dfrac{b^2 t_1 + (1/2)(d - 2t_1)t_2^2}{A}$	$\dfrac{bd^3 - (b - t_2)(d - 2t_1)^3}{12}$	$\dfrac{2t_1 b^3 - (d - 2t_1)t_2^3}{12}$ $+2bt_1\left(\dfrac{b}{2} - C_z\right)^2$ $+t_2(d - 2t_1)\left(C_z - \dfrac{t_2}{2}\right)^2$	$\dfrac{t^3(d + 2b)}{3}$
	$bt_1 + (d - t_1)t_2$	$\dfrac{bt_1(d - (t_1/2)) + (1/2)(d - t_1)^2 t_2}{A}$	$\dfrac{b}{2}$	$\dfrac{bt_1^3 + t_2(d - t_1)^3}{12}$ $+ bt_1\left(d - C_y - \dfrac{t_1}{2}\right)^2$ $+ t_2(d - t_1)\left(C_y - \dfrac{d - t_1}{2}\right)^2$	$\dfrac{t_1 b^3 - (d - t_1)t_2^3}{12}$	$\dfrac{t_1^3 b + t_2^3 d}{3}$
	$dt_2 + (b - t_2)t_1$	$\dfrac{d^2 t_2 + (b - t_2)^2 t_1}{2A}$	$\dfrac{dt_2^2 + bt_1(b - t_2)}{2A}$	$\dfrac{t_2 d^3 + (b - t_2)t_1^3}{12}$ $+ dt_2\left(\dfrac{d}{2} - C_y\right)^2$ $+ (b - t_2)\left(C_y - \dfrac{t_1}{2}\right)^2$	$\dfrac{t_1 b^3 + (d - t_1)t_2^3}{12}$ $+ bt_1\left(\dfrac{b}{2} - C_z\right)^2$ $+ (d - t_1)\left(C_z - \dfrac{t_2}{2}\right)^2$	$\dfrac{t_1^3 b + t_2^3 d}{3}$

Elastic modulus, $Z = I/y$, plastic modulus, S = sum of first moments of area about central axis, the shape factor = S/Z.

Parallel axis theorem

$$y = \frac{\sum A_i y_i}{\sum A} \qquad\qquad I_{xx} = \sum A_i (y - y_i)^2 + \sum I_c$$

where
y_i neutral axis depth of element from datum
y depth of the whole section neutral axis from the datum
A_i area of element
A area of whole section
I_{xx} moment of inertia of the whole section about the x-x axis
I_c moment of inertia of element

Composite sections

A composite section made of two materials will have a strength and stiffness related to the properties of these materials. An equivalent stiffness must be calculated for a composite section. This can be done by using the ratios of the Young's moduli to 'transform' the area of the weaker material into an equivalent area of the stronger material:

$$\alpha_E = \frac{E_1}{E_2} \quad \text{For concrete to steel, } \alpha_E \cong 15$$

$$\text{For timber to steel, } \alpha_E \cong 35$$

Typically the depth of the material (about the axis of bending) should be kept constant and the breadth should be varied: $b_1 = \alpha_E b_2$. The section properties and stresses can then be calculated based on the transformed section in the stronger material.

Material properties

Homogeneous: Same elastic properties throughout. *Isotropic*: Same elastic properties in all directions. *Anisotropic*: Varying elastic properties in two different directions. *Orthotropic*: Varying elastic properties in three different directions. All properties are given for a temperature of 20°C.

Properties of selected metals

Material	Specific weight γ (kN/m³)	Modulus of elasticity E (kN/mm²)	Shear modulus of elasticity G (kN/mm²)	Poisson's ratio ν	Proof or yield stress f_y (N/mm²)	Ultimate strength[a] $f_{y\ ult}$ (N/mm²)	Elongation at failure (%)
Aluminium pure	27	69	25.5	0.34	<25	<58	30–60
Aluminium alloy	27.1	70	26.6	0.32	130–250		
Aluminium bronze	77	120	46	0.30			
AB1					170–200	500–590	18–40
AB2					250–360	640–700	13–20
Copper	89	96	38	0.35	60–325	220–385	
Brass	84.5	102	37.3	0.35	290–300	460–480	
Naval brass (soft-hard)	84	100	39	0.34	170–140	410–590	30–15
Bronze	82–86	96–120	36–44	0.34	82–690	200–830	5–60
Phosphor bronze	88	116	43	0.33	–	410	15
Mild steel	78.5	205	82.2	0.3	275–355	430–620	20–22
Stainless steel 304 L	78–80	180	76.9	0.3	210	520–720	45
Stainless steel duplex 2205	78–80	180	76.9	0.3	460	640–840	20
Grey cast iron	72	130	48	–	–	150/600c	–
Blackheart cast iron	73.5	170	68	0.26	180	260/780C	10–14
Wrought iron	74–78	190	75	0.3	210	340	35

Note:
[a]   Ultimate tensile strength labelled c which denotes ultimate compressive stress.

Properties of selected stone, ceramics and composites

Material	Specific weight γ (kN/m³)	Modulus of elasticity E (kN/mm²)	Poisson's ratio ν	Characteristic crushing strength f_{cu} (N/mm²)	Ultimate tensile strength $f_{y\ ult}$ (N/mm²)
Carbon fibre (7.5 mmθ)	20	415			1750
Concrete	24	17–31	0.1–0.2	10–70	
Concrete blocks	5–20			3–20	
Clay brick	22.5–28	5–30		10–90	
Fibre glass	15	10		150	100
Glass (soda)	24.8	74	0.22	1000	30–90
Glass (float)	25–25.6	70–74	0.2–0.27	1000	45 annealed 120–150 toughened
Granite	26	40–70	0.2–0.3	70–280	
Limestone	20–29	20–70	0.2–0.3	20–200	
Marble	26–29	50–100	0.2–0.3	50–180	

Properties of selected timber[a]

Material	Specific weight γ (kN/m³)	Modulus of elasticity E (kN/mm²)	Ultimate tensile strength $f_{y\ ult}$ (N/mm²)	Ultimate compressive strength $f_{cu\ ult}$ (N/mm²)	Ultimate shear strength $f_{y\ ult}$ (N/mm²)
Ash	6.5	10	60	48	10
Beech	7.4	10	60–110	27–54	8–14
Birch	7.1	15	85–90	67–74	13–18
English elm	5.6	11	40–54	17–32	8–11
Douglas fir	4.8–5.6	11–13	45–73	49–74	7.4–8.8
Mahogany	5.4	8	60	45	6
Oak	6.4–7.2	11–12	56–87	27–50	12–18
Scots pine	5.3	8–10	41.8	21–42	5.2–9.7
Poplar	4.5	7	40–43	20	4.8
Spruce	4.3	7–9	36–62	18–39	4.3–8
Sycamore	6.2	9–14	62–106	26–46	8.8–15

Note:
[a] These values are ultimate values. See the chapter on timber for softwood and hardwood design stresses.

Properties of selected polymers and plastics

Material	Specific weight γ (kN/m³)	Modulus of elasticity E (kN/mm²)	Ultimate tensile strength $f_{y\ ult}$ (N/mm²)	Elongation at failure (–%)
Polythene HD	9–14	0.55–1	20–37	20–100
PVC	13–14	2.4–3.0	40–60	200
PVC plasticised	13–14	0.01	150	
Polystyrene	10–13	3–3.3	35–68	3
Perspex	12	3.3	80–90	6
Acrylic	11.7–12	2.7–3.2	50–80	2–8
PTFE	21–22	0.3–0.6	20–35	100
Polycarbonate	12	2.2–4	50–60	100–130
Nylon	11.5	2–3.5	60–110	50
Rubber	9.1	0.002–0.1	7–20	100–800
Epoxy resin	16–20	20	68–200	4
Neoprene		0.7–20	3.5–24	
Carbon fibre		240	3500	1.4
Kevlar 49		125	3000	2.8
Polyester fabric + PVC coat	14	14	900	14–20

Coefficients of linear thermal expansion

Amount of linear thermal expansion, $l_{thermal} = \alpha(t_{max}-t_{min})/_{overall}$. A typical internal temperature range for the United Kingdom might be: $-5-35°C$. Externally this might be more like $-15-60°C$ to allow for frost, wind chill and direct solar gain.

Material	α $10^{-6}/°C$
Aluminium	24
Aluminium bronze	17
Brass	18–19
Bronze	20
Copper	17
Float glass	8–9
Cast iron	10–11
Wrought iron	12
Mild steel	12
Stainless steel – austenitic	18
Stainless steel – ferritic	10
Lead	29
Wood – parallel to the grain	3
Wood – perpendicular to the grain	30
Zinc	26
Stone – granite	8–10
Stone – limestone	3–4
Stone – marble	4–6
Stone – sandstone	7–12
Concrete – dense gravel aggregate	10–14
Concrete – limestone aggregate	7–8
Plaster	18–21
Clay bricks	5–8
Concrete blocks	6–12
Polycarbonate	60–70
GRP (polyester/glass fibre)	18–25
Rigid PVC	42–72
Nylon	80–100
Asphalt	30–80

Coefficients of friction

The frictional force, $F = \mu N$, where N is the force normal to the frictional plane.

Materials	Coefficient of sliding friction (μ)
Metal on metal	0.15–0.60
Metal on hardwood	0.20–0.60
Wood on wood	0.25–0.50
Rubber on paving	0.70–0.90
Nylon on steel	0.30–0.50
PTFE on steel	0.05–0.20
Metal on ice	0.02
Masonry on masonry	0.60–0.70
Masonry on earth	0.50
Earth on earth	0.25–1.00

Sign conventions

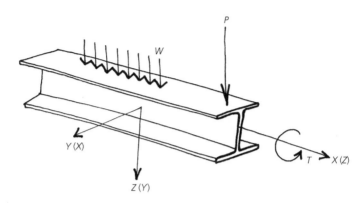

The axes convention shown is based on Eurocode standards. The axes notation in brackets relates to the convention traditionally used by the UK engineers prior to adoption of the Eurocodes as the 'default' codes of practice for structural design.

When members are cut into sections for the purpose of analysis, the cut section can be assumed to be held in equilibrium by the internal forces. A consistent sign convention like the following should be adopted:

A positive bending moment, M, results in tension in the bottom of the beam, causing the upper face of the beam to be concave. Therefore this is called a sagging moment. A negative bending moment is called a hogging moment. A tensile axial force, N, is normally taken as positive. Shear force, S, 'couples' are normally considered positive when they would result in a clockwise rotation of the cut element. A positive torque, T, is generally in an anti-clockwise direction.

Beam bending theory

$$\text{Moment: } M = -EI\frac{d^2y}{dx^2}$$

$$\text{Shear: } Q = \frac{dM}{dx}$$

Elastic constants

Hooke's law defines Young's modulus of elasticity, $E = \sigma/\varepsilon$. Young's modulus is an elastic constant to describe linear elastic behaviour, where a is stress and σ is the resulting strain. Hooke's law in shear defines the shear modulus of elasticity, $G = \tau/\gamma$, where τ is the shear stress and γ is the shear strain. Poisson's ratio, $v = \varepsilon_{lateral}/\varepsilon_{axiail}$, relates lateral strain over axial strain for homogeneous materials. The moduli of elasticity in bending and shear are related by: $G = E/(2(1 + v))$ for elastic isotropic materials. As v is normally from 0 to 1.5, G is normally between 0.3 and 0.5 of E.

Elastic bending relationships

$$\frac{M}{I} = \frac{\sigma}{y} = \frac{E}{R}$$

Where M is the applied moment, I is the section moment of inertia, σ is the fibre bending stress, y is the distance from the neutral axis to the fibre and R is the radius of curvature. The section modulus, $Z = I/y$ and the general equation can be simplified so that the applied bending stress, $\sigma = M/Z$.

Horizontal shear stress distribution

$$\tau = \frac{QAy}{bI}$$

Where τ is the horizontal shear stress, Q is the applied shear, b is the breadth of the section at the cut line being considered and A is the area of the segment above the cut line; I is the second moment of area of whole section and y is the distance from centre of area above the cut line to centroid of whole section.

Horizontal shear stresses have a parabolic distribution in a rectangular section. The average shear stress is about 60% of the peak shear (which tends to occur near the neutral axis).

Beam bending and deflection formulae

P is a point load in kN, W is the total load in kN on a span of length L and w is a distributed load in kN/m.

Loading condition	Reactions	Maximum moments	Maximum deflection
	$R_A = R_B = \dfrac{P}{2}$	$M_{midspan} = \dfrac{PL}{4}$	$\delta_{midspan} = \dfrac{PL^3}{48EI}$
	$R_A = \dfrac{Pb}{L}$ $R_B = \dfrac{Pa}{L}$	$M_C = \dfrac{Pab}{L}$	When $a > b$, $\delta_x = \dfrac{Pab(L + b)}{27EIL}\sqrt{3a(L + b)}$ at $x = \sqrt{\dfrac{a(L + b)}{3}}$ from A
	$R_A = R_B = P$ Thirdpoints: $a = \dfrac{L}{3}$	$M_C = Pa$ $M = \dfrac{PL}{3}$	$\delta_{midspan} = \dfrac{PL^3}{6EI}\left(\dfrac{3a}{4L} - \left(\dfrac{a}{L}\right)^3\right)$ $\delta = \dfrac{23PL^3}{648EI}$

continued

(continued) Beam bending and deflection formulae

Loading condition	Reactions	Maximum moments	Maximum deflection
	$R_A = R_B = \dfrac{W}{2}$	$M_{midspan} = \dfrac{WL}{8}$	$\delta_{midspan} = \dfrac{5WL^3}{384EI}$
	$R_A = \dfrac{W}{L}\left(\dfrac{b}{2} + c\right)$ $R_B = \dfrac{W}{L}\left(\dfrac{b}{2} + a\right)$	$M_{max} = \dfrac{W}{b}\left(\dfrac{x_1^2 - a^2}{2}\right)$ When $x_1 = a + \dfrac{R_A b}{W}$	$\delta_{max} = \dfrac{W}{384EI}(8L^3 - 4Lb^2 + b^3)$
	$R_A = R_B = \dfrac{W}{2}$	$M_{midspan} = \dfrac{WL}{6}$	$\delta_{midspan} = \dfrac{WL^3}{60EI}$

$$R_A = \frac{W}{3}$$

$$R_B = \frac{2W}{3}$$

$$M_x = \frac{Wx(L^2 - x^2)}{3L^2}$$

maximum at $x = 0.5774L$

$$\delta_{midspan} = \frac{5WL^3}{384EI}$$

$$\delta_{max} = \frac{0.01304WL^3}{EI}$$

when $x = 0.5193L$

$$R_A = \frac{W}{2a}(a^2 - b^2)$$

$$R_B = \frac{W}{2a}(a + b)^2$$

$$M_B = \frac{wb^2}{2}$$

$$M_C = \frac{w(a + b)^2(a + b)^2}{8a^2}$$

maximum at $x = \frac{a}{2}\left(1 - \frac{b^2}{a^2}\right)$

$$\delta_c = \frac{W}{24EI}\left(x^4 - 2ax^3 + \frac{2b^2}{a}x^3 + a^3x - 2ab^2x\right)$$

$$\delta_{free\,tip} = \frac{wb}{24EI}(3b^3 + 4ab^2 - a^3)$$

$$R_A = \frac{Pb^2(L + 2a)}{L^3}$$

$$R_B = \frac{Pb^2(L + 2b)}{L^3}$$

$$M_A = \frac{-Pab^2}{L^2}$$

$$M_B = \frac{-Pba^2}{L^2}$$

$$M_C = \frac{2Pa^2b^2}{L^3}$$

$$\delta_{max} = \frac{2Pa^3b^2}{3EI(L + 2a)^2}$$

when $x = \frac{L^2}{3L - 2a}$

$$\delta_c = \frac{Pa^3b^3}{3EIL}$$

continued

(continued) Beam bending and deflection formulae

Loading condition	Reactions	Maximum moments	Maximum deflection
	$R_A = R_B = \dfrac{W}{2}$	$M_A = M_B = \dfrac{-WL}{12}$ $M_C = \dfrac{WL}{24}$	$\delta_{midspan} = \dfrac{WL^3}{384EI}$
	$R_A = P$	$M_A = -Pa$	$\delta_{tip} = \dfrac{Pa^3}{3EI}\left(1 + \dfrac{3b}{2a}\right)$ $\delta_B = \dfrac{Pa^3}{3EI}$
	$R_A = W$	$M_A = \dfrac{-WL}{2}$	$\delta_B = \dfrac{WL^3}{8EI}$

	$R_A = P - R_B$ $$R_B = \frac{Pa^2(2L + b)}{2L^3}$$	$M_A = \dfrac{-Pb(L^2 - b^2)}{2L^2}$ $$M_C = \frac{Pb}{2}\left(2 - \frac{3b}{L} + \frac{b^3}{L^3}\right)$$	$\delta_C = \dfrac{Pa^3b^2}{12EIl^3}(4L - a)$
	$R_A = \dfrac{5W}{8}$ $$R_B = \frac{3W}{8}$$	$M_A = \dfrac{-WL}{8}$ $$M_D = \frac{9WL}{128} \text{ at } 0.62L \text{ from } A$$	$\delta_{max} = \dfrac{WL^3}{185EI} \text{ at } 0.58L \text{ from } A$
	$R_A = \dfrac{-3Pb}{2a}$ $$R_B = \frac{P}{a}\left(a + \frac{3b}{2}\right)$$	$M_A = \dfrac{Pb}{2}$ $$M_B = -Pb = -2M_A$$	$\delta_D = \dfrac{Pb^2}{4EI}\left(a + \frac{4b}{3}\right)$ $$\delta_D = \frac{-Pa^2b}{27EI} \text{ at } 0.66a$$

Clapeyron's equations of three moments

Clapeyron's equations can be applied to continuous beams with three supports, or to two-span sections of longer continuous beams.

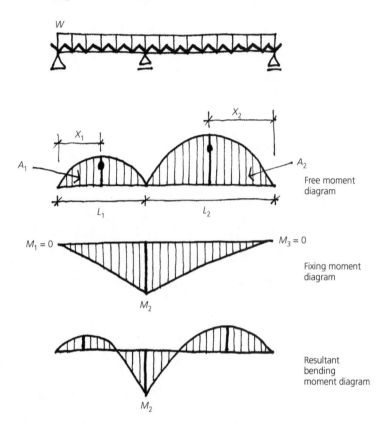

Free moment diagram

Fixing moment diagram

Resultant bending moment diagram

General equation

$$\frac{M_1 L_1}{I_1} + 2M_2\left(\frac{L_1}{I_1} + \frac{L_2}{I_2}\right) + \frac{M_3 L_2}{I_2} = 6\left(\frac{A_1 \bar{x}_1}{L_1 I_1} + \frac{A_2 \bar{x}_2}{L_2 I_2}\right) + 6E\left(\frac{y_2}{L_1} + \frac{(y_2 - y_3)}{L_2}\right)$$

where
- M bending moment
- A area of 'free' moment diagram if the span is treated as simply supported
- L span length
- $\bar{x}$ distance from support to centre of area of the moment diagram
- I second moment of area
- y deflections at supports due to loading

Usual case: level supports and uniform moment of area

$$M_1L_1 + 2M_2(L_1 + L_2) + M_3L_2 = 6\left(\frac{A_1\bar{x}_1}{L_1} + \frac{A_2\bar{x}_2}{L_2}\right)$$

where $y_1 = y_2 = y_3 = 0$ and $l_1 = l_2 = l_3$

M_1 and M_3 are either: unknown for fixed supports, zero for simple supports or known cantilever end moments, and can be substituted into the equation to provide a value for M_2.

Free ends: $M_1 = M_3 = 0$

$$2M_2(L_1 + L_2) = 3\left(\frac{A_1\bar{x}_1}{L_1} + \frac{A_2\bar{x}_2}{L_2}\right)$$

which can be further simplified to

$$M_2 = \frac{W(L_1^3 + L_2^3)}{8(L_1 + L_2)}$$

where $w = $ kN/m

Multiple spans

The general case can be applied to groups of three supports for longer continuous beams with n spans. This will produce $(n - 2)$ simultaneous equations which can be resolved to calculate the $(n - 2)$ unknown bending moments.

Continuous beam bending formulae

Moments of inertia are constant and all spans of L metres are equal. W is the total load on one span (in kN) from either distributed or point loads.

Reaction = coefficient $\times W$

Moment = coefficient $\times W \times L$

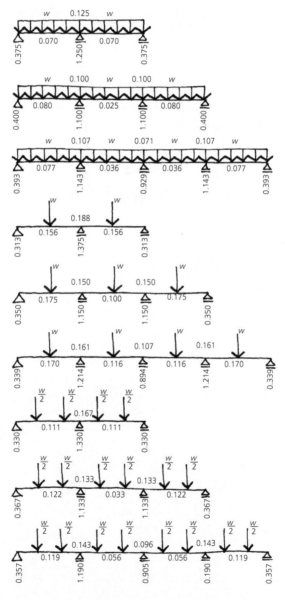

Struts

The critical buckling load of a strut is the applied axial load which will cause the strut to buckle elastically with a sideways movement. There are two main methods of determining this load: Euler's theory which is simple to use or the Perry–Robertson theory which forms the basis of the buckling tables in BS 449.

Effective length

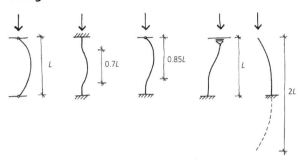

Euler

Euler critical buckling load:

$$P_E = \frac{\pi^2 EI}{L_e^2}$$

Euler critical buckling stress:

$$\sigma_e = \frac{P_E}{A} = \frac{\pi^2 E r_y^2}{L_e^2} = E \left(\frac{\pi r_y}{L_e} \right)^2$$

r_y and I are both for the weaker axis or for the direction of the effective length L_e under consideration.

Perry–Robertson

Perry–Robertson buckling load:

$$P_{PR} = A\left[\frac{(\sigma_c + \sigma_e(K + 1))}{2} - \sqrt{\left(\frac{\sigma_c + \sigma_e(K + 1)}{2}\right)^2}\right] - \sigma_c\sigma_e$$

where

$$K = 0.3\left(\frac{L_e}{100r_y}\right)^2$$

σ_e is the Euler critical stress as calculated above and crc is the yield stress in compression.

Pinned strut with uniformly distributed lateral load

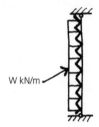

W kN/m

Maximum bending moment, where $\alpha = \sqrt{P/EI}$

$$M_{max} = \frac{wEI}{P}\left(\sec\left(\frac{\alpha L}{2}\right) - 1\right)$$

Maximum compressive stress

$$\sigma_{C\,max} = \frac{My}{I} + \frac{P}{A}$$

Maximum deflection

$$\delta_{max} = \frac{-M}{P} + \frac{wL^2}{8P}$$

Rigid frames under lateral loads

Rigid or plane frames are generally statically indeterminate. A simplified method of analysis can be used to estimate the effects of lateral load on a rigid frame based on its deflected shape, and assumptions about the load, shared between the columns. The method assumes notional pinned joints at expected points of contraflexure, so that the equilibrium system of forces can be established by statics. The vertical frame reactions as a result of the lateral loads are calculated by taking moments about the centre of the frame.

The following methods deal with lateral loads on frames, but similar assumptions can be made for vertical analysis (such as treating beams as simply supported) so that horizontal and vertical moments and forces can be superimposed for use in the sizing and design of members.

Rigid frame with infinitely stiff beam

It is assumed that the stiffness of the top beam will spread the lateral load evenly between the columns. From the expected deflected shape, it can be reasonably assumed that each column will carry the same load. Once the column reactions have been assumed, the moments at the head of the columns can be calculated by multiplying the column height by its horizontal base reaction. As the beam is assumed to be infinitely stiff, it is assumed that the columns do not transfer any moment into the beam.

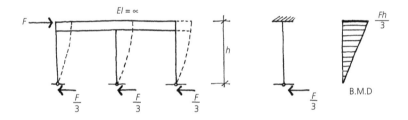

B.M.D

Rigid frame constant stiffness (*EI*)

As the top beam is not considerably stiffer than the columns, it will tend to flex and cause a point of contraflexure at mid span, putting extra load on the internal columns. It can be assumed that the internal columns will take twice the load (and therefore moment) of the external columns. As before, the moments at the head of columns can be calculated by multiplying the column height by its horizontal base reaction. The maximum moment in the beam due to horizontal loading of the frame is assumed to equal the moment at the head of the external columns.

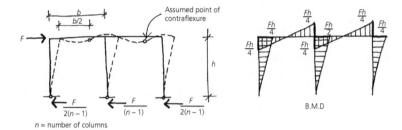

n = number of columns

Multi-storey frame with beams and columns of constant stiffness

For a multi-storey frame, points of contraflexure can be assumed at mid-points on beams and columns. Each storey is considered in turn as a separate subframe between the column points of contraflexure. The lateral shears are applied to the subframe columns in the same distribution as the single storey frames, so that internal columns carry twice the load of the external columns. As analysis progresses down the building, the total lateral shear applied to the top of each subframe should be the sum of the lateral loads applied above the notional point of contraflexure. The shears are combined with the lateral load applied to the subframe, to calculate lateral shear reactions at the bottom of each subframe. The frame moments in the columns due to the applied lateral loads increase towards the bottom of the frame. The maximum moments in the beam due to lateral loading of the frame are assumed to equal the difference between the moments at the external columns.

(continued) Multi-storey frame

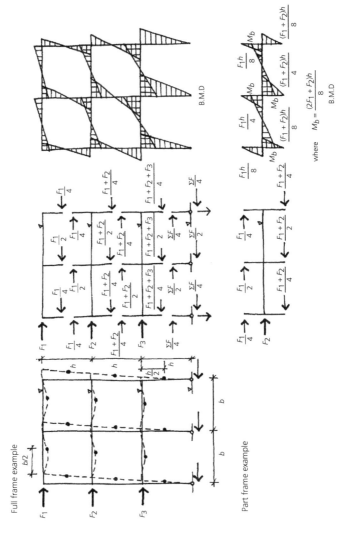

Full frame example

Part frame example

B.M.D

where $M_b = \dfrac{(2F_1 + F_2)h}{8}$

B.M.D

Plates

Johansen's yield line theory studies the ultimate capacity of plates. Deflection needs to be considered in a separate elastic analysis. Yield line analysis is a powerful tool which should not be applied without background reading and a sound understanding of the theory.

The designer must try to predict a series of failure crack patterns for yield line analysis by numerical or virtual work methods. Crack patterns relate to the expected deflected shape of the slab at collapse. For any one slab problem, there may be many potential modes of collapse which are geometrically and statically possible. All of these patterns should be investigated separately. It is possible for the designer to inadvertently omit the worst case pattern for analysis which could mean that the resulting slab might be designed with insufficient strength. Crack patterns can cover whole slabs, wide areas of slabs or local areas, such as failure at column positions or concentrated loads. Yield line moments are typically calculated as kNm/m width of slab.

The theory is most easily applied to isotropic plates which have the same material properties in both directions. An isotropic concrete slab is of constant thickness and has the same reinforcement in both directions. The reinforcement should be detailed to suit the assumptions of yield line analysis. Anisotropic slabs can be analysed if the 'degree of anisotropy' is selected before a standard analysis. As in the analysis of laterally loaded masonry panels, the results of the analysis can be transformed on completion to allow for the anisotropy.

The simplest case to consider is the isotropic rectangular slab:

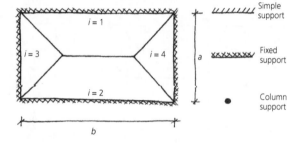

The designer must decide on the amount of fixity, i, at each support position. Generally $i = 0$ for simple support and $i = 1$ for fixed or encastre supports. The amount of fixity determines how much moment is distributed to the top of the slab m', where $m' = im$ and m is the moment in the bottom of the slab.

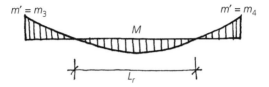

Fixed supports reduce the sagging moments, m, in the bottom of the slab. The distance between the points of zero moment can be considered as a 'reduced effective length', L_r:

$$L_r = \frac{2L}{\left[\sqrt{(1 + i_1)} + \sqrt{(1 + i_2)}\right]} \quad \text{and} \quad \text{where } L_r < L : M = \frac{wL_r^2}{8}$$

For the rectangular slab L_r should be calculated for both directions:

$$a_r = \frac{2a}{\left[\sqrt{(1 + i_1)} + \sqrt{(1 + i_2)}\right]} \quad b_r = \frac{2b}{\left[\sqrt{(1 + i_3)} + \sqrt{(1 + i_4)}\right]}$$

So that the design moment is:

$$M = \frac{wa_r b_r}{8(1 + (a_r/b_r) + (b_r/a_r))}$$

For fixity on all sides of a square slab (where $a = b = L$) the design moment, $M = wL^2/24$ kNm/m. For a point load or column support, $M = P/2\pi$ kNm/m.

Selected yield line solutions

These patterns are some examples of those which need to be considered for a given slab. Yield line analysis must be done on many different crack patterns to try to establish the worst case failure moment. Both top and bottom steel should be considered by examining different failure patterns with sagging and hogging crack patterns.

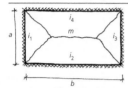

$$a_r = \frac{2a}{\sqrt{1 + i_2} + \sqrt{1 + i_4}} \qquad b_r = \frac{2b}{\sqrt{1 + i_1} + \sqrt{1 + i_3}}$$

$$m = \frac{wa_rb_r}{8(1 + (a_r/b_r) + (b_r/a_r))}$$

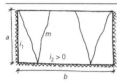

$$b_r = \frac{2b}{\sqrt{1 + i_1} + \sqrt{1 + i_3}} \qquad m = \frac{wa_rb_r}{3 + 12(a/b_r) + 2i_2(1 + (b_r/a))}$$

$$a \le b_r \qquad\qquad \text{Top steel also required.}$$

$$b_r = \frac{b}{\sqrt{1 + i_1}} \qquad m = \frac{wa_rb_r}{(3/2) + 3(a/b_r) + i_2(1 + 2(b_r/a))}$$

For opposite case, exchange a and b, l_1 and l_2.

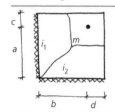

$$F = 0.6\frac{(a + c)i_1 + (b + d)i_2}{a + b + c + d} \qquad m_0 = \frac{3wab}{8(2 + (a/b) + (b/a))}$$

$$m = \frac{m_0 - 0.15wcd}{1 + F} \qquad a \le b \le 2a$$

$$m' = \frac{w}{6}(c^2 + d^2)$$

Bottom steel required for main span.

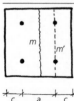

if $c = 0.35a$,

$$m = m' = \frac{wa^2}{16}$$

continued

(continued)

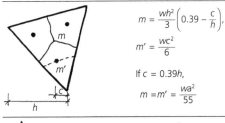

$$m = \frac{wh^2}{3}\left(0.39 - \frac{c}{h}\right),$$

$$m' = \frac{wc^2}{6}$$

If $c = 0.39h$,

$$m = m' = \frac{wa^2}{55}$$

$$m = \frac{wh^2}{2}\left(0.33 - \left(\frac{c}{2h}\right)^{\frac{2}{3}}\right)$$

| Point or concentrated load | $m = \frac{P}{2\pi}$ |

All moments are in kNm/m.

Torsion

Elastic torsion of circular sections:

$$\frac{T}{J} = \frac{\tau}{r} = \frac{G\phi}{L}$$

where T is the applied torque, J is the polar moment of inertia, τ is the torsional shear stress, r is the radius, ϕ is the angle of twist, G is the shear modulus of elasticity and L is the length of the member.

The shear strain, γ, is constant over the length of the member and $r\phi$ gives the displacement of any point along the member. Materials yield under torsion in a similar way to bending. The material has a stress/strain curve with gradient G up to a limiting shear stress, beyond which the gradient is zero.

The torsional stiffness of a member relies on the ability of the shear stresses to flow in a loop within the section shape which will greatly affect the polar moment of area, which is calculated from the relationship $J = \int r^2 dA$. This can be simplified in some closed loop cases to $J = I_{zz} = I_{xx} + I_{yy}$.

Therefore for a solid circular section, $J = \pi d^4/32$ for a solid square bar, $J = 5d^4/36$ and for thin-walled circular tubes, $J = \pi t(d^4_{outer} - r^4_{inner})/32$ or $J = 2\pi r^3 t$ and the shear stress, $\tau = (T/2At)$ where t is the wall thickness and A is the area contained within the tube.

Thin-walled sections of arbitrary and open cross-sections have less torsional stiffness than solid sections or tubular thin-walled sections which allow shear to flow around the section. In thin-walled sections the shear flow is only able to develop within the thickness of the walls and so the torsional stiffness comes from the sum of the stiffness of its parts: $J = (1/3)\int_{section} t^3 ds$. This can be simplified to $J \cong \Sigma(bt^3/3)$, where $\tau = Tt/J$.

J for thick open sections are beyond the scope of this book, and must be calculated empirically for the particular dimensions of a section. For non-square and circular shapes, the effect of the warping of cross-sections must be considered in addition to the elastic effects set out above.

Taut wires, cables and chains

The cables are assumed to have significant self-weight. Without any externally applied loads, the horizontal component of the tension in the cable is constant and the maximum tension will occur where the vertical component of the tension reaches a maximum. The following equations are relevant where there are small deflections relative to the cable length.

L	Span length
h	Cable sag
A	Area of cable
ΔLs	Cable elongation due to axial stress
C	Length of cable curve
E	Modulus of elasticity of the cable
$S = h/L$	Sag ratio
W	Applied load per unit length
V	Vertical reaction
Y	Equation for the deflected shape
D	Height of elevation
H	Horizontal reaction
T_{max}	Maximum tension in cable
x	Distance along cable

Uniformly loaded cables with horizontal chords

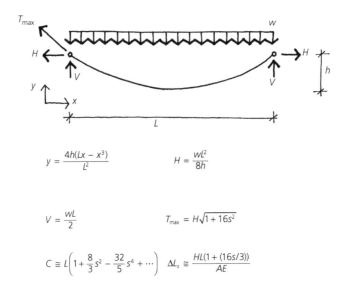

$$y = \frac{4h(Lx - x^3)}{L^2} \qquad H = \frac{wL^2}{8h}$$

$$V = \frac{wL}{2} \qquad T_{max} = H\sqrt{1 + 16s^2}$$

$$C \cong L\left(1 + \frac{8}{3}s^2 - \frac{32}{5}s^4 + \cdots\right) \quad \Delta L_s \cong \frac{HL(1 + (16s/3))}{AE}$$

Uniformly loaded cables with inclined chords

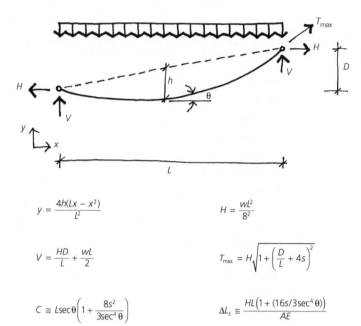

$$y = \frac{4h(Lx - x^2)}{L^2}$$

$$H = \frac{wL^2}{8^2}$$

$$V = \frac{HD}{L} + \frac{wL}{2}$$

$$T_{max} = H\sqrt{1 + \left(\frac{D}{L} + 4s\right)^2}$$

$$C \cong L\sec\theta\left(1 + \frac{8s^2}{3\sec^4\theta}\right)$$

$$\Delta L_s \cong \frac{HL\left(1 + (16s/3\sec^4\theta)\right)}{AE}$$

Vibration

When using long spans and lightweight construction, vibration can become an important issue. Human sensitivity to vibration has been shown to depend on frequency, amplitude and damping. Vibrations can detract from the use of the structure or can compromise the structural strength and stability.

Vibrations can be caused by wind, plant, people, adjacent building works, traffic, earthquakes or wave action. Structures will respond differently depending on their mass and stiffness. Damping is the name given to the ability of the structure to dissipate the energy of the vibrations – usually by friction in structural and non-structural components. While there are many sources of advice on vibrations in structures, assessment is not straightforward. In simple cases, structures should be designed so that their natural frequency is greater than 4.5 Hz to help prevent the structure from being dynamically excitable. Special cases may require tighter limits.

A simplified method of calculating the natural frequency of a structure (f in Hz) is related to the static dead load deflection of the structure, where g is the acceleration due to gravity, δ is the static dead load deflection estimated by normal elastic theory, k is the stiffness ($k = EI/L$), m is a UDL and M is a concentrated load. E is the modulus of elasticity, I is the moment of inertia and L is the length of the member. This method can be used to check the results of more complex analysis.

Member		Estimate of natural frequency, α_f
General rule for structures with concentrated mass		$f = \dfrac{1}{2\pi}\sqrt{\dfrac{g}{\delta}}$
General rule for most structures with distributed loads		$f = \dfrac{1}{2\pi}\sqrt{\dfrac{k}{m}}$
Simplified rule for most structures		$f = \dfrac{18}{\sqrt{\delta}}$
Simply supported, mass concentrated in the centre		$f = \dfrac{1}{2\pi}\sqrt{\dfrac{48EI}{ML^3}}$
Simply supported, sagging, mass and stiffness distributed		$f = \dfrac{\pi}{2}\sqrt{\dfrac{EI}{mL^4}}$
Simply supported, contraflexure, mass and stiffness distributed		$f = 2\pi\sqrt{\dfrac{EI}{ML^4}}$
Cantilever, mass concentrated at the end		$f = \dfrac{1}{2\pi}\sqrt{\dfrac{3EI}{ML^3}}$
Cantilever, mass and stiffness distributed		$f = 0.56\sqrt{\dfrac{EI}{mL^4}}$
Fixed ends, mass and stiffness distributed		$f = 3.56\sqrt{\dfrac{EI}{mL^4}}$

Note: For normal floors with span/depth ratios of 25 or less, there are unlikely to be any vibration problems. Typically problems are encountered with steel and lightweight floors with spans over about 8 m.

Source: Bolton, A. 1978.

5
Eurocodes

Eurocode background

The Eurocodes are the most technically advanced suite of design codes in the world and probably the most wide-ranging codification of structural design ever experienced. Eurocodes are intended to be a CEN (Comité Europeen de Normalisation) standard for buildings as a 'passport' for designers between EU countries. From March 2010 they are to be used as the principal codes to demonstrate compliance with the UK Building Regulations and the requirements of other public authorities. There are 10 codes, with each being split into a number of subsections (not listed here):

BS EN 1990	Basis of design
1991	Actions
1992	Concrete
1993	Structural steelwork
1994	Composite steel and concrete
1995	Timber
1996	Masonry
1997	Foundations
1998	Seismic
1999	Aluminium

These codes cover the principles, rules and recommended values for ultimate limit state structural design. However, safety, economy and durability have been 'derogated' to the member states, and for each Eurocode there is a supplementary National Annexe which must be consulted in parallel. Allowing for regional variations, the UK National Annexes have adjusted Eurocode values and methods to bring these into line with standard UK practice. These changes are termed Nationally Determined Parameters (NDPs). The British Standards Institute cannot change any text in the core CEN document nor publish a version of the CEN document with the values of the NDPs pasted in, but the National Annex may include reference to non-contradictory complimentary information (NCCI), such as other non-Eurocode national standards or guidance documents, for example information from Access Steel.

An additional complication is that while British Standards included design guidance alongside materials and workmanship requirements, the Eurocodes rely on separate standards for materials and workmanship. Also several topics typically covered in significant detail in British Standards have been omitted, so that the provision of supplementary technical information is essential in some areas. It is not considered safe or appropriate to adopt a 'pick and mix' design due to fundamental differences in philosophy, safety factors, underlying assumptions and construction detailing issues. Structures should be designed to either Eurocodes, or British Standards, not to a mixture of both.

To help keep the documents concise, Eurocodes are divided into 'Principles' and 'Application Rules'. Principles are normally general statements, definitions, models and requirements for which no alternative is permitted. This is indicated by the use of 'P' after the clause number. Applications rules give users design methods which follow the requirements described by the Principles. While it is possible to use alternative rules to demonstrate compliance with Eurocode Principles and produce a safe and economic design, any design which has not used the Application Rules cannot be called 'Eurocode Compliant' or obtain the equivalent of a CEN. This may be relevant to those who are contracted to provide Eurocode building designs under EU Public Procurement Directive rules or at their client's request.

Preparation of the Eurocode suite has been at glacial pace, with the whole process having taken over 25 years to get to the March 2010 implementation date. From this date the British Standards were 'Withdrawn' (which is not the same as outlawed) and the Eurocodes became current. In addition to design codes, engineers rely on a large amount of support documentation (e.g. textbooks, software, proprietary products) and it will take some time for this information to become fully available. In 2014, this process is not yet complete, with coverage being patchy when designs other than the most straightforward cases are considered.

Engineers need to refer to several Eurocode documents in place of one British Standard: the Eurocode, the National Annexe, the NCCI and the British Standard for materials and workmanship. British Standard users will have to familiarise themselves with a fairly daunting array of changes to both design vocabulary and notation. Some of these changes appear to stem from philosophical changes in the design approach and standardisation of symbols across Europe, but other (less obvious ones) appear to stem the need for precision when translating between other European languages. Some of the most common alterations are as follows:

Eurocode	English translation
Action	Load phenomena (including temperature, inertia, etc.)
Effect of actions	Stress/strain/deflection/rotation
Permanent action	Dead load
Variable action	Live load
Execution	Construction process
Auxiliary construction works	Temporary works
' + '	To be combined with
,	Decimal point

The cumulative effect of the changes in language, notation, technical approach and layout is quite considerable, particularly for those familiar with British Standards. It is therefore likely that it will take quite some time for Eurocodes to be fully adopted by established designers – especially when the new codes produce similar answers as British Standards and limit the use of quick/simple hand calculations. While many argue that the Eurocodes offer better nuance and scope for international work, others cite commercial reasons and limited benefits for the slow adoption. Whatever view is taken, the example of BS449 for steelwork design (still in use by practising engineers nearly 20 years after the introduction of BS5950) is probably a fair indicator of how speedy the full Eurocode transition will be.

For safety and simplicity, this book will use the UK convention of '.' for the decimal point throughout instead of using the Eurocode convention of ','. This is in line with the convention used in about half of the UK National Annexes.

Eurocodes and European public procurement rules

Under the Public Procurement rules, European Member States must accept designs to the Eurocodes for all public works contracts. Contracts and legal documents (such as the Building Regulations) are required to reference the appropriate European Standards. The same set of rules requires all publicly funded contracts above a certain value to be put out to competitive tender in the Official Journal of the European Union. Projects without public funding will be able to choose their own technical standards and procurement route in conjunction with their professional advisors.

Eurocode-supporters state that as British Standards will no longer be maintained, there will be little option but to use the Eurocodes. While this may ring true in the long term, Eurocode-sceptics highlight that it is not currently mandatory to design to the Eurocodes and that a designer may propose alternative standards for public contracts if they can demonstrate that the alternative is of 'technical equivalence' to a Eurocode solution.

Since 2004 the public procurement rules were set out in paragraph 29 of 2004/18/EC DIRECTIVE OF THE EUROPEAN PARLIAMENT AND OF THE COUNCIL on the coordination of procedures for the award of public works contracts, public supply contracts and public service contracts. An extract is:

> *'The technical specifications drawn up by public purchasers need to allow public procurement to be opened up to competition. To this end, it must be possible to submit tenders which reflect the diversity of technical solutions. Accordingly, it must be possible to draw up the technical specifications in terms of functional performance and requirements, and, where reference is made to the European standard or, in the absence thereof, to the national standard, tenders based on equivalent arrangements must be considered by contracting authorities. To demonstrate equivalence, tenderers should be permitted to use any form of evidence. Contracting authorities must be able to provide a reason for any decision that equivalence does not exist in a given case....'*

However, it is expected that this will be amended at some point in 2013 by paragraph 27 of 2011/0438 (COD) Proposal for a DIRECTIVE OF THE EUROPEAN PARLIAMENT AND OF THE COUNCIL on public procurement as follows:

> *'The technical specifications drawn up by public purchasers need to allow public procurement to be opened up to competition. To that end, it must be possible to submit tenders that reflect the diversity of technical solutions so as to obtain a sufficient level of competition. Consequently, technical specifications should be drafted in such a way to avoid artificially narrowing down competition through requirements that favour a specific economic operator by mirroring key characteristics of the supplies, services or works habitually offered by that economic operator. Drawing up the technical specifications in terms of functional and performance requirements generally allows this objective to be achieved in the best way possible and favours innovation. Where reference is made to a European standard or, in the absence thereof, to a national standard, tenders based on equivalent arrangements must be considered by contracting authorities. To demonstrate equivalence, tenderers can be required to provide third party verified evidence; however, other appropriate means of proof such as a technical dossier of the manufacturer should also be allowed where the economic operator concerned has no access to such certificates or test reports, or no possibility of obtaining them within the relevant time limits.'*

It is interesting to note that the 2004 Directive allows *'any form of evidence'* to demonstrate equivalence, while the amendment appears to be more restrictive: *'third party verified evidence'* or *'other appropriate means of proof'*.

For the foreseeable future it will probably be difficult to show that a design which complies with British Standard codes of practice does not also comply with the relevant Eurocodes, but this situation will probably change over time. It will presumably be down to lawyers, rather than engineers, to test this out in future. In the meantime, uptake of the Eurocodes for work in the private sector is on a voluntary basis.

Limit state philosophy and Eurocode partial safety factors

One of the main differences between the Eurocodes and British Standards is the use of different partial safety factors and the option to refine/reduce load using combination load factors when different load cases are combined. Load combinations are expressed by Equation 6.10 as follows:

$$\Sigma_{j \geq 1} \, \gamma_{G,j} \, G_{k,j} \; ' + ' \; \gamma_p \, P \; ' + ' \; \gamma_{Q,1} \, Q_{k,1} \; ' + ' \; \Sigma_{i>1} \, \gamma_{Q,i} \, \psi_{0,i} \, Q_{k,i}$$

In English this expression means:

Permanent ' + ' Prestress ' + ' Leading variable ' + ' Accompanying variable
Actions   Actions   Actions

Design values of actions – partial load factor combinations

Limit state	Description	Permanent actions Unfavourable (γ_{sup})	Permanent actions Favourable (γ_{inf})	Variable actions[c,d] Leading (γ)	Accompanying ($\gamma\psi$) Main	Accompanying ($\gamma\psi$) Others
ULS 'Set A'	EQU – Static equilibrium, overall stability	$1.1G_k$	$0.9G_k$	$1.5Q_k$		$1.5\psi_0 Q_k$
	EQU – Stability requiring resistance of structural members	$1.35G_k$	$1.15G_k$	$1.5Q_k$		$1.5\psi_0 Q_k$
ULS 'Set B'	STR (or less commonly GEO) – Resistance of structural member not involving geotechnical action	$1.35G_k$[b] Or worst of: $1.35G_k$ $1.25G_k$[a]	$1.0G_k$ $1.0G_k$ $1.0G_k$	$1.5Q_k$ $1.5Q_k$	$1.5\psi_0 Q_k$	$1.5\psi_0 Q_k$ $1.5\psi_0 Q_k$
ULS 'Set C'	GEO (or less commonly STR) – resistance of the ground (or structure)	$1.0G_k$	$1.0G_k$	$1.3Q_k$		$1.3\psi_0 Q_k$
ULS	Accidental	$1.1G_k$	$1.0G_k$	A_d	$\psi_1 Q_k$	$\psi_2 Q_k$
SLS	**Characteristic** – 'irreversible' limit state. Full values a rare occurrence: Function of structure, damage to structural or non-structural elements, etc.	$1.0G_k$	$1.0G_k$	$1.0Q_k$		$1.0\psi_0 Q_k$
	Frequent – 'reversible' limit state. Full values a frequent occurrence: User comfort, operation of machinery, water ponding, etc.	$1.0G_k$	$1.0G_k$		$1.0\psi_1 Q_k$	$1.0\psi_2 Q_k$
	Quasi-permanent – 'long-term effects' and appearance. Full values likely to be present most of the time: visual effects, etc.	$1.0G_k$	$1.0G_k$		$1.0\psi_2 Q_k$	$1.0\psi_2 Q_k$

Notes:

[a] $\xi = 0.925$, therefore $\gamma = 0.925 \times 1.35$.

[b] In most cases it is possible to use the first Equation 6.10 only, but more detailed analysis can be carried out using the two subsequent variations on this (6.10a and 6.10b).

[c] Where variable actions are favourable, take $Q_k = 0$.

[d] Variable actions include live, snow, wind and temperature and execution effects.

[e] Although BS EN 1990 does not define which variable actions should be considered as separate within load combinations, it would seem sensible to consider all normal gravity imposed loads as a single action, with wind load, earthquakes, etc. considered as separate 'actions'.

Source: BS EN 1990:2002 + A1:2005, Table A.12 (A, B, C).
NA to BS EN 1990: 2002 + A1:2005, 2.2.

Combination factors

Action		Combination value (ψ_0)	Frequent value (ψ_1)	Quasi-permanent value (ψ_2)
Imposed loads				
Category	A: domestic, residential areas	0.7	0.5	0.3
	B: office areas	0.7	0.5	0.3
	C: congregation areas	0.7	0.7	0.6
	D: shopping areas	0.7	0.7	0.6
	E: storage areas	1.0	0.9	0.8
	F: traffic area, vehicle weight ≤ 30 kN	0.7	0.7	0.6
	G: traffic area, 30 kN < vehicle weight ≤ 160 kN	0.7	0.5	0.3
	H: roofs	0.7	0	0
Snow loads				
– sites at altitude > 1000 m above sea level		0.7	0.5	0.2
– sites at altitude ≤ 1000 m above sea level		0.5	0.2	0
Wind loads		0.5	0.2	0
Temperature (non-fire)		0.6	0.5	0

Source: NA to BS EN 1990:2002 + A1:2005, Table NA.A1.1.

Summary of combined factors: Persistent situations

| Limit state | Permanent actions | | Variable actions | | | | | |
| | Unfavourable (γ_{sup}) | Favourable (γ_{inf}) | Imposed | | Wind/snow | | Temperature induced | |
			Leading (γ)	Accompanying ($\gamma\psi_0$)	Leading (γ)	Accompanying ($\gamma\psi_0$)	Leading (γ)	Accompanying ($\gamma\psi_0$)
Static equilibrium	1.10	0.90	1.50	1.05[c]	1.50	0.75	1.50	0.90
Structural strength	1.35[d]	1.00	1.50	1.05[c]	1.50	0.75	1.50	0.90
Geotechnical strength	1.00	1.00	1.30	0.91[c]	1.30	0.65	1.30	0.78
Serviceability	1.00	1.00	1.00	0.70	1.00	0.50	1.00	0.60

Notes:

[a] Table assumes $\psi_0 = 0.7$ for imposed loads, $\psi_0 = 0.5$ for wind & snow and $\psi_0 = 0.6$ for temperature.

[b] For favourable (restoring) variable actions, $\gamma = 0$.

[c] For storage areas $\psi_0 = 1.0$, therefore $\gamma\psi_0 = 1.5$ for static equilibrium and structural strength or $\gamma\psi_0 = 1.3$ for strength of the ground or $\gamma\psi_0 = 1.0$ for serviceability instead of the tabulated values above.

[d] $\gamma = 1.35$ assumes the use of only Equation 6.10. Equation 6.10b (with 6.10a considered separately) would use $\xi = 0.925$, therefore $\gamma = 0.925 \times 1.35 = 1.25$.

Comparison of BS and Eurocode partial load factors

For one variable action (imposed or wind):

British Standards: $1.4G_k + (1.4 \text{ or } 1.6)Q_k$
Eurocodes: $1.35G_k + 1.5Q_k$

For one variable action (imposed or wind) with restoring permanent action:

British Standards: $1.0G_k + (1.4 \text{ or } 1.6)Q_k$ for steel and concrete
                      $0.9G_k + (1.4 \text{ or } 1.6)Q_k$ for masonry
Eurocodes: $0.9G_k + 1.5Q_k$ for equilibrium
                      $1.0G_k + 1.5Q_k$ for structural strength

For two or more variable actions (imposed and wind):

British Standards: $1.2G_k + 1.2Q_{kl} + 1.2Q_{ka}$
Eurocodes: $1.35G_k + 1.5Q_{kl} + 0.75Q_{ka}$

Design life

Design working life category	Indicative design working life (years)	Examples
1	10	Temporary structures
2	10 to 30	Replaceable structural parts, e.g. gantry girders, bearings
3	15 to 25	Agricultural and similar structures
4	50	Building structures and other common structures
5	120	Monumental building structures, bridges, and other civil engineering structures

Note: Structures or parts of structures that can be dismantled with a view to being re-used should not be considered as temporary.

Source: NA to BS EN 1990:2002 + A1:2005, Table NA.2.1.

6
Actions on Structures

Permanent actions

Material	Description	Thickness/ quantity of unit	Unit load (kN/m²)	Bulk density (kN/m³)
Aggregate				16
Aluminium	Cast alloy			27
	Longstrip roofing	0.8 mm	0.022	
Aluminium bronze				76
Asphalt	Roofing – 2 layers	25 mm	0.58	
	Paving			21
Ballast	see Gravel			
Balsa wood				1
Bituminous felt roofing	3 layers and vapour barrier		0.11	
Bitumen				11–13
Blockboard	Sheet	18 mm	0.11	
Blockwork	Lightweight – dense			10–20
Books	On shelves			7
	Bulk			8–11
Brass	Cast			85
Brickwork	Blue			24
	Engineering			22
	Fletton			18
	London stock			19
	Sand lime			21
Bronze	Cast			83
Cast stone				23
Cement				15
Concrete	Aerated			10
	Lightweight aggregate			18
	Normal reinforced			24
Coal	Loose lump			9
Chalk				22
Chipboard				7

continued

(continued)

Material	Description	Thickness/ quantity of unit	Unit load (kN/m²)	Bulk density (kN/m³)
Chippings	Flat roof finish	1 year	0.05	
Clay	Undisturbed			19
Copper	Cast			87
	Longstrip roofing	0.6 mm	0.05	
Cork	Granulated			1
Double decker bus	see Vehicles			
Elephants	Adult group		3.2	
Felt	Roofing underlay		0.015	6
Glass	Crushed/refuse			16
Glass wool	Quilt	100 mm	0.01	
Gold				194
Gravel	Loose			16
	Undisturbed			21
Hardboard				6–8
Hardcore				19
Hardwood	Greenheart			10
	Oak			8
	Iroko, teak			7
	Mahogany			6
Hollow clay pot slabs	Including ribs and mortar but excluding topping	300 mm thick overall		12
		100 mm thick overall		15
Iron	Cast			72
	Wrought			77
Ivory				19
Lead	Cast			114
	Sheet	1.8 mm	0.21	
	Sheet	3.2 mm	0.36	
Lime	Hydrate (bags)			6
	Lump/quick (powder)			10
	Mortar (putty)			18
Linoleum	Sheet	3.2 mm	0.05	
Macadam	Paving			21
Magnesium	Alloys			18
MDF	Sheet			8
Mercury				136
Mortar				12–23
Mud				17–20
Partitions	Plastered brick	102 + 2 × 13 mm	2.6	21
	Medium dense plastered block	100 + 2 × 13 mm	2.0	16

(continued)

Material	Description	Thickness/ quantity of unit	Unit load (kN/m²)	Bulk density (kN/m³)
	Plaster board on timber stud	100 + 2 × 13 mm	0.35	3
Patent glazing	Single glazing		0.26–0.3	25
	Double glazed		0.52	
Pavement lights	Cast iron or concrete framed	100 mm	1.5	
Perspex	Corrugated sheets		0.05	12
Plaster	Lightweight	13 mm	0.11	9
	Wallboard and skim coat	13 mm	0.12	
	Lath and plaster	19 mm	0.25	
	Traditional lime plaster			20
	Traditional lath + plaster ceiling		0.5	
Plywood	Sheet			7
Polystyrene	Expanded sheet			0.2
Potatoes				7
Precast concrete planks	Beam and block plus 50 mm topping	150–225 mm	1.8–33	
	Hollowcore plank	150 mm	2.4	
	Hollowcore plank	200 mm	2.7	
	Solid plank and 50 mm topping	75–300 mm	3.7–7.4	
Quarry tiles	Including mortar bedding	12.5 mm	0.32	
Roofing tiles	Clay – plain		0.77	19
	Clay pantile		0.42	19
	Concrete		0.51	24
	Slate		0.30	28
Sand	Dry, loose			16
	Wet, compact			19
Screed	Sand/cement			22
Shingle	Coarse, graded, dry			19
Slate	Slab			28
Snow	Fresh		minimum 0.6	1
	Wet, compacted		minimum 0.6	3
Softwood	Battens for slating and tiling	C16–C24	0.03	3.7–4.2
	25 mm tongued and grooved boards on 100 × 50 timber joists at 400 c/c	Glulam	0.23	3.7–4.4
	25 mm tongued and grooved boards on 250 × 50 timber joists at 400 c/c		0.33	
Soils	Loose sand and gravels			16
	Dense sand and gravels			22
	Soft/firm clays and silts			18
	Stiff clays and silts			21

continued

(continued)

Material	Description	Thickness/ quantity of unit	Unit load (kN/m²)	Bulk density (kN/m³)
Stainless steel roofing	Longstrip	0.4 mm	0.05	78
Steel	Mild			78
Stone				
Granite	Cornish (Cornwall)			26
	Rublislaw (Grampian)			25
Limestone	Bath (Wiltshire)			21
	Mansfield (Nottinghamshire)			22
	Portland (Dorset)			22
Marble	Italian			27
Sandstone	Bramley Fell (West Yorkshire)			22
	Forest of Dean (Gloucestershire)			24
	Darley Dale or Kerridge (Derbyshire)			23–25
Slate	Welsh			28
Terracotta				18
Terrazzo	Paving	20 mm	0.43	22
Thatch	Including battens	305 mm	0.45	
Timber	See Hardwood or Softwood			
Vehicles	London bus	73.6 kN		
	New Mini Cooper	11.4 kN		
	Rolls Royce	28.0 kN		
	Volvo estate	17.8 kN		
Water	Fresh			10
	Salt			10–12
Woodwool slabs				6
Zinc	Cast			72
	Longstrip roofing	0.8 mm	0.06	

Variable actions: Imposed floor loads

The following table from BS EN 1991–1 gives the normally accepted minimum floor loadings. Clients can consider sensible reductions in these loads if it will not compromise future flexibility. A survey by Arup found that office loadings very rarely even exceed the values quoted for domestic properties.

The gross live load on columns and/or foundations from sections A to D in the table can be reduced in relation to the number of floors or floor area carried to BSEN 1991–1. Live load reductions are not permitted for loads from storage and/or plant, or where exact live loadings have been calculated.

Type of activity/ occupancy for part of the building or structure	Examples of specific use	Uniformity distributed load q_k (kN/m²)	Concen- trated load Q_k (kN)
A. Domestic and residential activities (Also see category C)	A1 All usages within self-contained dwelling units (a unit occupied by a single family or a modular student accommodation unit with a secure door comprising not more than six single bedrooms and an internal corridor). Communal areas (including kitchens) in blocks of flats with limited use (see Note a). For communal areas in other blocks of flats, see A5, A6 and C3	1.5	2.0
	A2 Bedrooms and dormitories except those in self-contained single family dwelling units and in hotels and motels	1.5	2.0
	A3 Bedrooms in hotels and motels; hospital wards; toilet areas	2.0	2.0
	A4 Billiard/snooker rooms	2.0	2.7
	Balconies		
	A5 Single family dwelling units and communal areas in blocks of flats with limited use (see Note a)	2.5	2.0
	A6 Hostels, guest houses, residential clubs and communal areas in blocks of flats except those covered by Note a	Same as rooms to which they give access but with a minimum of 3.0	2.0 (concen- trated at the outer edge)
	A7 Hotels and motels	Same as rooms to which they give access but with a minimum of 4.0	2.0 (concen- trated at the outer edge)
B. Offices areas	B1 General use other than B2	2.5	2.7
	B2 At or below ground floor level	3.0	2.7
C. Areas where people may congregate (with the exception of areas defined under category A, B and D)	**C1 Areas with tables**		
	C11 Public, institutional and communal dining rooms and lounges, cafes and restaurants (see Note b)	2.0	3.0
	C12 Reading rooms with no book storage	2.5	4.0
	C13 Classrooms	3.0	3.0

continued

C2 Areas with fixed seats

(continued)

Type of activity/ occupancy for part of the building or structure	Examples of specific use	Uniformity distributed load q_k (kN/m²)	Concen-trated load Q_k (kN)
	C21 Assembly areas with fixed seating (see Note c)	4.0	3.6
	C22 Places of worship	3.0	2.7
	C3 Areas without obstacles for moving people		
	C31 Corridors, hallways, aisles in institutional type buildings not subjected to crowds or wheeled vehicles, hostels, guest houses, residential clubs, and communal areas in blocks of flats not covered by Note a	3.0	4.5
	C32 Stairs, landings in institutional type buildings not subjected to crowds or wheeled vehicles, hostels, guest houses, residential clubs, and communal areas in blocks of flats not covered by Note a	3.0	4.0
	C33 Corridors, hallways, aisles in all buildings not covered by C31 and C32, including hotels and motels and in institutional type buildings subjected to crowds	4.0	4.5
	C34 Corridors, hallways, aisles in all buildings not covered by C31 and C32, including hotels and motels and in institutional type buildings subjected to wheeled vehicles, including trolleys	5.0	4.5
	C35 Stairs, landings in all buildings not covered by C31 and C32, including hotels and motels, and institutional buildings subjected to crowds	4.0	4.0
	C36 Walkways – Light duty (access suitable for one person, walkway width approx. 600 mm)	3.0	2.0
	C37 Walkways – General duty (regular two-way pedestrian traffic)	5.0	3.6
	C38 Walkways – Heavy duty (high-density pedestrian traffic including escape routes)	7.5	4.5
	C39 Museum floors and art galleries for exhibition purposes	4.0	4.5
C. Areas where people may congregate (with the exception of areas defined under category A, B and D)	**C4 Areas with possible physical activities**		
	C41 Dance halls and studios, gymnasia, stages (see Note e)	5.0	3.6
	C42 Drill halls and drill rooms (see Note e)	5.0	7.0
	C5 Areas susceptible to large crowds		
	C51 Assembly areas without fixed seating, concert halls, bars and places of worship (see Note d and Note e)	5.0	3.6
	C52 Stages in public assembly areas (see Note e)	7.5	4.5

(continued)

Type of activity/ occupancy for part of the building or structure	Examples of specific use	Uniformity distributed load q_k (kN/m²)	Concentrated load Q_k (kN)
D. Shopping areas	**D1** Areas in general retail shops	4.0	3.6
	D2 Areas in department stores		
E1 Areas susceptible to accumulation of goods including access areas	**E11** General areas for static equipment not specified elsewhere (institutional and public buildings)	2.0	1.8
	E12 Reading rooms with book storage, e.g. libraries	4.0	4.5
	E13 General storage other than those specified (but designers are encouraged to determine project specific values)	2.4 per metre of storage height	7.0
	E14 File rooms, filing and storage space (offices)	5.0	4.5
	E15 Stack rooms (books)	2.4 per metre of storage height (6.5 kN/m² min)	7.0
	E16 Paper storage for printing plants and stationery stores	4.0 per metre of storage height	9.0
	E17 Dense mobile stacking (books) on mobile trolleys, in public and institutional buildings	4.8 per metre of storage height (9.6 kN/m² min)	7.0
	E18 Dense mobile stacking (books) on mobile trucks, in warehouses	4.8 per metre of storage height (15 kN/m² min)	7.0
	E19 Cold storage	5.0 per metre of storage height (15 kN/m² min)	9.0
E2 Industrial use (See PD 6688-1-1 for imposed loads on floors for areas of industrial use. The following loads are minimum suggested)	**E201** Communal kitchens except those covered by occupancy class A	3.0	4.5
	E202 Operating theatres, x-ray rooms, utility rooms	2.0	4.5
	E203 Work rooms (light industrial) without storage	2.5	1.8
	E204 Kitchens, laundries, laboratories	3.0	4.5
	E205 Rooms with mainframe computers or similar equipment	3.5	4.5
	E206 Machinery halls, circulation spaces therein	4.0	4.5
	E207 Cinematographic projection rooms	5.0	To be determined for specific use
	E208 Factories, workshops and similar buildings (general industrial)	5.0	4.5
	E209 Foundries	20.0	To be determined for specific use

continued

(continued)

Type of activity/ occupancy for part of the building or structure	Examples of specific use	Uniformity distributed load q_k (kN/m²)	Concen-trated load Q_k (kN)
	E210 Catwalks	–	1.0 at 1 m centres
	E211 Fly galleries (i.e. access structures used in theatres to hang scenery, curtains, etc.)	4.5 kN/m	–
	E212 Ladders	–	1.5 rung load
	E213 Plant rooms, boiler rooms, fan rooms, etc., including weight of machinery	7.5	4.5
F. Traffic areas	**F** Traffic areas vehicle weight ≤30 kN	2.5	9.0
G. Traffic areas	**G** Traffic areas 30 kN < vehicle weight ≤160 kN	5.0	To be determined for specific use
H. Roofs	**H** Roofs not accessible except for normal maintenance and repair	0.6	0.9
I. Roofs	**I** Roofs accessible – occupancy according to categories A to G	As category	As category
K. Roofs	**K** Roofs accessible for special services, such as helicopter landing areas	To be determined for specific use	To be determined for specific use

Notes:

[a] Communal areas in blocks of flats with limited are blocks of flats not more than three storeys in height and with not more than four self-contained dwelling units per floor accessible from one staircase.

[b] Where the areas described by C11 might be subjected to loads due to physical activities or overcrowding, e.g. a hotel dining room used as a dance floor, imposed loads should be based on occupancy C4 or C5 as appropriate. Reference should also be made to Note e.

[c] Fixed seating is seating where its removal and use of the space for other purposes is improbable.

[d] For grandstands and stadia, reference should be made to the requirements of the appropriate certifying authority.

[e] For structures that might be susceptible to resonance effects, reference should be made to NA2.1.

[f] Museums, galleries and exhibition spaces often need more capacity than this, sometimes up to 10 kN/m².

Source: NA to BS EN 1991-1-1:2002, Tables NA.2, NA.3, NA.4, NA.5, NA.6, NA.7
BS EN 1991-1-1:2002, Table 6.9. PD 6688 1-1:2011, Table 1.

Typical unit floor and roof loadings

Permanent partitions shown on the floor plans should be considered as dead load. Flexible partitions which may be movable should be allowed for in imposed loads, with a minimum of 1 kN/m²

Timber floor	Live loading: domestic/office	1.5/2.5 (kN/m²)
	(Office partitions)	(1.0)
	Timber boards/plywood	0.15
	Timber joists	0.2
	Ceiling and services	0.15
	Domestic/office totals	**Total 2.0/4.0 kN/m²**
Timber flat roof	Snow and access	0.75 kN/m²
	Asphalt waterproofing	0.45
	Timber joists and insulation	0.2
	Ceiling and services	0.15
		Total 1.55 kN/m²
Timber-pitched roof	Snow	0.6 kN/m²
	Slates, timber battens and felt	0.55
	Timber rafters and insulation	0.2
	Ceiling and services	0.15
		Total 1.5 kN/m²
Internal RC slab	Live loading: office/classroom/ corridors, etc.	2.5/3.0/4.0 kN/m²
	Partitions	1.0 (minimum)
	50 screed/75 screed/raised floor	1.2/1.8/0.4
	Solid reinforced concrete slab	24t
	Ceiling and services	0.15
		Total – kN/m²
External RC slab	Live loading: snow and access/office/ bar	0.75/2.5/5.0 kN/m²
	Slabs/paving	0.95
	Asphalt waterproofing and insulation	0.45
	50 screed	1.2
	Solid reinforced concrete slab	24t
	Ceiling and services	0.15
		Total – kN/m²
Metal deck roofing	Live loading: snow/wind uplift	0.6/–1.0 kN/m²
	Outer covering, insulation and metal deck liner	0.3
	Purlins – 150 deep at 1.5 m c/c	0.1
	Services	0.1
	Primary steelwork: light beams/ trusses	0.5–0.8/0.7–2.4
		Total – kN/m²

Typical 'all up' loads

For very rough assessments of the loads on foundations, 'all up' loads can be useful. The best way is to 'weigh' the particular building, but very general values for small-scale buildings might be:

Steel clad steel frame	5–10 kN/m^2
Masonry clad timber frame	10–15 kN/m^2
Masonry walls and precast concrete floor slabs	15–20 kN/m^2
Masonry clad steel frame	15–20 kN/m^2
Masonry clad concrete frame	20–25 kN/m^2

Variable actions: Wind loading

BS EN 1991-1-4 give methods for determining the peak gust wind loads on buildings and their components. Structures susceptible to dynamic excitation fall outside the scope of the guidelines. While the methods in theory allow for a very site-specific study of the many design parameters, it does mean that grossly conservative values can be calculated if the 'path of least resistance' is taken through the code.

As wind loading relates to the size and shape of the building, the size and spacing of surrounding structures, altitude and proximity to the sea or open stretches of country, it is difficult to summarise the design methods. The following dynamic pressure values have been calculated (on a whole building basis) for an imaginary building 20×20 m^2 in plan and 10 m tall (with equal exposure conditions and no dominant openings) in different UK locations. The following values should not be taken as prescriptive, but as an idea of an 'end result' to aim for. Taller structures will tend to have slightly higher values and where buildings are close together, funnelling should be considered. Small buildings located near the bases of significantly taller buildings are unlikely to be sheltered as the wind speed around the bases of tall buildings tends to increase.

Typical values of peak velocity pressure, q_p (z) in kN/m^2

Building location	Maximum q for prevailing south westerly wind (kN/m^2)	Minimum q for north easterly wind (kN/m^2)	Arithmetic mean q (kN/m^2)
Scottish mountaintop	3.40	1.81	2.60
Dover clifftop	1.69	0.90	1.30
Rural Scotland	1.14	0.61	0.87
Coastal Scottish town	1.07	0.57	0.82
City of London high rise	1.03	0.55	0.80
Rural northern England	1.02	0.54	0.78
Suburban southeast England	0.53	0.28	0.45
Urban northern Ireland	0.88	0.56	0.72
Rural northern Ireland	0.83	0.54	0.74
Rural upland Wales	1.37	0.72	1.05
Coastal Welsh town	0.94	0.40	0.67
Conservative quick scheme value for most UK buildings	−	−	1.20

Note: These are typical values which do not account for specific exposure or topographical conditions.

Variable actions: Horizontal barrier loads

Minimum horizontal imposed loads for barriers, parapets and balustrades, etc.

Type of occupancy for part of the building or structure	Examples of specific use	Line load (kN/m)	UDL on infill (kN/m²)	Point load on infill (kN)
A	i. All areas within or serving exclusively one dwelling including stairs, landings, etc. but excluding external balconies and edges of roofs (see vii)	0.36	0.5	0.25
B and C1	ii. Residential area not covered by i	0.74	1.0	0.5
	iii. Areas not susceptible to overcrowding in office and institutional buildings, reading rooms and classrooms including stairs.	0.74	1.0	0.5
	iv. Restaurants and cafes	1.50	1.5	1.5
C2, C3, C4 and D	v. Areas having fixed seating within 530 mm of the barrier, balustrade or parapet	1.50	1.5	1.5
	vi. Stairs, landings, balustrades, corridors and ramps	0.74	1.0	0.5
	vii. External balconies and edges of roofs. Footways within building cartilage and adjacent to basement/sunken areas	0.74	1.0	0.5
	viii. All retail areas	1.50	1.5	1.5
C5	ix. Footways or pavements less than 3 m wide adjacent to sunken areas	1.50	1.5	1.5
	x. Theatres, cinemas, discotheques, bars, auditoria, shopping malls, assembly areas, studios. Footways or pavements greater than 3 m wide adjacent to sunken areas	3.0	1.5	1.5
	xi. Grandstands and stadia	See requirements of the appropriate certifying authority		
E	xii. Industrial; and storage buildings except as given by (xiii) and (xiv)	0.74	1.0	0.5
	xiii. Light pedestrian traffic routes in industrial and storage buildings except designated escape routes	0.36	0.5	0.25
	xiv. Light access stairs and gangways not more than 600 mm wide	0.22	N/A	N/A
F and G	xv. Pedestrian areas in car parks including stairs, landings, ramps, edges or internal floors, footways, edges of roofs	1.5	1.5	1.5
	xvi. Horizontal loads imposed by vehicles	See Annex B BS EN 1991-1-1:2002		

Note: Line load applied at 1.1 m above datum (a finished level on which people may stand: on a floor, roof, wide parapet, balcony, ramp, pitch line of stairs (nosings), etc.).

Source: NA to BS EN 1991-1-1:2002, Table NA.8. PD6688-1-1: 2011, Table 2.

Minimum barrier heights

Use	Position	Height (mm)
Single family dwelling	(a) Barriers in front of a window	800
	(b) Stairs, landings, ramps, edges of internal floors	900
	(c) External balconies, including Juliette balconies, edges of roofs	1100
All other uses	(d) Barrier in front of a window	800
	(e) Stairs	900
	(f) Balconies and stands, etc. having fixed	800[a]
	(g) Balconies and stands, etc. having fixed seating within 530 mm of the barrier, provided the sum of the barrier width and the barrier height is greater than 975 mm	750[a]
	(h) Other positions including Juliette balconies	1100

Note:
[a] Site lines should be considered as set out in clause 6.8 of BS 6180.

Source: BS 6180:2011.

Eurocode combinations of actions for serviceability limit states

Eurocodes BS EN 1990 to BS 1996 provide generic guidance on serviceability and deflection loadcases and criteria. In the absence of specific requirements, the following combinations of actions are suggested by the Institution of Structural Engineers:

- The *'characteristic combination'* should be used to verify function and damage to structural and non-structural elements (e.g. partition walls). If the functioning of the structure or damage to finishes or non-structural members (e.g. partition walls, claddings) is being considered, account should be taken of those effects of permanent and variable actions that occur after the execution of the member or finish concerned.
- The *'frequent combination'* should be used to verify comfort criteria for users, criteria associated with the use of machinery, or criteria for the avoidance of water ponding, etc.
- The *'quasi-permanent combination'* should be used for criteria associated with the appearance of the structure. Long-term deformations due to shrinkage, relaxation or creep should be considered where relevant, and calculated by using the effects of the permanent actions and quasi-permanent values of the variable actions.

Typical deflection limits

Typical vertical deflection limits

Total deflection	span/250
Live load deflection	span/360
Domestic timber floor joists	span/333 or 14 mm
Deflection of brittle elements	span/500
Cantilevers	span/180
Vertical deflection of crane girders due to static vertical wheel loads from overhead travelling cranes	span/600
Purlins and sheeting rails (dead load only)	span/200
Purlins and sheeting rails (worst case dead, imposed, wind and snow)	span/100

Typical horizontal deflection limits

Sway of single storey columns	height/300
Sway of each storey of multi-storey column	height/300
Sway of columns with movement sensitive cladding	height/500
Sway of portal frame columns (no cranes)	to suit cladding
Sway of portal frame columns (supporting crane runways)	to suit crane runway
Horizontal deflection of crane girders (calculated on the top flange properties alone) due to horizontal crane loads	span/500
Curtain wall mullions and transoms (single glazed)	span/175
Curtain wall mullions and transoms (double glazed)	span/250

Stability, robustness and disproportionate collapse

While stability, solidity and robustness are essential qualities in a structure able to cope with uncertainties, they are abstract qualities which are not easy to codify. Although British Standards provided a considerable amount of advice for designers, the Eurocodes are much less prescriptive. As a result, the BSi have published a series of 'PD' documents, which provide the 'missing' information as NCCI for the UK use.

Stability

Stability of a structure must be achieved in two orthogonal directions. Circular structures should also be checked for rotational failure. The positions of movement and/or acoustic joints should be considered and each part of the structure should be designed to be independently stable and robust. Lateral loads can be transferred across the structure and/or down to the foundations by using any of the following methods:

- Cross bracing which carries the lateral forces as axial load in diagonal members.
- Diaphragm action of floors or walls which carry the forces by panel/plate/shear action.
- Frame action with 'fixed' connections between members and 'pinned' connections at the supports.
- Vertical cantilever columns with 'fixed' connections at the foundations.
- Buttressing with diaphragm, chevron or fin walls.

Stability members must be located on the plan so that their shear centre is aligned with the resultant of the overturning forces. If an eccentricity cannot be avoided, the stability members should be designed to resist the resulting torsion across the plan.

Robustness, accidental damage and disproportionate collapse

Structures are expected to be safe under normal service conditions, as well as in 'accidental' situations (e.g. Vehicle impact, fire, explosion, human error, minor alterations, etc.) to BS EN 1991-1-7. The Building Regulations have specific requirements to limit disproportionate collapse of buildings. These requirements can have a significant effect on the design and detailing of structures, so robustness should be considered early in the design process.

The 2004 revision of the Building Regulations made substantial alterations to part A3. The requirements of the regulations and various material codes of practice are summarised in the following table. Eurocode and NCCI references have been included. Where possible, Eurocode requirements for notional horizontal loads can generally be found in clauses titled 'geometric imperfections'.

BS EN 1990 clause 2.1 sets on the principle that potential damage shall be avoided or limited by appropriate choice of one of more of the following:

- Avoiding, eliminating or reducing hazards to which the structure can be subjected
- Selecting a structural form which has low sensitivity to the hazards considered
- Selecting a structural form and design that can survive adequately the accidental removal of an individual member or a limited part of the structure, or the occurrence of acceptable localised damage
- Avoiding as far as possible structure systems that can collapse without warning
- Tying the structural members together

Disproportionate collapse requirements with British Standard clause references

Building class	Building type and occupancy	Building regulations/ general Eurocode requirements	British standard material references and summary outline guidance				
			BS EN 1992 – concrete	BS EN 1993 – steel	BS EN 1995 – timber	BS EN 1996 – masonry
1	Houses not exceeding 4 storeys. Agricultural buildings. Buildings into which people rarely go (provided no part of the building is closer than 1.5 times its height to occupied areas)	Basic requirements as BS EN 1990 to 1999. (BS EN 1990 CL 2.1 applies to all building classes)	Structures should be constructed so that no collapse should be disproportionate to the cause and reduce the risk of localised damage spreading – but that permanent deformation of members/connections is acceptable.			
			CL 9.10.2 plus allowance for 'out of verticality' loading CL 5.2[f]	Basic requirements: generic advice from BS EN 1991-1-7. Annex A	No specific mention but 'out of verticality' implied[g]	No specific guidance beyond the generic advice in BS EN 1991-1-7 Annex A.[h]
2A	5 storey single occupancy house. Hotels not exceeding 4 storeys. Flats, apartments and other residential buildings not exceeding 4 storeys. Industrial buildings not exceeding 3 storeys. Retailing premises not exceeding 3 storeys and less than 2000 m² at each storey.	Option 1: Effective anchorage of suspended floors to walls	PD6687 CL 2.26.2 BS EN 1991-1-7, CL A.5.2	Generally N/A But refer to PD6687 for precast concrete building elements.	BS EN 1991-1-7 CL A.5.2 and as class 1	BS EN 1991-1-7 CL A.5.2 and as class 1
		Option 2: Provision of horizontal ties	As class 1 plus BS EN 1991-1-7, CL A.5.1	BS EN 1991-1-7, CL A.5.1 and as class 1	BS EN 1991-1-7 CL A.5.1 and as class 1 or proof by component test	BS EN 1991-1-7, CL A.5.1 and as class 1.

continued

(continued)

Building class	Building type and occupancy	Building regulations/ general Eurocode requirements	British standard material references and summary outline guidance			
			BS EN 1992 – concrete	BS EN 1993 – steel	BS EN 1995 – timber	BS EN 1996 – masonry
2B	Hotels, flats, apartments and other residential buildings greater than 4 storeys but not exceeding 15 storeys. Educational buildings greater than 1 storey but less than 15 storeys. Retailing premises greater than 3 storeys but less than 15 storeys. Hospitals not exceeding 3 storeys. Offices greater than 4 storeys but less than 15 storeys. All buildings to which members of the public are admitted which contain floor areas exceeding 2000 m² but are less than 5000 m² at each storey. Car parking not exceeding 6 storeys.	Option 1: Provision of horizontal and vertical ties	As class 1 plus NA A.1.3 and PD6687 CL 2.26.1 and BS EN 1991-1-7, CL A.6 + A.5.2.1	BS EN 1991-1-7, CL A.5.1 plus A.6	Generally impractical	Generally impractical
		Option 2: Check notional removal of load-bearing elements	As class 2B option 1 plus BS EN 1991-1-7, CL A.7	BS EN 1991-1-7 and as class 2B option 1	BS EN 1991-1-7, CL A.7 and as class 2A	BS EN 1991-1-7, CL A.7 and as class 2A
			Check notional removal of load-bearing elements such that for removal of any element the building remains stable and that the area of floor at any storey at risk of collapse is less than the lesser of 70 m² or 15% of the floor area of that storey. The nominal length of load-bearing wall should be the distance between vertical lateral restraints (not exceeding 2.25H for reinforced concrete walls or internal walls of masonry, timber or steel stud). If catenary action is assumed, allowance should be made for the necessary horizontal reactions.			

	Option 3: Key element design	As class 2B option 1 plus BS EN 1991-1-7, CL A.8	BS EN 1991-1-7, CL A.8 and as class 2B option 1	BS EN 1991-1-7, CL A.8 plus A.1.3.2 and Table A.1.3 plus supporting National Annex	BS EN 1991-1-7, CL A.8 and as class 2A.
		Design of key elements to be capable of withstanding 34 kN/m² applied one direction at a time to the member and attached components subject to the limitations of their strength and connections, such accidental loading should be considered to act simultaneously with full dead loading and 1/3 of all normal wind/imposed loadings unless permanent storage loads, etc. Where relevant, partial load factors of 1.05 or 0.9 should be applied for overturning and restoring loads, respectively. Elements providing stability to key elements should be designed as key elements themselves.			
3	Systematic risk assessment of the building should be undertaken taking into account all the normal hazards that may be reasonably forseen, together with any abnormal hazards.	Information on risk assessment methods in Annex B, BS EN 1991-1-7: 2006. Class 2B assumed as a minimum requirement			

Row 3 first description column: All buildings defined above as Class 2A and 2B that exceed the limits of area or number of storeys. Stadia accommodating more than 5000 spectators. Buildings containing hazardous substances and/or processes. All buildings to which member of the public are admitted in significant numbers

Notes:

[a] Refer to the detailed British Standard clauses for full details of design and detailing requirements.

[b] Where provided, horizontal and vertical ties should be safeguarded against damage and corrosion.

[c] Key elements may be present in any class of structure and should be designed accordingly.

[d] The construction details required by Class 2B can make buildings with load-bearing walls difficult to justify economically.

[e] In Class 2B and 3 buildings, precast concrete elements not acting as ties should be effectively anchored (C1.5.1.8.3), such anchorage being capable of carrying the dead weight of the member.

[f] This allowance is in addition to wind actions and replaces the BS 8110 notional load allowance of 1.5% of design dead loads.

[g] Not directly mentioned in BS EN 1995-1-1 but principle 'out of verticality' within the Eurocode suite would suggest a notional horizontal long-term load of 2.5% design load.

[h] Guidance is only available in BS5628 until BSi publish the extracted information as NCCI. See clause 16 for class 1, clause 33.4 and Table 12 for class 2A; check notional removal of bearing elements for class 2B option 2, or clause 33.5 and Table 12 for class 2B option 3 key element design.

Source: Adapted from Table 11, Part A3 Approved Document A, HMSO. and Table A.1, Annex A, BS EN 1991-1-7: 2006.

7
Reinforced Concrete

The Romans are thought to have been the first to use the binding properties of volcanic ash in mass concrete structures. The art of making concrete was then lost until Portland Cement was discovered in 1824 by Joseph Aspedin from Leeds. His work was developed by two Frenchmen, Monier and Lambot, who began experimenting with reinforcement. Deformed bars were developed in America in the 1870s, and the use of reinforced concrete has developed worldwide since 1900–1910. Concrete consists of a paste of aggregate, cement and water that can be reinforced with steel bars, or occasionally fibres, to enhance its flexural strength. Concrete constituents are as follows:

Cement: Limestone and clay fired to temperatures of about 1400°C and ground to a powder. Grey is the standard colour but white can be used to change the mix appearance. The cement content of a mix affects the strength and finished surface appearance.

Aggregate: Coarse aggregate (10–20 mm) and sand make up about 75% of the mix volume. Coarse aggregate can be natural dense stone or lightweight furnace by-products.

Water: Water is added to create the cement paste that coats the aggregate. The water/cement ratio must be carefully controlled as the addition of water to a mix will increase workability and shrinkage, but will reduce strength if cement is not added. A typical water:cement ratio is about 0.4–0.45 by volume.

Reinforcement: Reinforcement normally consists of deformed steel bars. Traditionally the main bars were typically high yield steel ($f_y = 460$ N/mm²) and the links mild steel ($f_y = 250$ N/mm²). However, the new standards on bar bending now allow small diameter high yield bars to be bent to the same small radii as mild steel bars. This may mean that the use of mild steel links will reduce. The bars can be loose, straight or shaped, or as high yield welded mesh. Less commonly steel, plastic or glass fibres can be added (1–2% by volume) instead of bars to improve impact and cracking resistance, but this is generally only used for ground bearing slabs.

Admixtures: Workability, durability and setting time can be affected by the use of admixtures.

Formwork: Generally designed by the contractor as part of the temporary works (now referred to as 'auxiliary construction works' by the Eurocodes) this is the steel, timber or plastic mould used to keep the liquid concrete in place until it has hardened. Formwork can account for up to half the cost of a concrete structure and should be kept simple and standardised where possible.

Summary of material properties

Density: 17–24 kN/m³ depending on the density of the chosen aggregate.

Compressive strength: Design strengths have a good range. Cube strengths: $F_{cu} = 7$–60 N/mm²; cylinder strengths: $f_{ck} = 5.6$–48 N/mm² (about 80% of cube strengths).

Tensile strength: Poor at about 6–12% of f_{ck} (8–15% of F_{cu}). Reinforcement provides flexural strength.

Modulus of elasticity: This varies with the mix design strength, reinforcement content and age. Typical short-term (28 days) values are: 24–32 kN/mm². Long-term values are about 30–50% of the short-term values.

Linear coefficient of thermal expansion: 8–12×10^{-6}°C.

Shrinkage: As water is lost in the chemical hydration reaction with the cement, the concrete section will shrink. The amount of shrinkage depends on the water content, aggregate properties and section geometry. Normally, a long-term shrinkage strain of 0.03% can be assumed, of which 90% occurs in the first year.

Creep: Irreversible plastic flow will occur under sustained compressive loads. The amount depends on the temperature, relative humidity, applied stress, loading period, strength of concrete, allowed curing time and size of the element. It can be assumed that about 40% and 80% of the final creep occurs in 1 month and 30 months, respectively. The final (30-year) creep value is estimated from $\sigma\phi/E$, where σ is the applied stress, E is the modulus of elasticity of the concrete at the age of loading and ϕ is the creep factor which varies between about 1.0 and 3.2 for the United Kingdom concrete loaded at 28 days.

Concrete mixes

Concrete mix design is not an absolute science. The process is generally iterative, based on an initial guess at the optimum mix constituents, followed by testing and mix adjustments on a trial-and-error basis.

There are different ways to specify concrete. A 'prescribed mix' is where the purchaser specifies the mix proportions and is responsible for ensuring (by testing) that these proportions produce a suitable mix. A 'designed mix' is where the engineer specifies the required performance, the concrete producer selects the mix proportions and concrete cubes are tested in parallel with the construction to check the mix compliance and consistency. Grades of designed mixes are prefixed by C. Special proprietary mixes such as self-compacting or waterproof concrete can also be specified, such as Pudlo or Caltite.

The majority of the concrete in the United Kingdom is specified on the basis of strength, workability and durability as a designated mix to BS EN 206 and BS 8500. This means that the ready mix companies must operate third-party accredited quality assurance (to BS EN ISO 9001), which substantially reduces the number of concrete cubes which need to be tested. Grades of designated mixes are prefixed by GEN, FND or RC depending on their proposed use.

Cementitious content

Cement is a single powder (containing, for example, OPC and fly ash) supplied to the concrete producer and denoted 'CEM'. Where a concrete producer combines cement, fly ash, ggbs and/or limestone fines at the works, the resulting cement combination is denoted 'C' and must set out the early day strengths as set out in BS 8500-2. As a result mixes can have the same constituents and strength (e.g. cement combination type IIA), with their labels indicating whether they are as a result of processes by the cement manufacturers (e.g. CEM II/A) or producers (e.g. CIIA). However, use of broad designations is suitable for most purposes, with cement and cement combinations being considered the same for specification purposes.

Cement combination types

CEM 1	Portland cement (OPC)
SRPC	Sulphate-resisting Portland cement
IIA	OPC + 6–20% fly ash, ggbs or limestone fines
IIB-S	OPC + 21–35% ggbs
IIB-V	OPC + 21–35% fly ash
IIIA	OPC + 36–65% ggbs (low early strength)
IIIB	OPC + 66–80% ggbs (low early strength)
IVB-V	OPC + 36–55% fly ash

Designated concrete mixes to BS 8500 and BS EN 206

Designated mix	Compressive strength (N/mm²)	Strength class	Typical application	Min cement content (kg/m³)	Max free water/ cement Ratio	Typical slump (mm)	Consistence class
GEN 0	7.5	C6/8	Kerb bedding and backing	120	N/A	Nominal 10	S1
GEN 1	10	C8/10	Blinding and mass concrete fill	180	N/A	75	S3
			Drainage works	180	N/A	10–50	S1
			Oversite below suspended slabs	180	N/A	75	S3
GEN 3	20	C16/20	Mass concrete foundations	220	N/A	75	S3
			Trench fill foundations	220	N/A	125	S4
FND 2, 3, 4 or 4A	35	C28/35	Foundations in sulphate conditions: 2, 3, 4 or 4A	320–380	0.5–0.35	75	S3
RC 30	30	C25/30	Reinforced concrete	260	0.65	50–100	S3
RC 35	35	C28/35		280	0.60	50–100	S3
RC 40	40	C32/40		300	0.55	50–100	S3
RC 50	50	C 40/50		340	0.45	50–100	S3

Note: Where strength class C28/35 indicates that the minimum characteristic cylinder strength is 28 N/mm² and cube strength 35 N/mm².

Source: BS 8500: Part 1:2002, Table A.7 adapted.

Traditional prescribed mix proportions

Concrete was traditionally specified on the basis of prescribed volume proportions of cement to fine and coarse aggregates. This method cannot allow for variability in the mix constituents, and as a result mix strengths can vary widely. This variability means that prescribed mixes batched by volume are rarely used for anything other than small works where the concrete does not need to be of a particularly high quality.

Typical volume batching ratios and the probable strengths achieved (with a slump of 50–75 mm) are:

Typical prescribed mix volume batching proportions cement: sand : 20 mm aggregate	Probable characteristic crushing strength, F_{cu} (N/mm²)
1:1.5:3	40
1:2:4	30
1:3:6	20

Concrete cube testing for strength of designed mixes

Concrete cube tests should be taken to check compliance of the mix with the design and specification. The amount of testing will depend on how the mix has been designed or specified. If the concrete is a designed mix from a ready mix plant BS 5328 gave the following minimum rates for sampling:

1 Sample per	Maximum quantity of concrete at risk under any one decision	Examples of applicable structures
10 m³ or 10 batches	40 m³	Masts, columns, cantilevers
20 m³ or 20 batches	80 m³	Beams, slabs, bridges, decks
50 m³ or 50 batches	200 m³	Solid rafts, breakwaters

This seems clearer that BS EN 206 (which is the current standard) suggests three 'samples' in the first 50 m³ and then one 'sample' per 150–400 m³ depending on certification and stage of production.

At least one 'sample' should be taken, for each type of concrete mix on the day it is placed, prepared to the requirements of BS EN 12350. If the above table is not used, 60 m³ should be the maximum quantity of concrete represented by four consecutive test results. Higher rates of sampling should be adopted for critical elements.

A sample consists of two concrete cubes for each test result. Where results are required for 7- and 28-day strengths, four cubes should be prepared. The concrete cubes are normally cured under water at a minimum of 20°C ± 2°C. If the cubes are not cured at this temperature, their crushing strength can be seriously reduced.

Cube results must be assessed for validity using the following rules for 20 N/mm² concrete or above:

- A cube test result is said to be the mean of the strength of two cube tests. Any individual test result should not be more than 3 N/mm² below the specified characteristic compressive strength.
- When the difference between the two cube tests (i.e. four cubes) divided by their mean is greater than 15%, the cubes are said to be too variable in strength to provide a valid result.
- If a group of test results consist of four consecutive cube results (i.e. eight cubes), the mean of the group of test results should be at least 3 N/mm² above the specified characteristic compressive strength.

Separate tests are required to establish the conformity of the mix on the basis of workability, durability and so forth.

Source: BS 5328: Part 1:1997; BS EN 206-1: 2000.

Reinforcement

The ultimate design strength is $f_y = 250$ N/mm² for mild steel and $f_y = 500$ N/mm² high yield reinforcement.

Weight of reinforcement bars by diameter (kg/m)

6 mm	8 mm	10 mm	12 mm	16 mm	20 mm	25 mm	32 mm	40 mm
0.222	0.395	0.616	0.888	1.579	2.466	3.854	6.313	9.864

Reinforcement area (mm²) for groups of bars

Number of bars	Bar diameter (mm)								
	6	8	10	12	16	20	25	32	40
1	28	50	79	113	201	314	491	804	1257
2	57	101	157	226	402	628	982	1608	2513
3	85	151	236	339	603	942	1473	2413	3770
4	113	201	314	452	804	1257	1963	3217	5027
5	141	251	393	565	1005	1571	2454	4021	6283
6	170	302	471	679	1206	1885	2945	4825	7540
7	198	352	550	792	1407	2199	3436	5630	8796
8	226	402	628	905	1608	2513	3927	6434	10053
9	254	452	707	1018	1810	2827	4418	7238	11310

Reinforcement area (mm²/m) for different bar spacing

Spacing (mm)	Bar diameter (mm)								
	6	8	10	12	16	20	25	32	40
50	565	1005	1571	2262	4021	6283	9817	–	–
75	377	670	1047	1508	2681	4189	6545	10723	–
100	283	503	785	1131	2011	3142	4909	8042	12566
125	226	402	628	905	1608	2513	3927	6434	10053
150	188	335	524	754	1340	2094	3272	5362	8378
175	162	287	449	646	1149	1795	2805	4596	7181
200	141	251	393	565	1005	1571	2454	4021	6283
225	126	223	349	503	894	1396	2182	3574	5585
250	113	201	314	452	804	1257	1963	3217	5027

Reinforcement mesh to BS 4483

Fabric reference	Longitudinal wires			Cross wires			Mass (kg/m²)
	Diameter (mm)	Pitch (mm)	Area (mm²/m)	Diameter (mm)	Pitch (mm)	Area (mm²/m)	
Square mesh – High tensile steel							
A393	10	200	393	10	200	393	6.16
A252	8	200	252	8	200	252	3.95
A193	7	200	193	7	200	193	3.02
A142	6	200	142	6	200	142	2.22
A98	5	200	98	5	200	98	1.54
Structural mesh – High tensile steel							
B131	12	100	131	8	200	252	10.90
B785	10	100	785	8	200	252	8.14
B503	8	100	503	8	200	252	5.93
B385	7	100	385	7	200	193	4.53
B283	6	100	283	7	200	193	3.73
B196	5	100	196	7	200	193	3.05
Long mesh – High tensile steel							
C785	10	100	785	6	400	70.8	6.72
C636	9	100	636	6	400	70.8	5.55
C503	8	100	503	5	400	49	4.34
C385	7	100	385	5	400	49	3.41
C283	6	100	283	5	400	49	2.61
Wrapping mesh – Mild steel							
D98	5	200	98	5	200	98	1.54
D49	2.5	100	49	2.5	100	49	0.77

Stock sheet size	Longitudinal wires	Cross wires	Sheet area
	Length 4.8 m	Width 2.4 m	11.52 m²

Source: BS 4486: 2005, Table A.1.

Shear link reinforcement areas

| No. of link legs | Shear Link Area, A_{sv} (mm²) | | | | Shear link area/link bar spacing, A_{sv}/S_v (mm²/mm) | | | | | | | | |
| | Link diameter (mm) | | | | Link spacing, S_v (mm) | | | | | | | | |
	6	8	10	12	100	125	150	175	200	225	250	275	300
2	56				0.560	0.448	0.373	0.320	0.280	0.249	0.224	0.204	0.187
		100			1.000	0.800	0.667	0.571	0.500	0.444	0.400	0.364	0.333
			158		1.580	1.264	1.053	0.903	0.790	0.702	0.632	0.575	0.527
				226	2.260	1.808	1.507	1.291	1.130	1.004	0.904	0.822	0.753
3	84				0.840	0.672	0.560	0.480	0.420	0.373	0.336	0.305	0.280
		150			1.500	1.200	1.000	0.857	0.750	0.667	0.600	0.545	0.500
			237		2.370	1.896	1.580	1.354	1.185	1.053	0.948	0.862	0.790
				339	3.390	2.712	2.260	1.937	1.695	1.507	1.356	1.233	1.130
4	112				1.120	0.896	0.747	0.640	0.560	0.498	0.448	0.407	0.373
		200			2.000	1.600	1.333	1.143	1.000	0.889	0.800	0.727	0.667
			316		3.160	2.528	2.107	1.806	1.580	1.404	1.264	1.149	1.053
				452	4.520	3.616	3.013	2.583	2.260	2.009	1.808	1.644	1.507
6	168				1.680	1.344	1.120	0.960	0.840	0.747	0.672	0.611	0.560
		300			3.000	2.400	2.000	1.714	1.500	1.333	1.200	1.091	1.000
			474		4.740	3.792	3.160	2.709	2.370	2.107	1.896	1.724	1.580
				678	6.780	5.424	4.520	3.874	3.390	3.013	2.712	2.465	2.260

Durability and fire resistance

Concrete durability

Concrete durability requirements based on BS 8500 (also known as BS EN 206-1) are based on the expected mode of deterioration, therefore all relevant exposure classes should be identified to establish the worst case for design. Concrete cover is also specified on the basis of a minimum cover, with the addition of a margin for fixing tolerance (normally between 5 and 15 mm) based on the expected quality control and possible consequences of low cover. In severe exposure conditions (e.g. marine structures, foundations) and/or for exposed architectural concrete, stainless steel reinforcement can be used to improve corrosion resistance, and prevent spalling and staining. Austenitic is the most common type of stainless steel used in reinforcing bars.

Carbonation

Carbon dioxide from the atmosphere combines with water in the surface of concrete elements to form carbonic acid. This reacts with the alkaline concrete to form carbonates which reduces the concrete pH and its passive protection of the steel rebar. With further air and water, the steel rebar can corrode (expanding in volume by two to four times) causing cracking and spalling. Depth of carbonation is most commonly identified by the destructive phenolphthalein test which turns from clear to magenta where carbonation has not occurred. In practice, corrosion due to carbonation is a minor risk compared with chloride-related problems, due to the way concrete is typically used and detailed in the United Kingdom.

Chlorides

Chlorides present in air or water from de-icing salts or a marine environment can enter the concrete through cracks, diffusion, capillary action or hydrostatic head to attack steel reinforcement. Chloride ions destroy any passive oxide layer on the steel, leaving it at risk of corrosion in the presence of air and water, resulting in cracking and spalling of the concrete. Although lower water/cement ratios and higher cover are thought to be the main way that will reduce the risk of chloride-related corrosion, these measures are largely unsuccessful if not also accompanied by good workmanship and detailing to help keep moisture out of the concrete – particularly on horizontal surfaces.

Freeze thaw

Structures such as dams, spillways, tunnel inlets and exposed horizontal surfaces are likely to be saturated for much of the time. In low temperatures, ice crystals can form in the pores of saturated concrete (expanding by up to eight times compared with the original water volume) and generate internal stresses which break down the concrete structure. Repeated cycles have a cumulative effect. Air-entrainment has been found to provide resistance to freeze–thaw effects as well as reduce the adverse effects of chlorides.

Aggressive chemicals

All concretes are vulnerable to attack by salts present in solution and to attack by acids. Such chemicals can be present in natural ground and groundwater, as well as contaminated land, so precautions are often necessary to protect buried concrete. Although sulphates are relatively common in natural clay soils, sulphate attack is believed to be a relatively rare cause of deterioration (with acid attack being even rarer). Protection is usually provided by using a sulphate-resisting cement or use of slag as a cement replacement. Flowing water is also more aggressive than still water.

Exposure classification and recommendations for resisting corrosion of reinforcement

Likely risk identified by designer	Class	Exposure condition	Recommended strength class, max. w/c ratio, min. cement content (kg/m³)[a]							
			Nominal Cover to Reinforcement							
			15 + Δc	20 + Δc	25 + Δc	30 + Δc	35 + Δc	40 + Δc	45 + Δc	50 + Δc
No risk	X0	Completely dry	Recommended that this exposure is not applicable to reinforced concrete							
Carbonation-induced corrosion of rebar	XC1	Dry or permanently wet / Inside buildings with low humidity	C20/25 0.7 240	—	—	—	—	—	—	—
	XC2	Wet, rarely dry / Water retaining structures, many foundations	X	X	C25/30 0.65 260	—	—	—	—	—
	XC3[a]	Moderate humidity / Internal concrete or external sheltered from rain	X	C40/50 0.45 340	C32/40 0.55 300	C28/35 0.6 280	C25/30 0.65 260	—	—	—
	XC4[a]	Cyclic wet and dry / More severe than XC3 exposure								
Chloride-induced corrosion of rebar excl. seawater chlorides	XD1	Moderate humidity / Exposure to direct spray	X	X	C40/50 0.45 360	C32/40 0.55 320	C28/35 0.6 300	—	—	—

	XD2	Wet, rarely dry	X	X	X	X	X	X	C28/35 0.55 320	—	—
		Swimming pools or industrial water contact									
	XD3	Cyclic wet and dry	X	X	X	X	X	X	C45/55 0.35 380	C40/50 0.4 380	C35/45 0.45 360
		Parts of bridges, pavements, car parks									
Seawater-induced corrosion of rebar	XS1	Airborne salts but no direct contact	X	X	X	X	X	X	C35/45 0.5 340	—	—
		Near to, or on, the coast									
	XS2	Wet, rarely dry	X	X	X	X	X	X	C28/35 0.55 320	—	—
		Parts of marine structures									
	XS3	Tidal, splash and spray zones	X	X	X	X	X	X	X	C45/55 0.35 380	C40/50 0.4 380
		Parts of marine structures									
Freeze–thaw attack on concrete	XF	Damp or saturated surfaces subject to freeze–thaw degradation	Refer to BS 8500-1 Table A.8								
Aggressive Chemical attack on concrete & rebar	DS	Surfaces exposed to aggressive chemicals present in natural soils or contaminated land	Refer to BS 8500-2 Table 1								

Notes:

a   For all cement/combination types unless noted[a] indicating IVB.
b   Design life of 50 years with normal weight concrete and 20 mm aggregate. See BS 8500 for other variations.
c   More economic to follow BS 8500 if specific cement/combination is to be specified.
d   BS 8500 sets a minimum standard for cement content not to be reduced, therefore – indicates minimum standard specified in cell to left.
e   Where Δc is fixing tolerance, normally 5–15 mm.

Source: BS 8500-1:2006. Tables A.6 and A.13 adapted.

Fire resistance

BS EN 1992-1-2 Structural fire design gives several methods for determining the fire resistance of concrete elements. However, this can also be established using tabulated values for simplicity.

The concept of 'minimum cover' as used in British standards has been abandoned. In its place is 'nominal axis distance' (NAD) represented by the letter, a. This is the distance from the centre or a main reinforcement bar to the surface of the member.

$$a \geq c_{nom} + \phi_{link} + \frac{\phi_{bar}}{2}$$

Within the Eurocodes, fire rating is referred to as 'REI' as there are three standard fire exposive conditions to be satisfied:

R mechanical resistance for load bearing
E integrity of compartment separation
I Insulation

The acronym REI is usually combined with the number of minutes fire performance for classification purposes e.g. REI 60, REI 120 etc. to indicate overall fire rating.

Preliminary sizing of concrete elements

Typical span/depth ratios

Element	Typical spans (m)	Overall depth or thickness		
		Simply supported	Continuous	Cantilever
One-way spanning slabs	5–6	$L/22$–30	$L/28$–36	$L/$–10
Two-way spanning slabs	6–11	$L/24$–35	$L/34$–40	–
Flat slabs	4–8	$L/27$	$L/36$	$L/$–10
Close centre-ribbed slabs (ribs at 600 mm c/c)	6–14	$L/23$	$L/31$	$L/9$
Coffered slabs (ribs at 900–1500 mm slabs)	8–14	$L/15$–20	$L/18$–24	$L/7$
Post-tensioned flat slabs	9–10	$L/35$–40	$L/38$–45	$L/10$–12
Rectangular beams (width > 250 mm)	3–10	$L/12$	$L/15$	$L/6$
Flanged beams	5–15	$L/10$	$L/12$	$L/6$
Columns	2.5–8	$H/10$–20	$H/10$–20	$H/10$
Walls	2–4	$H/30$–35	$H/45$	$H/15$–18
Retaining wall	2–8	–	–	$H/10$–14

Note: 125 mm is normally the minimum concrete floor thickness for fire resistance.

Preliminary sizing

Beams: Although the span/depth ratios are a good indication, beams tend to need more depth to fit sufficient reinforcement into the section in order to satisfy deflection requirements. Check the detailing early – especially for clashes with steel at column/beam junctions. The shear stress should be limited to 2 N/mm^2 for preliminary design.

Solid slabs: Two-way spanning slabs are normally about 90% of the thickness of one-way spanning slabs.

Profiled slabs: Obtain copies of proprietary mould profiles to minimise shuttering costs. The shear stress in ribs should be limited to 0.6 N/mm^2 for preliminary design.

Columns: A plain concrete section with no reinforcement can take an axial stress of about 0.36 f_{ck} or 0.45/F_{CU}. The minimum column dimensions for a stocky braced column = clear column height/17.5.

A simple allowance for moment transfer in the continuous junction between slab and column can be made by factoring up the load from the floor immediately above the column being considered (by 1.25 for interior, 1.50 for edge and 2.00 for corner columns). The column design load is this factored load plus any other column loads.

For stocky columns, the column area (A_c) can be estimated by: $A_c = N/15$, $N/18$ or $N/21$ for columns in C28/35 concrete containing 1%, 2% or 3% high yield steel, respectively.

Minimum dimensions and nominal axis distance (NAD a) for fire resistance periods

Member	Requirements	Fire rating, REI (minutes)					
		30	60	90	120	180	240
Columns fully exposed to fire	Minimum column width	200–300	250–350	350–450[g]	350–450[g]	450[g]	–[h]
	NAD	32–27	46–40	53–40	57–51	40	–
Walls (2 sides exposed)	Minimum wall thickness	120	140	170	220	270	350
	NAD	10[i]	10[i]	25	25	55	60
Beams	Minimum beam width	160–200	160–300	200–400	240–500	300–600	500–700
	NAD for simply supported	15–15[i]	35–25	45–35	60–50	70–60	75–70
	NAD for continuous	12[i]–12[i]	25–12[i]	30–25	40–30	55–40	60–50
Slabs with plain soffits (one or two way)	Minimum slab thickness	60	80	100	120	150	175
	NAD for simply supported	10[i]	20	30	40	55	65
Ribbed slabs (open soffit and no stirrups)[j]	Minimum top slab thickness	80	80	100	120	150	175
	Minimum rib width	80	100–>200	120–>250	190>300	260–>410	280–>500
	NAD for simply supported	15[i]	35–15[i]	45–30	55–40	70–60	90–70
Flat slabs	Minimum slab thickness	150	180	200	200	200	200
	NAD	10[i]	15[i]	25	35	45	50

Notes:

a   REI indicates fire exposure conditions (R = mechanical resistance for load bearing, E = Integrity of separation and I = Insulation). NAD = nominal axis distance, that is centre of bar to member surface, so greater than 'cover' specified under BS 8110.

b   Table contains selected ranges of member sizes and axis distances for simplified cases using method A. See original Eurocode text for full range available.

c   Effective length of braced column should be ≤3 m. Effective length may be taken as 50% of actual length for intermediate floors and 50–70% for the upper floor column.

d   Assumptions: First-order eccentricity order fire conditions should be ≤0.15b. Otherwise use method B (see T5.26 BS EN 1992) for more slender conditions.

e   The reinforcement area (outside lap locations) does not exceed 4% of concrete cross-section.

f   The ratio of design axial load to design resistance of the column or wall (μ_n) may be conservatively taken as 0.7.

g   Minimum of eight bars.

h   Method B may be used (see T5.2b BS EN 1992).

i   The practical minimum cover is 20 mm, but corrosion protection cover will probably govern.

j   Lateral cover is NAD + 10 mm.

Source: BS EN1992-1-2:2004. (T5.2a, 5.4, 5.5, 5.6, 5.8, 5.9).

Reinforced concrete design to BS EN 1992

Bending

BS EN 1992 expects engineers to derive bending formulae based on recommended values and data from the National Annexe. This has been collated for the UK situation below for ease of reference.

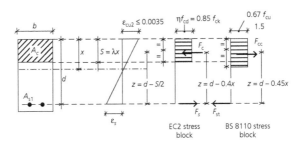

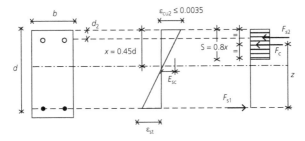

where:

$\lambda = 0.8$

$\eta = 1.0$

$F_{CD} = a_{cc} \dfrac{f_{ck}}{\gamma_c}$ where $\gamma_c = 1.5$ and $a_{cc} = 0.85$ (for flexure)

$\varepsilon_{CU2} \leq 3.5\%$ (i.e. 0.0035) for $f_{ck} \leq 50$ N/mm^2

and if the width of the compression zone decreases by the extreme compression fibre (e.g. circular column), ηF_{Cd} should be reduced by 10%;

and the depth of the neutral axis, $x \leq 0.45d$ for $f_{ck} \leq 50$ N/mn^2 (equivalent to BS $x \leq 0.5d$)

and for doubly reinforced beams:

$\dfrac{d_2}{x} \leq 0.38$

$k_o' = 0.167$, therefore $\dfrac{z}{d} = 0.82$

Source: BS EN 1992-1-1:2004: Figure 3.5 adapted by clauses 3.17(3), 3.16(1), T3.1 in 3.1.3, T2.IN in 2.4.2.4 5.6.3 and TNAI in NA to BS EN 1992-1-1: 2004.

Ultimate moment capacity of beam section from derived formulae – $M_{RD} = 0.167 \, f_{ck} \, bd^2$ where there is no moment redistribution and $M_{RD} = 0.142 \, f_{ck} \, bd^2$ where there is less than 10% moment redistribution.

Factors for lever arm (z/d) and neutral axis (x/d) depth

$k_o = \dfrac{M}{f_{ck} bd^2}$	0.043	0.050	0.070	0.090	0.110	0.130	0.0145	0.156
z/d	0.950	0.941	0.915	0.887	0.857	0.825	0.798	0.777
x/d	0.13	0.15	0.19	0.25	0.32	0.39	0.45	0.50

where z = lever arm and x = neutral axis depth.

$\dfrac{z}{d} = 0.5 + \sqrt{0.25 - (3k_o/3.4)}$ (≤ 0.95 although this B.S. limit is not recommended in BS EN 1992 it is sensible to retain this) and $x = (d - z)/0.4$.

Area of tension reinforcement for rectangular beams from derived formulae – If the applied moment M is less than M_{RD}, then the area of tension reinforcement,

$$A_{s1} = M / \left[0.87 \left(\frac{z}{d} \right) f_{yk} d \right].$$

If the applied moments M is greater than M_{RD}, then the area of compression steel is $A_{s2} = (M - M_{RD})/0.87 f_{yk} \, (d - d_2)$ and the area of tension reinforcement is $A_{s1} = (M_{RD}/(0.87 \, f_{yk} \, (z/d)d)) + A_{s2}$ if redistribution is less than 10%.

Approximate equivalent breadth and depth of neutral axis for flanged beams

Flanged beams	Simply supported	Continuous	Cantilever
T beam	$b_w + L/5$	$b_w + L/7$	b_w
L beam	$b_w + L/10$	$b_w + L/13$	b_w

Note: Where b_w = breadth of web, L = actual flange width or beam spacing, h_1 is the depth of the flange.

Calculate k_o using b_w. From k, calculate $0.9x$ from the tabulated values of the neutral axis depth, x/d.

If $0.8x < h_f$, the neutral axis is in the beam flange and steel areas can be calculated as rectangular beams.

If $0.8x > h_f$, the neutral axis is in the beam web and steel areas should be calculated as flanged beams in accordance with BS EN 1992-1-2.

Shear

Although BS EN 1992 generally works with shear forces, calculations are more readily simplified using shear stresses.

The applied shear stress is $v_{Ed} = \dfrac{V_{Ed}}{0.9b_w d}$

Basic shear capacity of concrete – The shear capacity of concrete, $v_{Rd,c}$ relates to the section size, effective depth and percentage reinforcement. $V_{Rd,c} = v_{Rd,c}\, b_w d$.

Shear reinforcement – variable strut inclination method (N/mm²)

Strength class	f_{ck}	$V_{Rd\ max\ (cot\ \theta = 2.5)}$ ($\theta = 21.8°$)	$V_{Rd\ max\ (cot\ \theta = 1.0)}$ ($\theta = 45°$)
C20/25	20	2.54	3.68
C25/30	25	3.10	4.50
C28/35	28	3.43	4.97
C30/38	30	3.64	5.28
C32/40	32	3.84	5.58
C35/45	35	4.15	6.02
C40/50	40	4.63	6.72
C45/55	45	5.08	7.38
C50/60	50	5.51	8.00

Value of applied shear stress v_{Ed} (N/mm²)	Form of shear reinforcement to be provided	Area of shear reinforcement to be provided A_{sw} as vertical links (mm²)
$v_{Ed} < v_{Rd.c}$	Minimum links should normally be provided other than in elements of minor importance such as lintels	Suggested minimum : $$\frac{A_{sw}}{s} \geq \frac{0.08 b_w \sqrt{f_{ck}}}{f_{yk}}$$
$v_{Rd,c} < v_{Ed} < v_{Rd\ max\ (cot\ \theta = 2.5)}$	Links for whole length of beam to provide all shear resistance	Conservatively assume $\cot\theta = 2.5$: $$\frac{A_{sw}}{s} \geq \frac{v_{Ed} b_w}{2.5 f_{yk}}$$
$v_{Rd\ max\ (cot\theta = 2.5)} < v_{Ed} < v_{Rd\ max\ (cot\theta = 1.0)}$	Links for whole length of beam to provide all shear resistance	$$\theta = 0.5\sin^{-1}\left[\frac{v_{Ed}}{0.2 f_{ck}\left(1 - (f_{ck}/250)\right)}\right]$$ $$\frac{A_{sw}}{s} \geq \frac{v_{Ed} b_w}{f_{yk} \cot\theta}$$

Note:

[a] Max shear link spacing, $s < 0.75d$ for vertical links.

[b] If $v_{Ed} > v_{Rd\ max\ (cot\ \theta = 1.0)}$, then change to a deeper section or use stronger concrete.

Source: BS EN 1992-1-2: 2004: Clause 9.2.2.

Punching shear

Design stress

$$v_{Ed} = \frac{\beta \, V_{Ed}}{U_i d}$$

For braced structures where spans do not differ by 25%

where
  $\beta = 1.0$ for columns with no eccentricity/continuity
  $\beta = 1.15$ for internal columns
  $\beta = 1.4$ for edge columns
  $\beta = 1.5$ for corner columns
  U_i = shear perimeter under consideration
  d = effective depth

Punching shear checks – Check shear at column face, U_o to ensure $V_{Ed} \leq V_{Rd,\,max}$.
Check basic control perimeter, U_1 at $2d$ from column face to ensure $V_{Ed} \leq V_{Rd,c}$.
If shear stresses exceed the basic concrete shear capacity (i.e. $V_{Rd,c} > V_{Ed}$), then shear reinforcement should be provided and the shear stresses checked against the slab's reinforced capacity (i.e. $v_{Ed} < v_{Rd,cs}$)

$$v_{Rd,cs} = 0.75 \, v_{Rd,c} + 1.5 \frac{d}{S_r} A_{sw} f_{yk\ ef} \left(\frac{\sin\alpha}{Ud} \right)$$

where
  A_{sw} = area of one perimeter of reinforcement around the column (mm^2)
  S_r = radial spacing of shear reinforcement perimeters
  $f_{yk\ ef}$ = design strength of the shear reinforcement = $250 + 0.25d \leq f_{yk}$
  d = effective depth
  α = angle of shear reinforcement ($\sin \alpha = 1.0$ for vertical links/bars)

Source: BS EN 1992-1-1:2004; clause 6.4.3.

Axial load

Vertical elements (of clear height, l, and dimensions, $b \times h$) are considered as columns if $h \leq 4b$, otherwise they should be considered as walls. Generally, the clear column height between restraints should be less than $60b$. It must be established early in the design whether the columns will be in a braced frame where stability is to be provided by shear walls or cores, or whether the columns will be unbraced, meaning that they will maintain the overall stability for the structure. This has a huge effect on the effective length of columns, $l_e = Fl$, as the design method for columns depends on their slenderness, l_{ex}/b or l_{ey}/h. BS EN 1992 has detailed methods for assessment of slenderness and additional moments generated by slenderness/second-order effects in clause 5.8. For simplicity, the method outlined below deals only with stocky columns in braced frames.

Effective length factor (F) for columns – For columns in braced frames where the stiffness of any vertically adjacent column does not vary significantly effective column length factors can be taken from the table below. Otherwise, more detailed calculation from BS EN 1992 clause 5.8.3 should be used.

End condition at top of column		End condition at base of column		
		Condition 1	**Condition 2**	**Condition 3**
Condition 1	'Moment' connection to a beam or foundation which is at least as deep as the column dimension[a]	0.75	0.80	0.90
Condition 2	'Moment' connection to a beam or foundation which is shallower than the column dimension [a]	0.80	0.85	0.95
Condition 3	'Pinned' connection	0.90	0.95	1.00

Note:
[a] Column dimensions measured in the direction under consideration.

Source: BS 8110: Part 1:1997. Table 3.19; BS EN 1992-1-1:2004. BSI. Clause 5.8.3.

Slenderness – The slenderness (λ) of a column is defined by the ratio of its effective height ($l_o = Fl$) to the radius of gyration (i) depending on the axis being considered.

$$\text{For a rectangular section } i = h/3.46, \quad \text{therefore } \lambda = \frac{3.46 l_o}{h}$$

$$\text{For a circular section } i = \lambda/4.0, \quad \text{therefore } \lambda = \frac{4.0\, l_o}{d}$$

Slenderness limit – If the slenderness (λ) of a column is less than the calculated slenderness limit (λ_{lim}), then the second-order effects can be ignored:

$$\lambda_{lim} = \frac{20\,ABC}{\sqrt{n}} \equiv 12.6c\sqrt{\frac{A_c\, f_{ck}}{N_{ed}}}$$

Derived from:

$A = 0.7$

$B = 1.1$

$C = 1.7 - r_m$, where $r_m = (m_{o1}/m_{o2})$. C can be taken as 0.7 but as this factor has the greatest influence over the value of λ_{lim} it is worth calculating.

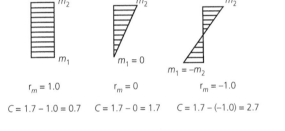

$$r_m = 1.0 \qquad\qquad r_m = 0 \qquad\qquad r_m = -1.0$$

$$C = 1.7 - 1.0 = 0.7 \quad C = 1.7 - 0 = 1.7 \quad C = 1.7 - (-1.0) = 2.7$$

$$n = \frac{N_{ed}}{A_c f_{cd}} = \frac{N_{ed}}{0.67 bh f_{ck}} \quad \text{as } f_{cd} = \frac{\alpha_{cc} f_{ck}}{\gamma_c} \quad \text{where } \gamma_c = 1.5 \text{ and } \alpha_{cc} = 1.0 \text{ for axial load}$$

N_{ed} is the axial design load

A_c is the section area of concrete

F_{ck} is the characteristic concrete strength

Source: BS EN 1992-1-1: 2004. Clause 5.8.3.1, NA 3.1.6(1)P.

Framing moments transferred to columns

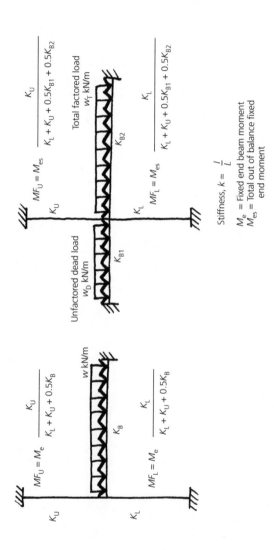

Stiffness, $k = \dfrac{I}{L}$

M_e = Fixed end beam moment
M_{es} = Total out of balance fixed
            end moment

Column design methods – Column design charts must be used where the column has to resist axial and bending stresses. Stocky columns need only normally be designed for the maximum design moment about one axis. The minimum design moment is the axial load multiplied by the greater of the eccentricity or $h/20$ in the plane being considered.

If a full frame analysis has not been carried out, the effect of moment transfer can be approximated by using column subframes or by using increasing axial loads by 10% for symmetrical simply supported loads. Alternately, column moments can be obtained using subframes subject to minimum suggested design moments.

Slender columns can be designed in the same way as short columns, but must resist an additional moment due to eccentricity caused by the deflection of the column as set out in clause BS EN 1992-1-1:2004: Clause 5,8,7,3.

Biaxial bending in columns – Where BS8110 enhanced the moment in one direction to account to the effects of biaxial bending, BS EN 1992 advocates separate design checks in each direction, disregarding biaxial bending.

No further check is necessary if

$$0.5 \leq \frac{\lambda_y}{\lambda_z} \ \text{ or } \ \frac{\lambda_z}{\lambda_y} \ \leq 2.0 \ \text{ and}$$

$$0.2 \geq \frac{(e_y/h_{eq})}{(e_z/b_{eq})} \ \text{ or } \ \frac{(e_z/b_{eq})}{(e_y/h_{eq})} \ \geq 5.0$$

where eccentricity, $e_z = (M_{Edz}/N_{Ed})$ and $e_y = (M_{Edy}/N_{Ed})$ and equivalent dimensions, $b_{eq} = i_y\sqrt{12}$ (= b for rectangular sections) and $h_{eq} = i_z\sqrt{12}$ (= h for rectangular sections).

For rectangular sections $(e_y/h_{eq})/(e_z/b_{eq}) = (M_{Edy}b/M_{Edz}h)$.

Otherwise, biaxially bent columns should be designed to satisfy:

$$\left(\frac{M_{EdZ}}{M_{RdZ}}\right)^a + \left(\frac{M_{Edy}}{M_{Rdy}}\right)^a \leq 1.0$$

where
    a = 2.0 for circular or elliptical sections or
    a = 1.0 for $N_{Ed}/N_{Rd} = 0.1$ for rectangular section (interpolate between)
    a = 1.5 for $N_{Ed}/N_{Rd} = 0.7$
    a = 2.0 for $N_{Ed}/N_{Rd} = 1.0$

Reinforced concrete column design charts

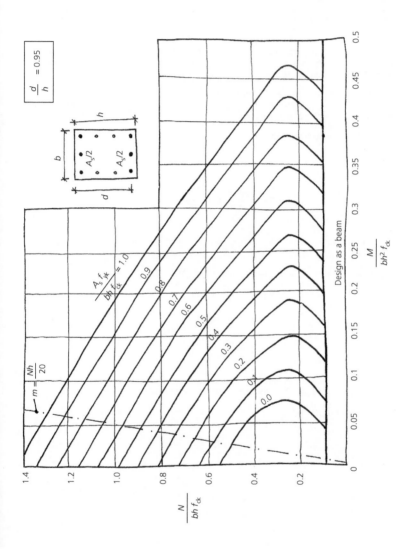

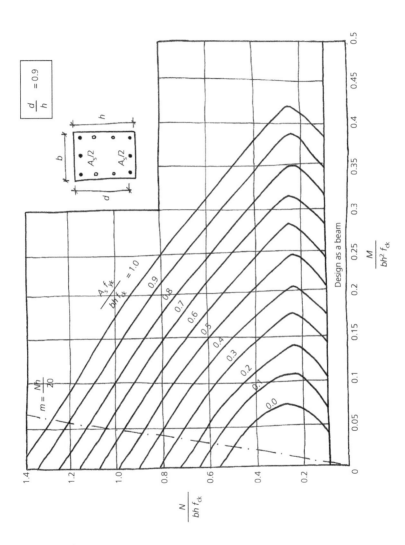

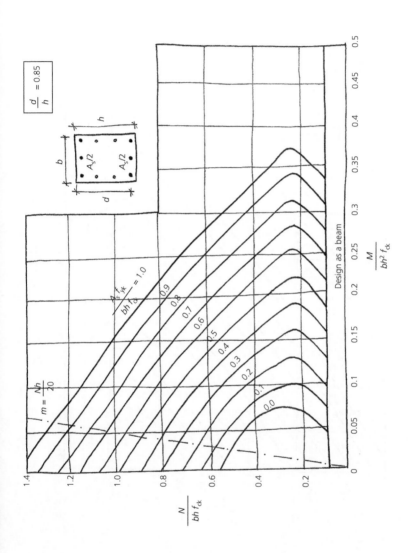

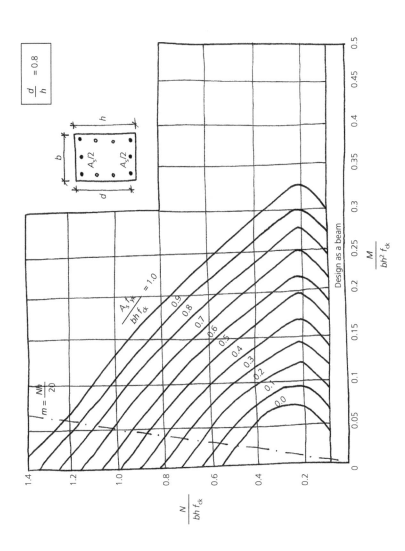

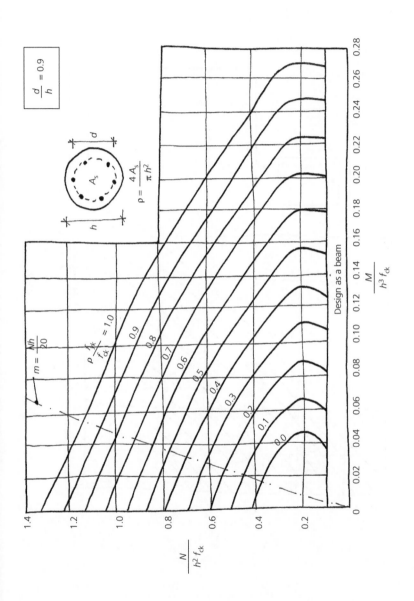

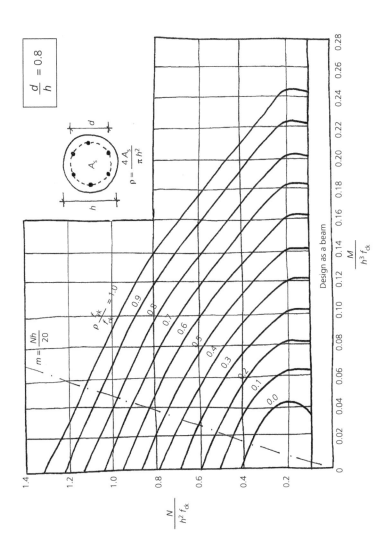

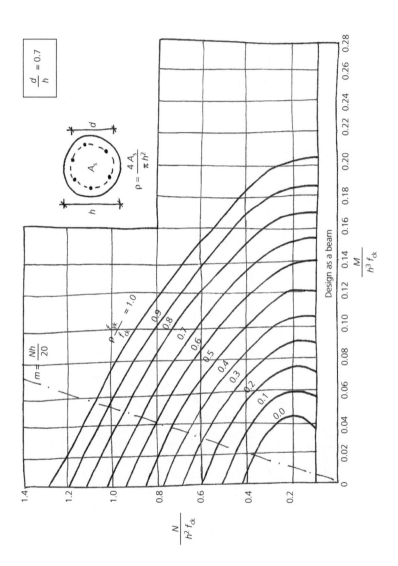

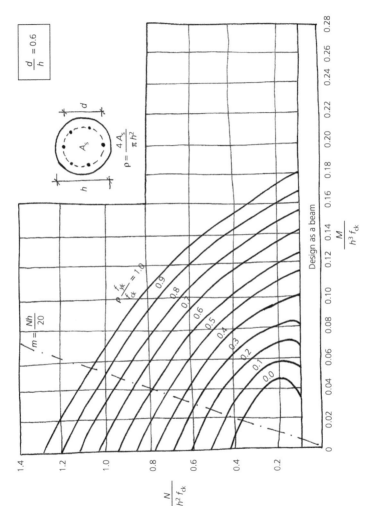

Stiffness and deflection – BS EN 1992-1 provides engineers the option of calculating deflections or using basic span/depth ratios which limit the total deflection to span/250 and live load and creep deflection to span/500 for spans up to 10 m.

Basic span/depth ratios for beams

Support conditions	Concrete highly stressed ($p \geq 1.5\%$)	Concrete lightly stressed ($p \leq 0.5\%$)
Simply supported	14	20
End span of continuous structure	18	26
Interior span of continuous structure	20	30
Flat slab	17	24
Cameleer	6	8

Notes:
[a] Span/depth ratios are conservative when compared with calculated results.
[b] For two-way spanning slab, check deflections on shorter span.
[c] For flat slabs, check deflections on longer span.
[d] Span depth varies need detailed consideration if formwork is to be struck early.
[e] If $f_{yk} > 500$ N/mm² multiply the tabulated values by $500/f_{yk}$.
[f] $\rho = (A_s/bd)$ calculated for the maximum span moment.
[g] Values for C30/37 concrete.

Allowable span/depth ratio
Allowable span/depth $= F_1 \times F_2 \times F_3 \times F_4 \times$ Basic span/depth ratio

where
    F_1 modification factor for ribbed slabs

$$F_1 = 1 - 0.1\left(\frac{b_f}{b_w} - 1\right) \geq 0.8 \text{ or } F_1 = 1.0$$

F_2 modification for flat slabs >8.5 m and brittle partitions

$$F_2 = 8.5/L_{eff} \leq 1.0$$

F_3 modification for stress in reinforcement

$$F_3 = 310/6_s \equiv \frac{500 A_{s\,prov}}{f_{yk}} A_{s\,req} \leq 1.5$$

Source: NA to BS EN 1992-1-1:2004: T NA.5.

Selected detailing rules for high yield reinforcement to BS EN 1992-1-1 – Generally, no more than four bars should be arranged in contact at any point. Minimum percentages of reinforcement are intended to control cracking and maximum percentages are intended to ensure that concrete can be placed and adequately compacted around the reinforcement. A_c is the area of the concrete section. Simplified rules are given below.

BS EN 1992 detailing rules differ from those in BS8110 and therefore the rules are incompatible. If designing to BS EN 1992, ensure the reinforcement detailing and scheduling follows Eurocode rules to avoid design failures.

Minimum percentages of reinforcement – Elements containing less reinforcement than A_s min should be considered unreinforced

For tension reinforcement in rectangular beams/slabs in bending	A_s min = 0.13% A_c
For compression reinforcement (if required) in rectangular beams in bending	A_s min = 0.2% A_c
For compression reinforcement in columns	A_s min = 0.4% A_c
For compression reinforcement in walls	A_s min = 0.2% A_c
For lateral reinforcement in walls	A_s min = 0.1% A_c

Maximum percentages of reinforcement

For beams	A_s max = 4% A_c
For vertically cast columns and walls	A_s max = 4% A_c
For horizontally cast columns	A_s max = 6% A_c
At lap positions in vertically or horizontally cast columns	A_s max = 8 = 4% A_c

Typical bond lengths
Bond is the friction and adhesion between the concrete and the steel reinforcement. It depends on the properties of the concrete and steel, as well as the position of the bars within the concrete. Bond forces are transferred through the concrete rather than relying on contact between steel bars. Deformed Type 2 high yield bars are the most commonly used. For a bar diameter, ϕ, typical bond lengths for tension and compression laps are 46ϕ, 43ϕ, or 39ϕ, for C25/30, C28/35 or C32/40, respectively. In 'poor' bond conditions these lengths should be increased by a factor of (at least) 1.5.

Source: BS EN 1992-1-1:2004; Section 9.

Reinforcement bar bending to BS 8666

BS 8666 sets down the specification for the scheduling, dimensioning, bending and cutting of steel reinforcement for concrete.

Minimum scheduling radius, diameter and bending allowances for reinforcement bars (mm)

Nominal bar diameter, D	Minimum radius for schedule, R	Minimum diameter of bending former, M	Minimum end dimension P	
			Bend ≥ 150° (min 5d straight)	Bend <150° (min 10d straight)
6	12	24	110	110
8	16	32	115	115
10	20	40	120	130
12	24	48	125	160
16	32	64	130	210
20	70	140	190	290
25	87	175	240	365
32	112	224	305	465
40	140	280	380	580
50	175	350	475	725

Note:

[a] Grade 250 bars are no longer commonly used.

[b] Grade H bars (formerly known as T) denote high yield Type 2 deformed bars $f_y = 500N/mm^2$. Ductility grades A, B and C are available within the classification, with B being most common, C being used where extra ductility is required (e.g. earthquake design) and A for bars (12 mm diameter and less) being bent to tight radii where accuracy is particularly important.

[c] Due to 'spring back' the actual bend radius will be slightly greater than half of the bending former diameter.

Source: BS 8666:2005.

Bar bending shape codes to BS 8666

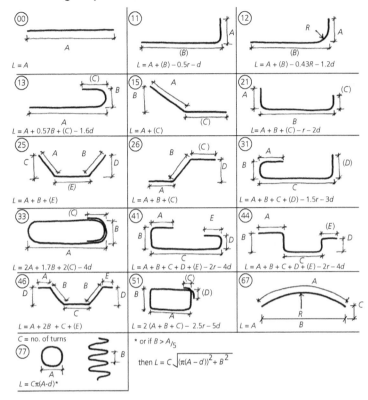

Source: BS 8666:2005.

Summary of differences with BS 8110

	BS EN 1992	BS 8110
Concrete strength	Cylinder strength, f_{ck}, is used throughout as a measure of concrete strength. Great care must be taken to use cylinder strengths in Eurocode calculations to avoid erroneously overestimating member strength!	Cube strength, F_{cu}, is used throughout as a measure of concrete strength. Cylinder strength is about 80% of the cube strength, which are now both reflected in concrete class references. For example: C28/35, where the 28 N/mm² is the cylinder strength and 35 N/mm² is the cube strength
Concrete strength	High strength concrete (up to C90/105) can be used in designs	Methodology based on typical concrete strengths being in the C25/30 to C40/50 range
Layout	Set out in sections based on load phenomenon, e.g. bending, shear, deflection, etc. For new users this can make it difficult initially to locate relevant design clauses. For example, axial loading (i.e. column, wall, pile design, etc.) is considered in Section 5.8 under 'Analysis of second order effects with axial load' quite separate from all the other 'Ultimate limit states' covered in Section 6	Set out in sections based on building element
Notional horizontal loads	Considered in addition to lateral loads	Generally considered only if greater than imposed lateral loads
Concrete cover	Nominal axis distance (NAD) refers to the dimension from the centre of the bar to the outside face of the element for checking fire resistance and durability. Expect larger dimensions than BS8110 and ensure detailing is completed appropriately on reinforcement schedules and drawings	Concrete cover for durability and fire resistance purposes is based on the distance from the face of the concrete to the face of the reinforcement
Durability	Requirements explicit: $c_{min} + \Delta c_{,dev}$	Requirements slightly vague on how/what fixing tolerances should be included over and above minimum cover figures provided
Bending formulae	BS EN 1992 provides stress block diagrams and expects engineers to derive their own bending formulae based on a myriad of clauses for factors and subfactors, in both the core Standard and National Annexe.	BS8110 provides stress block diagrams and the bending formulae which have been derived from these
Shear convention	Proposes the use of shear forces, but in practice the formulae provided are more manageable for hand calculation if shear stresses are compared	Proposes the use of shear stresses

continued

(continued)

	BS EN 1992	**BS 8110**
Maximum shear	No maximum shear capacity proposed. Possibly good practice to adhere to the BS8110 limits in the absence of other guidance.	Maximum shear capacity proposed is $0.8\sqrt{F_{cu}}$ or 5 N/mm²
Shear links	BS EN 1992 shear reinforcement should be designed to provide all of the shear capacity when the applied shear exceeds the basic shear resistance of the concrete	BS 8110 permits shear capacity to be provided by a combination of concrete and reinforcement strengths
Detailing for shear resistance	BS EN 1992 shear reinforcement calculations result in the use of less shear reinforcement, but require the main longitudinal bars to be extended further towards the supports. Care must therefore be taken to check that any rebar curtailment and detailing matches the Eurocode, rather than British Standard, rules	BS 8110 shear reinforcement calculations allow curtailment of the main bars on the basis that more shear reinforcement is used
Punching shear	Punching shear perimeters are calculated using rounded corners and maximum radii for any shear reinforcement is easier to calculate than in BS 8110	Punching shear perimeters are calculated using sharp corners
Deflection calculations	May be carried out using two methods: basic span/depth ratios or by theoretical calculation	Carried out using basic span/depth ratios only
Reinforcement strengths	Design methods based on the use of high strength reinforcement only	Design methods for plain or mild steel included alongside high strength reinforcement

Reinforcement Estimates

'Like fountain pens, motor cars and wives, steel estimates have some personal features. It is difficult to lay down hard and fast rules and one can only provide a guide to the uninitiated'. This marvellous (but now rather dated) quote was the introduction to an unpublished guide to better reinforcement estimates. These estimates are difficult to get right and the best estimate is based on a proper design and calculations.

DO NOT: Give a reinforcement estimate to anyone without an independent check by another engineer.

DO: Remember that you use more steel than you think and that although you may remember to be generous, you will inevitably omit more than you overestimate. Compare estimates with similar previous projects. Try to keep the QS happy by differentiating between mild and high tensile steel, straight and bent bars, and bars of different sizes. Apply a factor of safety to the final estimate. Keep a running total of the steel scheduled during preparation of the reinforcement drawings so that if the original estimate starts to look tight, it may be possible to make the ongoing steel detailing more economical.

As a useful check on a detailed estimate, the following are typical reinforcement quantities found in different structural elements:

Slabs	80–110 kg/m³
RC pad footings	70–90 kg/m³
Transfer slabs	150 kg/m³
Pile caps/rafts	115 kg/m³
Columns	150–450 kg/m³
Ground beams	230 kg/m³
Beams	220 kg/m³
Retaining walls	110 kg/m³
Stairs	135 kg/m³
Walls	65 kg/m³

'All up' estimates for different building types:

Heavy industrial	125 kg/m³
Commercial	95 kg/m³
Institutional	85 kg/m³

Source: Price and Myers. 2001.

8
Structural Steel

The method of heating iron ore in a charcoal fire determines the amount of carbon in the iron alloy. The following three iron ore products contain differing amounts of carbon: cast iron, wrought iron and steel.

Cast iron involves the heat treatment of iron castings and was developed as part of the industrial revolution between 1800 and 1900. It has a high carbon content and is therefore quite brittle, which means that it has a much greater strength in compression than in tension. Typical allowable working stresses were 23 N/mm^2 tension, 123 N/mm^2 compression and 30 N/mm^2 shear.

Wrought iron has relatively uniform properties and, between the 1840s and 1900, wrought iron took over from cast iron for structural use, until it was in turn superseded by mild steel. Typical allowable working stresses were 81 N/mm^2 tension, 61 N/mm^2 compression and 77 N/mm^2 shear.

'Steel' can cover many different alloys of iron, carbon and other alloying elements to alter the properties of the alloys. Steel can be formed into structural sections by casting, hot rolling or cold rolling. Mild steel, which is now mostly used for structural work, was first introduced in the mid-nineteenth century.

Types of steel products

Cast steel

Castings are generally used for complex or non-standard structural components. The casting shape and moulding process must be carefully controlled to limit residual stresses. Sand casting is a very common method, but the lost wax method is generally used where a very fine surface finish is required.

Cold rolled

Cold rolling is commonly used for lightweight sections, such as purlins and wind posts. Work hardening and residual stresses caused by the cold working cause an increase in the yield strength but this is at the expense of ductility and toughness. Cold-rolled steel cannot be designed using the same method as hot-rolled steel and design methods are given in BS 5950: Part 5.

Hot rolled steel

Most steel in the United Kingdom is produced by continuous casting where ingots or slabs are preheated to about 1300°C and the working temperatures fall as processing continues through the intermediate stages. The total amount of rolling work and the finishing temperatures are controlled to keep the steel grain size fine, which gives a good combination of strength and toughness. Although hollow sections (RHS, CHS and SHS) are often cold bent into shape, they tend to be hot finished and are considered 'hot rolled' for design purposes. This pocket book deals only with hot-rolled steel.

Summary of hot-rolled steel material properties

Density	78.5 kN/m^3
Tensile strength	275–460 N/mm^2 yield stress and 430–550 N/mm^2 ultimate strength
Poisson's ratio	0.3
Modulus of elastiity, _E_	205–210 kN/mm^2
Modulus of rigidity, _G_	80–81 kN/mm^2
Linear coefficient of thermal expansion	12×10^{-6}/°C

Mild steel section sizes and tolerances

Fabrication tolerances

BS 4 covers the dimensions of many of the hot-rolled sections produced by Corus. Selected rolling tolerances for different sections are covered by the following standards:

UB and UC sections: BS EN 10034

Section height (mm)	$h \leq 180$	$180 < h \leq 400$	$400 < h \leq 700$	$700 < h$
Tolerance (mm)	+3/–2	+4/–2	+5/–3	±5

Flange width (mm)	$b \leq 110$	$110 < b < 210$	$210 < b \leq 325$	$325 < b$
Tolerance (mm)	+4/ –1	+4/ –2	±4	+6/–5

Out of squareness for flange width (mm)	$b \leq 110$	$110 < b$
Tolerance (mm)	1.5	2% of b up to maximum 6.5 mm

Straightness for section height (mm)	$80 < h \leq 180$	$180 < h \leq 360$	$360 < h$
Tolerance on section length (mm)	$0.003L$	$0.0015L$	$0.001L$

RSA sections: BS EN 10056-2

Leg length (mm)	$h \leq 50$	$50 < h \leq 100$	$100 < h \leq 150$	$150 < h \leq 200$	$200 < h$
Tolerance (mm)	±1	±2	±3	±4	+6/–4

Straightness for section height	$h \leq 150$	$h \leq 200$	$200 < h$
Tolerance along section length (mm)	$0.004L$	$0.002L$	$0.001L$

PFC sections: BS EN 10279

Section height (mm)	$h \leq 65$	$65 < h \leq 200$	$200 < h \leq 400$	$400 < h$
Tolerance (mm)	±1.5	±2	±3	±4

Out of squareness for flange width		$b \leq 100$	$100 < b$
Tolerance (mm)		2.0	2.5% of b

Straightness	$h \leq 150$	$150 < h \leq 300$	$300 < h$
Tolerance across flanges (mm)	0.005L	0.003L	0.002L
Tolerance parallel to web (mm)	0.003L	0.002L	0.0015L

Hot finished RHS[a] SHS and CHS sections: BS EN 10210-2

Straightness	0.2%L and 3 mm over any 1 m
Depth, breadth of diameter	± 1% (minimum ±0.5 mm and maximum ±10 mm)
Squareness of side for SHS and RHS	90° ± 1°
Twist for SHS and RHS	2 mm + 0.5 mm per m maximum
Twist for EHS	4 mm + 1.0 mm per m maximum

Examples of minimum bend radii for selected steel sections

The minimum radius to which any section can be curved depends on its metallurgical properties, particularly its ductility, cross-sectional geometry and end use (the latter determines the standard required for the appearance of the work). It is therefore not realistic to provide a definitive list of the radii to which every section can be curved due to the wide number of end uses, but a selection of examples is possible. Normal bending tolerances are about 8 mm on the radius. In cold rolling the steel is deformed in the yield stress range and therefore becomes work hardened and displays different mechanical properties (notably a loss of ductility). However, if the section is designed to be working in the elastic range, there is generally no significant difference to its performance.

Section	Typical bend radius for S275 steel (m)
610 × 305 UB 238	40.0
533 × 210 UB 122	30.0
305 × 165 UB 40	15.0
250 × 150 × 12.5 RHS	9.0
305 × 305 UC 118	5.5
300 × 100 PFC 46	4.6
150 × 150 × 12.5 SHS	3.0
254 × 203 RSJ 82	2.4
191 × 229 TEE 49	1.5
152 × 152 UC 37	1.5
125 × 65 PFC 15	1.0
152 × 127 RSJ 37	0.8

Source: Angle Ring Company Limited. 2002.

Hot rolled section tables

UK beams – dimensions and properties

Section designation	Mass per metre	Depth of section	Width of section	Thickness		Root radius	Depth between fillets	Ratios for local buckling		Buckling ratio
		h	b	Web t_w	Flange t_f	r	d or c_w	Web c_w/t_w	Flange c_f/t_f	h/t_f or
	kg/m	mm	mm	mm	mm	mm	mm			D/T
British Standard Notation		h	b	t	T	r	d	d/t	–	d/t
$1016 \times 305 \times 487^a$	486.7	1036.3	308.5	30	54.1	30	868.1	28.9	2.02	19
$1016 \times 305 \times 437^a$	437	1026.1	305.4	26.9	49	30	868.1	32.3	2.23	21
$1016 \times 305 \times 393^a$	392.7	1015.9	303	24.4	43.9	30	868.1	35.6	2.49	23
$1016 \times 305 \times 349^a$	349.4	1008.1	302	21.1	40	30	868.1	41.1	2.76	25
$1016 \times 305 \times 314^a$	314.3	999.9	300	19.1	35.9	30	868.1	45.5	3.08	28
$1016 \times 305 \times 272^a$	272.3	990.1	300	16.5	31	30	868.1	52.6	3.6	32
$1016 \times 305 \times 249^a$	248.7	980.1	300	16.5	26	30	868.1	52.6	4.3	38
$1016 \times 305 \times 222^a$	222	970.3	300	16	21.1	30	868.1	54.3	5.31	46
$914 \times 419 \times 388$	388	921	420.5	21.4	36.6	24.1	799.6	37.4	4.79	25
$914 \times 419 \times 343$	343.3	911.8	418.5	19.4	32	24.1	799.6	41.2	5.48	28
$914 \times 305 \times 289$	289.1	926.6	307.7	19.5	32	19.1	824.4	42.3	3.91	29
$914 \times 305 \times 253$	253.4	918.4	305.5	17.3	27.9	19.1	824.4	47.7	4.48	33
$914 \times 305 \times 224$	224.2	910.4	304.1	15.9	23.9	19.1	824.4	51.8	5.23	38
$914 \times 305 \times 201$	200.9	903	303.3	15.1	20.2	19.1	824.4	54.6	6.19	45
$838 \times 292 \times 226$	226.5	850.9	293.8	16.1	26.8	17.8	761.7	47.3	4.52	32
$838 \times 292 \times 194$	193.8	840.7	292.4	14.7	21.7	17.8	761.7	51.8	5.58	39
$838 \times 292 \times 176$	175.9	834.9	291.7	14	18.8	17.8	761.7	54.4	6.44	44
$762 \times 267 \times 197$	196.8	769.8	268	15.6	25.4	16.5	686	44	4.32	30
$762 \times 267 \times 173$	173	762.2	266.7	14.3	21.6	16.5	686	48	5.08	35
$762 \times 267 \times 147$	146.9	754	265.2	12.8	17.5	16.5	686	53.6	6.27	43
$762 \times 267 \times 134$	133.9	750	264.4	12	15.5	16.5	686	57.2	7.08	48
$686 \times 254 \times 170$	170.2	692.9	255.8	14.5	23.7	15.2	615.1	42.4	4.45	29
$686 \times 254 \times 152$	152.4	687.5	254.5	13.2	21	15.2	615.1	46.6	5.02	33
$686 \times 254 \times 140$	140.1	683.5	253.7	12.4	19	15.2	615.1	49.6	5.55	36
$686 \times 254 \times 125$	125.2	677.9	253	11.7	16.2	15.2	615.1	52.6	6.51	42
$610 \times 305 \times 238$	238.1	635.8	311.4	18.4	31.4	16.5	540	29.3	4.14	20
$610 \times 305 \times 179$	179	620.2	307.1	14.1	23.6	16.5	540	38.3	5.51	26
$610 \times 305 \times 149$	149.2	612.4	304.8	11.8	19.7	16.5	540	45.8	6.6	31
$610 \times 229 \times 140$	139.9	617.2	230.2	13.1	22.1	12.7	547.6	41.8	4.34	28
$610 \times 229 \times 125$	125.1	612.2	229	11.9	19.6	12.7	547.6	46	4.89	31
$610 \times 229 \times 113$	113	607.6	228.2	11.1	17.3	12.7	547.6	49.3	5.54	35
$610 \times 229 \times 101$	101.2	602.6	227.6	10.5	14.8	12.7	547.6	52.2	6.48	41
$610 \times 178 \times 100^a$	100.3	607.4	179.2	11.3	17.2	12.7	547.6	48.5	4.14	35

Second moment of area		Radius of gyration		Elastic modulus		Plastic modulus		Buckling parameter	Torsional index	Warping constant	Torsional constant	Area of section
y–y	z–z	y–y	z–z	y–y	z–z	y–y	z–z					
I	I	i	i	W_{el}	W_{el}	W_{pl}	W_{pl}	U	X	I_w	I_T	A
cm⁴	cm⁴	cm	cm	cm³	cm³	cm³	cm³			dm⁶	cm⁴	cm²
I_{x-x}	I_{y-y}	r_{x-x}	r_{y-y}	Z_{x-x}	Z_{y-y}	S_{x-x}	S_{y-y}	U	x	H	J	
1020000	26700	40.6	6.57	19700	1730	23200	2800	0.867	21.1	64.4	4300	620
910000	23400	40.4	6.49	17700	1540	20800	2470	0.868	23.1	56	3190	557
808000	20500	40.2	6.4	15900	1350	18500	2170	0.868	25.5	48.4	2330	500
723000	18500	40.3	6.44	14300	1220	16600	1940	0.872	27.9	43.3	1720	445
644000	16200	40.1	6.37	12900	1080	14800	1710	0.872	30.7	37.7	1260	400
554000	14000	40	6.35	11200	934	12800	1470	0.872	35	32.2	835	347
481000	11800	39	6.09	9820	784	11300	1240	0.861	39.9	26.8	582	317
408000	9550	38	5.81	8410	636	9810	1020	0.85	45.7	21.5	390	283
720000	45400	38.2	9.59	15600	2160	17700	3340	0.885	26.7	88.9	1730	494
626000	39200	37.8	9.46	13700	1870	15500	2890	0.883	30.1	75.8	1190	437
504000	15600	37	6.51	10900	1010	12600	1600	0.867	31.9	31.2	926	368
436000	13300	36.8	6.42	9500	871	10900	1370	0.865	36.2	26.4	626	323
376000	11200	36.3	6.27	8270	739	9530	1160	0.86	41.3	22.1	422	286
325000	9420	35.7	6.07	7200	621	8350	982	0.853	46.9	18.4	291	256
340000	11400	34.3	6.27	7980	773	9160	1210	0.869	35	19.3	514	289
279000	9070	33.6	6.06	6640	620	7640	974	0.862	41.6	15.2	306	247
246000	7800	33.1	5.9	5890	535	6810	842	0.856	46.5	13	221	224
240000	8170	30.9	5.71	6230	610	7170	958	0.869	33.1	11.3	404	251
205000	6850	30.5	5.58	5390	514	6200	807	0.865	38	9.39	267	220
169000	5460	30	5.4	4470	411	5160	647	0.858	45.2	7.4	159	187
151000	4790	29.7	5.3	4020	362	4640	570	0.853	49.8	6.46	119	171
170000	6630	28	5.53	4920	518	5630	811	0.872	31.8	7.42	308	217
150000	5780	27.8	5.46	4370	455	5000	710	0.871	35.4	6.42	220	194
136000	5180	27.6	5.39	3990	409	4560	638	0.87	38.6	5.72	169	178
118000	4380	27.2	5.24	3480	346	3990	542	0.863	43.8	4.8	116	159
209000	15800	26.3	7.23	6590	1020	7490	1570	0.886	21.3	14.5	785	303
153000	11400	25.9	7.07	4930	743	5550	1140	0.885	27.7	10.2	340	228
126000	9310	25.7	7	4110	611	4590	937	0.886	32.7	8.17	200	190
112000	4510	25	5.03	3620	391	4140	611	0.875	30.6	3.99	216	178
98600	3930	24.9	4.97	3220	343	3680	535	0.875	34	3.45	154	159
87300	3430	24.6	4.88	2870	301	3280	469	0.87	38	2.99	111	144
75800	2910	24.2	4.75	2520	256	2880	400	0.863	43	2.52	77	129
72500	1660	23.8	3.6	2390	185	2790	296	0.854	38.7	1.44	95	128

continued

(continued) UK beams – dimensions and properties

Section designation	Mass per metre	Depth of section	Width of section	Thickness		Root radius	Depth between fillets	Ratios for local buckling		Buck-ling ratio
				Web	Flange		*d* or	Web	Flange	*h*/*t*_f
		h	*b*	*t*_w	*t*_f	*r*	*c*_w	*c*_w/*t*_w	*c*_f/*t*_f	or
	kg/m	mm	mm	mm	mm	mm	mm			*D/T*
British Standard Notation		*h*	*b*	*t*	*T*	*r*	*d*	*d/t*	–	*d/t*
610 × 178 × 92[a]	92.2	603	178.8	10.9	15	12.7	547.6	50.2	4.75	40
610 × 178 × 82[a]	81.8	598.6	177.9	10	12.8	12.7	547.6	54.8	5.57	47
533 × 312 × 272[a]	273.3	577.1	320.2	21.1	37.6	12.7	476.5	22.6	3.64	15
533 × 312 × 219[a]	218.8	560.3	317.4	18.3	29.2	12.7	476.5	26	4.69	19
533 × 312 × 182[a]	181.5	550.7	314.5	15.2	24.4	12.7	476.5	31.3	5.61	23
533 × 312 × 150[a]	150.6	542.5	312	12.7	20.3	12.7	476.5	37.5	6.75	27
533 × 210 × 138[a]	138.3	549.1	213.9	14.7	23.6	12.7	476.5	32.4	3.68	23
533 × 210 × 122[a]	122	544.5	211.9	12.7	21.3	12.7	476.5	37.5	4.08	26
533 × 210 × 109[a]	109	539.5	210.8	11.6	18.8	12.7	476.5	41.1	4.62	29
533 × 210 × 101[a]	101	536.7	210	10.8	17.4	12.7	476.5	44.1	4.99	31
533 × 210 × 92[a]	92.1	533.1	209.3	10.1	15.6	12.7	476.5	47.2	5.57	34
533 × 210 × 82[a]	82.2	528.3	208.8	9.6	13.2	12.7	476.5	49.6	6.58	40
533 × 165 × 85[a]	84.8	534.9	166.5	10.3	16.5	12.7	476.5	46.3	3.96	32
533 × 165 × 74[a]	74.7	529.1	165.9	9.7	13.6	12.7	476.5	49.1	4.81	39
533 × 165 × 66[a]	65.7	524.7	165.1	8.9	11.4	12.7	476.5	53.5	5.74	46
457 × 191 × 161[a]	161.4	492	199.4	18	32	10.2	407.6	22.6	2.52	15
457 × 191 × 133[a]	133.3	480.6	196.7	15.3	26.3	10.2	407.6	26.6	3.06	18
457 × 191 × 106[a]	105.8	469.2	194	12.6	20.6	10.2	407.6	32.3	3.91	23
457 × 191 × 98	98.3	467.2	192.8	11.4	19.6	10.2	407.6	35.8	4.11	24
457 × 191 × 89	89.3	463.4	191.9	10.5	17.7	10.2	407.6	38.8	4.55	26
457 × 191 × 82	82	460	191.3	9.9	16	10.2	407.6	41.2	5.03	29
457 × 191 × 74	74.3	457	190.4	9	14.5	10.2	407.6	45.3	5.55	32
457 × 191 × 67	67.1	453.4	189.9	8.5	12.7	10.2	407.6	48	6.34	36
457 × 152 × 82	82.1	465.8	155.3	10.5	18.9	10.2	407.6	38.8	3.29	25
457 × 152 × 74	74.2	462	154.4	9.6	17	10.2	407.6	42.5	3.66	27
457 × 152 × 67	67.2	458	153.8	9	15	10.2	407.6	45.3	4.15	31
457 × 152 × 60	59.8	454.6	152.9	8.1	13.3	10.2	407.6	50.3	4.68	34
457 × 152 × 52	52.3	449.8	152.4	7.6	10.9	10.2	407.6	53.6	5.71	41
406 × 178 × 85[a]	85.3	417.2	181.9	10.9	18.2	10.2	360.4	33.1	4.14	23
406 × 178 × 74	74.2	412.8	179.5	9.5	16	10.2	360.4	37.9	4.68	26
406 × 178 × 67	67.1	409.4	178.8	8.8	14.3	10.2	360.4	41	5.23	29
406 × 178 × 60	60.1	406.4	177.9	7.9	12.8	10.2	360.4	45.6	5.84	32
406 × 178 × 54	54.1	402.6	177.7	7.7	10.9	10.2	360.4	46.8	6.86	37

Second moment of area		Radius of gyration		Elastic modulus		Plastic modulus		Buckling para-meter	Torsio-nal index	Warping constant	Torsional constant	Area of section
y–y	z–z	y–y	z–z	y–y	z–z	y–y	z–z					
I	I	i	i	W_{el}	W_{el}	W_{pl}	W_{pl}	U	X	I_w	I_T	A
cm^4	cm^4	cm	cm	cm^3	cm^3	cm^3	cm^3			dm^6	cm^4	cm^2
I_{x-x}	I_{y-y}	r_{x-x}	r_{y-y}	Z_{x-x}	Z_{y-y}	S_{x-x}	S_{y-y}	U	x	H	J	
64600	1440	23.4	3.5	2140	161	2510	258	0.85	42.7	1.24	71	117
55900	1210	23.2	3.4	1870	136	2190	218	0.843	48.5	1.04	48.8	104
199000	20600	23.9	7.69	6890	1290	7870	1990	0.891	15.9	15	1290	348
151000	15600	23.3	7.48	5400	982	6120	1510	0.884	19.8	11	642	279
123000	12700	23.1	7.4	4480	806	5040	1240	0.886	23.4	8.77	373	231
101000	10300	22.9	7.32	3710	659	4150	1010	0.885	27.8	7.01	216	192
86100	3860	22.1	4.68	3140	361	3610	568	0.874	24.9	2.67	250	176
76000	3390	22.1	4.67	2790	320	3200	500	0.878	27.6	2.32	178	155
66800	2940	21.9	4.6	2480	279	2830	436	0.875	30.9	1.99	126	139
61500	2690	21.9	4.57	2290	256	2610	399	0.874	33.1	1.81	101	129
55200	2390	21.7	4.51	2070	228	2360	355	0.873	36.4	1.6	75.7	117
47500	2010	21.3	4.38	1800	192	2060	300	0.863	41.6	1.33	51.5	105
48500	1270	21.2	3.44	1820	153	2100	243	0.861	35.5	0.857	73.8	108
41100	1040	20.8	3.3	1550	125	1810	200	0.853	41.1	0.691	47.9	95.2
35000	859	20.5	3.2	1340	104	1560	166	0.847	47	0.566	32	83.7
79800	4250	19.7	4.55	3240	426	3780	672	0.881	16.5	2.25	515	206
63800	3350	19.4	4.44	2660	341	3070	535	0.879	19.6	1.73	292	170
48900	2510	19	4.32	2080	259	2390	405	0.876	24.4	1.27	146	135
45700	2350	19.1	4.33	1960	243	2230	379	0.881	25.8	1.18	121	125
41000	2090	19	4.29	1770	218	2010	338	0.878	28.3	1.04	90.7	114
37100	1870	18.8	4.23	1610	196	1830	304	0.879	30.8	0.922	69.2	104
33300	1670	18.8	4.2	1460	176	1650	272	0.877	33.8	0.818	51.8	94.6
29400	1450	18.5	4.12	1300	153	1470	237	0.873	37.8	0.705	37.1	85.5
36600	1180	18.7	3.37	1570	153	1810	240	0.872	27.4	0.591	89.2	105
32700	1050	18.6	3.33	1410	136	1630	213	0.872	30.1	0.518	65.9	94.5
28900	913	18.4	3.27	1260	119	1450	187	0.868	33.6	0.448	47.7	85.6
25500	795	18.3	3.23	1120	104	1290	163	0.868	37.5	0.387	33.8	76.2
21400	645	17.9	3.11	950	84.6	1100	133	0.859	43.8	0.311	21.4	66.6
31700	1830	17.1	4.11	1520	201	1730	313	0.88	24.4	0.728	93	109
27300	1550	17	4.04	1320	172	1500	267	0.882	27.5	0.608	62.8	94.5
24300	1360	16.9	3.99	1190	153	1350	237	0.88	30.4	0.533	46.1	85.5
21600	1200	16.8	3.97	1060	135	1200	209	0.88	33.7	0.466	33.3	76.5
18700	1020	16.5	3.85	930	115	1050	178	0.871	38.3	0.392	23.1	69

continued

(continued) UK beams – dimensions and properties

Section designation	Mass per metre	Depth of section	Width of section	Thickness Web	Thickness Flange	Root radius	Depth between fillets d or	Ratios for local buckling Web	Ratios for local buckling Flange	Buckling ratio h/t_f
		h	b	t_w	t_f	r	c_w	c_w/t_w	c_f/t_f	or
	kg/m	mm	mm	mm	mm	mm	mm			D/T
British Standard Notation		h	b	t	T	r	d	d/t	–	d/t
$406 \times 140 \times 53$[a]	53.3	406.6	143.3	7.9	12.9	10.2	360.4	45.6	4.46	32
$406 \times 140 \times 46$	46	403.2	142.2	6.8	11.2	10.2	360.4	53	5.13	36
$406 \times 140 \times 39$	39	398	141.8	6.4	8.6	10.2	360.4	56.3	6.69	46
$356 \times 171 \times 67$	67.1	363.4	173.2	9.1	15.7	10.2	311.6	34.2	4.58	23
$356 \times 171 \times 57$	57	358	172.2	8.1	13	10.2	311.6	38.5	5.53	28
$356 \times 171 \times 51$	51	355	171.5	7.4	11.5	10.2	311.6	42.1	6.25	31
$356 \times 171 \times 45$	45	351.4	171.1	7	9.7	10.2	311.6	44.5	7.41	36
$356 \times 127 \times 39$	39.1	353.4	126	6.6	10.7	10.2	311.6	47.2	4.63	33
$356 \times 127 \times 33$	33.1	349	125.4	6	8.5	10.2	311.6	51.9	5.82	41
$305 \times 165 \times 54$	54	310.4	166.9	7.9	13.7	8.9	265.2	33.6	5.15	23
$305 \times 165 \times 46$	46.1	306.6	165.7	6.7	11.8	8.9	265.2	39.6	5.98	26
$305 \times 165 \times 40$	40.3	303.4	165	6	10.2	8.9	265.2	44.2	6.92	30
$305 \times 127 \times 48$	48.1	311	125.3	9	14	8.9	265.2	29.5	3.52	22
$305 \times 127 \times 42$	41.9	307.2	124.3	8	12.1	8.9	265.2	33.2	4.07	25
$305 \times 127 \times 37$	37	304.4	123.4	7.1	10.7	8.9	265.2	37.4	4.6	28
$305 \times 102 \times 33$	32.8	312.7	102.4	6.6	10.8	7.6	275.9	41.8	3.73	29
$305 \times 102 \times 28$	28.2	308.7	101.8	6	8.8	7.6	275.9	46	4.58	35
$305 \times 102 \times 25$	24.8	305.1	101.6	5.8	7	7.6	275.9	47.6	5.76	44
$254 \times 146 \times 43$	43	259.6	147.3	7.2	12.7	7.6	219	30.4	4.92	20
$254 \times 146 \times 37$	37	256	146.4	6.3	10.9	7.6	219	34.8	5.73	23
$254 \times 146 \times 31$	31.1	251.4	146.1	6	8.6	7.6	219	36.5	7.26	29
$254 \times 102 \times 28$	28.3	260.4	102.2	6.3	10	7.6	225.2	35.7	4.04	26
$254 \times 102 \times 25$	25.2	257.2	101.9	6	8.4	7.6	225.2	37.5	4.8	31
$254 \times 102 \times 22$	22	254	101.6	5.7	6.8	7.6	225.2	39.5	5.93	37
$203 \times 133 \times 30$	30	206.8	133.9	6.4	9.6	7.6	172.4	26.9	5.85	22
$203 \times 133 \times 25$	25.1	203.2	133.2	5.7	7.8	7.6	172.4	30.2	7.2	26
$203 \times 102 \times 23$	23.1	203.2	101.8	5.4	9.3	7.6	169.4	31.4	4.37	22
$178 \times 102 \times 19$	19	177.8	101.2	4.8	7.9	7.6	146.8	30.6	5.14	23
$152 \times 89 \times 16$	16	152.4	88.7	4.5	7.7	7.6	121.8	27.1	4.48	20
$127 \times 76 \times 13$	13	127	76	4	7.6	7.6	96.6	24.2	3.74	17

Note:
[a] TATA Sections produced in addition to the range of BS Sections.

Source: TATA Steel. 2008 – reproduced with the kind permission of TATA.

Second moment of area		Radius of gyration		Elastic modulus		Plastic modulus		Buckling parameter	Torsional index	Warping constant	Torsional constant	Area of section
y–y	z–z	y–y	z–z	y–y	z–z	y–y	z–z					
I	I	i	i	W_{el}	W_{el}	W_{pl}	W_{pl}	U	X	I_w	I_T	A
cm⁴	cm⁴	cm	cm	cm³	cm³	cm³	cm³			dm⁶	cm⁴	cm²
I_{x-x}	I_{y-y}	r_{x-x}	r_{y-y}	Z_{x-x}	Z_{y-y}	S_{x-x}	S_{y-y}	U	x	H	J	
18300	635	16.4	3.06	899	88.6	1030	139	0.87	34.1	0.246	29	67.9
15700	538	16.4	3.03	778	75.7	888	118	0.871	39	0.207	19	58.6
12500	410	15.9	2.87	629	57.8	724	90.8	0.858	47.4	0.155	10.7	49.7
19500	1360	15.1	3.99	1070	157	1210	243	0.886	24.4	0.412	55.7	85.5
16000	1110	14.9	3.91	896	129	1010	199	0.882	28.8	0.33	33.4	72.6
14100	968	14.8	3.86	796	113	896	174	0.881	32.1	0.286	23.8	64.9
12100	811	14.5	3.76	687	94.8	775	147	0.874	36.8	0.237	15.8	57.3
10200	358	14.3	2.68	576	56.8	659	89	0.871	35.2	0.105	15.1	49.8
8250	280	14	2.58	473	44.7	543	70.2	0.863	42.1	0.081	8.79	42.1
11700	1060	13	3.93	754	127	846	196	0.889	23.6	0.234	34.8	68.8
9900	896	13	3.9	646	108	720	166	0.89	27.1	0.195	22.2	58.7
8500	764	12.9	3.86	560	92.6	623	142	0.889	31	0.164	14.7	51.3
9570	461	12.5	2.74	616	73.6	711	116	0.873	23.3	0.102	31.8	61.2
8200	389	12.4	2.7	534	62.6	614	98.4	0.872	26.5	0.085	21.1	53.4
7170	336	12.3	2.67	471	54.5	539	85.4	0.872	29.7	0.073	14.8	47.2
6500	194	12.5	2.15	416	37.9	481	60	0.867	31.6	0.044	12.2	41.8
5370	155	12.2	2.08	348	30.5	403	48.4	0.859	37.3	0.035	7.4	35.9
4460	123	11.9	1.97	292	24.2	342	38.8	0.846	43.4	0.027	4.77	31.6
6540	677	10.9	3.52	504	92	566	141	0.891	21.1	0.103	23.9	54.8
5540	571	10.8	3.48	433	78	483	119	0.89	24.3	0.086	15.3	47.2
4410	448	10.5	3.36	351	61.3	393	94.1	0.879	29.6	0.066	8.55	39.7
4000	179	10.5	2.22	308	34.9	353	54.8	0.873	27.5	0.028	9.57	36.1
3410	149	10.3	2.15	266	29.2	306	46	0.866	31.4	0.023	6.42	32
2840	119	10.1	2.06	224	23.5	259	37.3	0.856	36.3	0.018	4.15	28
2900	385	8.71	3.17	280	57.5	314	88.2	0.882	21.5	0.037	10.3	38.2
2340	308	8.56	3.1	230	46.2	258	70.9	0.876	25.6	0.029	5.96	32
2100	164	8.46	2.36	207	32.2	234	49.7	0.888	22.4	0.015	7.02	29.4
1360	137	7.48	2.37	153	27	171	41.6	0.886	22.6	0.010	4.41	24.3
834	89.8	6.41	2.1	109	20.2	123	31.2	0.89	19.5	0.005	3.56	20.3
473	55.7	5.35	1.84	74.6	14.7	84.2	22.6	0.894	16.3	0.002	2.85	16.5

UK columns – dimensions and properties

Section designation	Mass per metre	Depth of section	Width of section	Thickness		Root radius	Depth between fillets	Ratios for local buckling		Buckling ratio
				Web	Flange		d or	Web	Flange	h/t_f
	metre	h	b	t_w	t_f	r	c_w	c_w/t_w	c_f/t_f	or
	kg/m	mm	mm	mm	mm	mm	mm			D/T
British Standard Notation		h	b	t	T	r	d	d/t	–	d/t
356 × 406 × 634	633.9	474.6	424	47.6	77	15.2	290.2	6.1	2.25	6
356 × 406 × 551	551	455.6	418.5	42.1	67.5	15.2	290.2	6.89	2.56	7
356 × 406 × 467	467	436.6	412.2	35.8	58	15.2	290.2	8.11	2.98	8
356 × 406 × 393	393	419	407	30.6	49.2	15.2	290.2	9.48	3.52	9
356 × 406 × 340	339.9	406.4	403	26.6	42.9	15.2	290.2	10.9	4.03	9
356 × 406 × 287	287.1	393.6	399	22.6	36.5	15.2	290.2	12.8	4.74	11
356 × 406 × 235	235.1	381	394.8	18.4	30.2	15.2	290.2	15.8	5.73	13
356 × 368 × 202	201.9	374.6	374.7	16.5	27	15.2	290.2	17.6	6.07	14
356 × 368 × 177	177	368.2	372.6	14.4	23.8	15.2	290.2	20.2	6.89	15
356 × 368 × 153	152.9	362	370.5	12.3	20.7	15.2	290.2	23.6	7.92	17
356 × 368 × 129	129	355.6	368.6	10.4	17.5	15.2	290.2	27.9	9.37	20
305 × 305 × 283	282.9	365.3	322.2	26.8	44.1	15.2	246.7	9.21	3	8
305 × 305 × 240	240	352.5	318.4	23	37.7	15.2	246.7	10.7	3.51	9
305 × 305 × 198	198.1	339.9	314.5	19.1	31.4	15.2	246.7	12.9	4.22	11
305 × 305 × 158	158.1	327.1	311.2	15.8	25	15.2	246.7	15.6	5.3	13
305 × 305 × 137	136.9	320.5	309.2	13.8	21.7	15.2	246.7	17.9	6.11	15
305 × 305 × 118	117.9	314.5	307.4	12	18.7	15.2	246.7	20.6	7.09	17
305 × 305 × 97	96.9	307.9	305.3	9.9	15.4	15.2	246.7	24.9	8.6	20
254 × 254 × 167	167.1	289.1	265.2	19.2	31.7	12.7	200.3	10.4	3.48	9
254 × 254 × 132	132	276.3	261.3	15.3	25.3	12.7	200.3	13.1	4.36	11
254 × 254 × 107	107.1	266.7	258.8	12.8	20.5	12.7	200.3	15.6	5.38	13
254 × 254 × 89	88.9	260.3	256.3	10.3	17.3	12.7	200.3	19.4	6.38	15
254 × 254 × 73	73.1	254.1	254.6	8.6	14.2	12.7	200.3	23.3	7.77	18
203 × 203 × 127[a]	127.5	241.4	213.9	18.1	30.1	10.2	160.8	8.88	2.91	8
203 × 203 × 113[a]	113.5	235	212.1	16.3	26.9	10.2	160.8	9.87	3.26	9
203 × 203 × 100[a]	99.6	228.6	210.3	14.5	23.7	10.2	160.8	11.1	3.7	10
203 × 203 × 86	86.1	222.2	209.1	12.7	20.5	10.2	160.8	12.7	4.29	11
203 × 203 × 71	71	215.8	206.4	10	17.3	10.2	160.8	16.1	5.09	12
203 × 203 × 60	60	209.6	205.8	9.4	14.2	10.2	160.8	17.1	6.2	15
203 × 203 × 52	52	206.2	204.3	7.9	12.5	10.2	160.8	20.4	7.04	16
203 × 203 × 46	46.1	203.2	203.6	7.2	11	10.2	160.8	22.3	8	18
152 × 152 × 51[a]	51.2	170.2	157.4	11	15.7	7.6	123.6	11.2	4.18	11
152 × 152 × 44[a]	44	166	155.9	9.5	13.6	7.6	123.6	13	4.82	12
152 × 152 × 37	37	161.8	154.4	8	11.5	7.6	123.6	15.5	5.7	14
152 × 152 × 30	30	157.6	152.9	6.5	9.4	7.6	123.6	19	6.98	17
152 × 152 × 23	23	152.4	152.2	5.8	6.8	7.6	123.6	21.3	9.65	22

Note:
[a] TATA Sections produced in addition to the range of BS Sections.

Source: TATA Steel. 2008 – reproduced with the kind permission of TATA.

Second moment of area		Radius of gyration		Elastic modulus		Plastic modulus		Buckling para-meter	Torsio-nal index	Warping constant	Torsional constant	Area of section
y–y	z–z	y–y	z–z	y–y	z–z	y–y	z–z					
I	I	i	i	W_{el}	W_{el}	W_{pl}	W_{pl}	U	X	I_w	I_T	A
cm⁴	cm⁴	cm	cm	cm³	cm³	cm³	cm³			dm⁶	cm⁴	cm²
I_{x-x}	I_{y-y}	r_{x-x}	r_{y-y}	Z_{x-x}	Z_{y-y}	S_{x-x}	S_{y-y}	U	x	H	J	
275000	98100	18.4	11	11600	4630	14200	7110	0.843	5.46	38.8	13700	808
227000	82700	18	10.9	9960	3950	12100	6060	0.841	6.05	31.1	9240	702
183000	67800	17.5	10.7	8380	3290	10000	5030	0.839	6.86	24.3	5810	595
147000	55400	17.1	10.5	7000	2720	8220	4150	0.837	7.86	18.9	3550	501
123000	46900	16.8	10.4	6030	2330	7000	3540	0.836	8.85	15.5	2340	433
99900	38700	16.5	10.3	5070	1940	5810	2950	0.835	10.2	12.3	1440	366
79100	31000	16.3	10.2	4150	1570	4690	2380	0.834	12.1	9.54	812	299
66300	23700	16.1	9.6	3540	1260	3970	1920	0.844	13.4	7.16	558	257
57100	20500	15.9	9.54	3100	1100	3460	1670	0.844	15	6.09	381	226
48600	17600	15.8	9.49	2680	948	2960	1430	0.844	17	5.11	251	195
40200	14600	15.6	9.43	2260	793	2480	1200	0.844	19.9	4.18	153	164
78900	24600	14.8	8.27	4320	1530	5110	2340	0.855	7.65	6.35	2030	360
64200	20300	14.5	8.15	3640	1280	4250	1950	0.854	8.74	5.03	1270	306
50900	16300	14.2	8.04	3000	1040	3440	1580	0.854	10.2	3.88	734	252
38700	12600	13.9	7.9	2370	808	2680	1230	0.851	12.5	2.87	378	201
32800	10700	13.7	7.83	2050	692	2300	1050	0.851	14.2	2.39	249	174
27700	9060	13.6	7.77	1760	589	1960	895	0.85	16.2	1.98	161	150
22200	7310	13.4	7.69	1450	479	1590	726	0.85	19.3	1.56	91.2	123
30000	9870	11.9	6.81	2080	744	2420	1140	0.851	8.49	1.63	626	213
22500	7530	11.6	6.69	1630	576	1870	878	0.85	10.3	1.19	319	168
17500	5930	11.3	6.59	1310	458	1480	697	0.848	12.4	0.898	172	136
14300	4860	11.2	6.55	1100	379	1220	575	0.85	14.5	0.717	102	113
11400	3910	11.1	6.48	898	307	992	465	0.849	17.3	0.562	57.6	93.1
15400	4920	9.75	5.5	1280	460	1520	704	0.854	7.38	0.549	427	162
13300	4290	9.59	5.45	1130	404	1330	618	0.853	8.11	0.464	305	145
11300	3680	9.44	5.39	988	350	1150	534	0.852	9.02	0.386	210	127
9450	3130	9.28	5.34	850	299	977	456	0.85	10.2	0.318	137	110
7620	2540	9.18	5.3	706	246	799	374	0.853	11.9	0.25	80.2	90.4
6130	2070	8.96	5.2	584	201	656	305	0.846	14.1	0.197	47.2	76.4
5260	1780	8.91	5.18	510	174	567	264	0.848	15.8	0.167	31.8	66.3
4570	1550	8.82	5.13	450	152	497	231	0.847	17.7	0.143	22.2	58.7
3230	1020	7.04	3.96	379	130	438	199	0.848	10.1	0.061	48.8	65.2
2700	860	6.94	3.92	326	110	372	169	0.848	11.5	0.0499	31.7	56.1
2210	706	6.85	3.87	273	91.5	309	140	0.848	13.3	0.0399	19.2	47.1
1750	560	6.76	3.83	222	73.3	248	112	0.849	16	0.0308	10.5	38.3
1250	400	6.54	3.7	164	52.6	182	80.1	0.84	20.7	0.0212	4.63	29.2

Rolled joists – dimensions and properties

Inside slope = 8°

Section designation	Mass per metre	Depth of section	Width of section	Thickness		Radius		Depth between fillets	Ratios for local buckling		Buckling ratio
				Web	Flange	Root	Toe	d or	Web	Flange	h/t_f
		h	b	t_w	t_f	r_1	r_2	c_w	c_w/t_w	c_f/t_f	or
	kg/m	mm	mm	mm	mm	mm	mm	mm			D/T
British Standard Notation		h	b	t	T	r_1	r_2	d	d/t	–	d/t
$254 \times 203 \times 82$	82	254	203.2	10.2	19.9	19.6	9.7	166.6	16.3	3.86	12.7638191
$254 \times 114 \times 37$	37.2	254	114.3	7.6	12.8	12.4	6.1	199.3	26.2	3.20	19.84375
$203 \times 152 \times 52$	52.3	203.2	152.4	8.9	16.5	15.5	7.6	133.2	15.0	3.41	12.31515152
$152 \times 127 \times 37$	37.3	152.4	127	10.4	13.2	13.5	6.6	94.3	9.1	3.39	11.54545455
$127 \times 114 \times 29$	29.3	127	114.3	10.2	11.5	9.9	4.8	79.5	7.8	3.67	11.04347826
$127 \times 114 \times 27$	26.9	127	114.3	7.4	11.4	9.9	5	79.5	10.7	3.82	11.14035088
$127 \times 76 \times 16$	16.5	127	76.2	5.6	9.6	9.4	4.6	86.5	15.4	2.70	13.22916667
$114 \times 114 \times 27$	26.9	114.3	114.3	9.5	10.7	14.2	3.2	60.8	6.4	3.57	10.68224299
$102 \times 102 \times 23$	23	101.6	101.6	9.5	10.3	11.1	3.2	55.2	5.8	3.39	9.86407767
$102 \times 44 \times 7$	7.5	101.6	44.5	4.3	6.1	6.9	3.3	74.6	17.3	2.16	16.6557377
$89 \times 89 \times 19$	19.5	88.9	88.9	9.5	9.9	11.1	3.2	44.2	4.7	2.89	8.97979798
$76 \times 76 \times 15$	15	76.2	80	8.9	8.4	9.4	4.6	38.1	4.3	3.11	9.071428571
$76 \times 76 \times 13$	12.8	76.2	76.2	5.1	8.4	9.4	4.6	38.1	7.5	3.11	9.071428571

Source: TATA Steel. 2008 – reproduced with the kind permission of TATA.

Second moment of area		Radius of gyration		Elastic modulus		Plastic modulus		Buckling parameter	Torsional index	Warping constant	Torsional constant	Area of section
y–y	z–z	y–y	z–z	y–y	z–z	y–y	z–z					
I	I	i	i	W_{el}	W_{el}	W_{pl}	W_{pl}	U	X	I_w	I_T	A
cm⁴	cm⁴	cm	cm	cm³	cm³	cm³	cm³			dm⁶	cm⁴	cm²
I_{x-x}	I_{y-y}	r_{x-x}	r_{y-y}	Z_{x-x}	Z_{y-y}	S_{x-x}	S_{y-y}	U	x	H	J	
12000	2280	10.7	4.67	947	224	1080	371	0.888	11	0.312	152	105
5080	269	10.4	2.39	400	47.1	459	79.1	0.885	18.7	0.0392	25.2	47.3
4800	816	8.49	3.5	472	107	541	176	0.89	10.7	0.0711	64.9	66.6
1820	378	6.19	2.82	239	59.6	279	99.8	0.867	9.33	0.0183	33.9	47.5
979	242	5.12	2.54	154	42.3	181	70.8	0.853	8.77	0.00807	20.8	37.4
946	236	5.26	2.63	149	41.3	172	68.2	0.868	9.31	0.00788	16.9	34.2
571	60.8	5.21	1.7	90	16	104	26.4	0.891	11.8	0.0021	6.73	21.1
736	224	4.62	2.55	129	39.2	151	65.8	0.839	7.92	0.000601	18.9	34.5
486	154	4.07	2.29	95.6	30.3	113	50.6	0.836	7.42	0.00321	14.2	29.3
153	7.82	4.01	0.907	30.1	3.51	35.4	6.03	0.872	14.9	0.000178	1.26	9.5
307	101	3.51	2.02	69	22.8	82.7	38	0.83	6.58	0.00158	11.5	24.9
172	60.9	3	1.78	45.2	15.2	54.2	25.8	0.82	6.42	0.0007	6.84	19.1
158	51.8	3.12	1.79	41.5	13.6	48.7	22.4	0.853	7.21	0.000595	4.6	16.2

UK parallel flange channels – dimensions and properties

Section designation	Mass per metre	Depth of section	Width of section	Thickness		Root radius	Depth between fillets	Ratios for local buckling		Distance
				Web	Flange		d or	Web	Flange	
	metre	h	b	t_w	t_f	r	c_w	c_w/t_w	c_f/t_f	e_0
	kg/m	mm	mm	mm	mm	mm	mm			cm
British Standard Notation		h	b	t	T	r	d	d/t	–	e
$430 \times 100 \times 64$	64.4	430	100	11	19	15	362	32.9	3.89	3.27
$380 \times 100 \times 54$	54	380	100	9.5	17.5	15	315	33.2	4.31	3.48
$300 \times 100 \times 46$	45.5	300	100	9	16.5	15	237	26.3	4.61	3.68
$300 \times 90 \times 41$	41.4	300	90	9	15.5	12	245	27.2	4.45	3.18
$260 \times 90 \times 35$	34.8	260	90	8	14	12	208	26	5	3.32
$260 \times 75 \times 28$	27.6	260	75	7	12	12	212	30.3	4.67	2.62
$230 \times 90 \times 32$	32.2	230	90	7.5	14	12	178	23.7	5.04	3.46
$230 \times 75 \times 26$	25.7	230	75	6.5	12.5	12	181	27.8	4.52	2.78
$200 \times 90 \times 30$	29.7	200	90	7	14	12	148	21.1	5.07	3.6
$200 \times 75 \times 23$	23.4	200	75	6	12.5	12	151	25.2	4.56	2.91
$180 \times 90 \times 26$	26.1	180	90	6.5	12.5	12	131	20.2	5.72	3.64
$180 \times 75 \times 20$	20.3	180	75	6	10.5	12	135	22.5	5.43	2.87
$150 \times 90 \times 24$	23.9	150	90	6.5	12	12	102	15.7	5.96	3.71
$150 \times 75 \times 18$	17.9	150	75	5.5	10	12	106	19.3	5.75	2.99
$125 \times 65 \times 15$	14.8	125	65	5.5	9.5	12	82	14.9	5	2.56
$100 \times 50 \times 10$	10.2	100	50	5	8.5	9	65	13	4.24	1.94

Source: TATA Steel. 2008 – reproduced with the kind permission of TATA.

Buckling ratio	Second moment of area		Radius of gyration		Elastic modulus		Plastic modulus		Buckling parameter	Torsional index	Warping constant	Torsional constant	Area of section
h/t_f	y–y	z–z	y–y	z–z	y–y	z–z	y–y	z–z					
or	I	I	i	i	W_{el}	W_{el}	W_{pl}	W_{pl}	U	X	I_w	I_T	A
D/T	cm^4	cm^4	cm	cm	cm^3	cm^3	cm^3	cm^3			dm^6	cm^4	cm^2
d/t	$I_{x\text{-}x}$	$I_{y\text{-}y}$	$r_{x\text{-}x}$	$r_{y\text{-}y}$	$Z_{x\text{-}x}$	$Z_{y\text{-}y}$	$S_{x\text{-}x}$	$S_{y\text{-}y}$	U	x	H	J	
23	21900	722	16.3	2.97	1020	97.9	1220	176	0.917	22.5	0.219	63	82.1
22	15000	643	14.8	3.06	791	89.2	933	161	0.932	21.2	0.15	45.7	68.7
18	8230	568	11.9	3.13	549	81.7	641	148	0.944	17	0.081	36.8	58
19	7220	404	11.7	2.77	481	63.1	568	114	0.934	18.4	0.058	28.8	52.7
19	4730	353	10.3	2.82	364	56.3	425	102	0.942	17.2	0.038	20.6	44.4
22	3620	185	10.1	2.3	278	34.4	328	62	0.932	20.5	0.020	11.7	35.1
16	3520	334	9.27	2.86	306	55	355	98.9	0.95	15.1	0.028	19.3	41
18	2750	181	9.17	2.35	239	34.8	278	63.2	0.947	17.3	0.015	11.8	32.7
14	2520	314	8.16	2.88	252	53.4	291	94.5	0.954	12.9	0.020	18.3	37.9
16	1960	170	8.11	2.39	196	33.8	227	60.6	0.956	14.8	0.011	11.1	29.9
14	1820	277	7.4	2.89	202	47.4	232	83.5	0.949	12.8	0.014	13.3	33.2
17	1370	146	7.27	2.38	152	28.8	176	51.8	0.946	15.3	0.008	7.34	25.9
13	1160	253	6.18	2.89	155	44.4	179	76.9	0.936	10.8	0.009	11.8	30.4
15	861	131	6.15	2.4	115	26.6	132	47.2	0.946	13.1	0.005	6.1	22.8
13	483	80	5.07	2.06	77.3	18.8	89.9	33.2	0.942	11.1	0.002	4.72	18.8
12	208	32.3	4	1.58	41.5	9.89	48.9	17.5	0.942	10	0.000	2.53	13

UK equal angles – dimensions and properties

Section designation		Mass per metre	Radius		Area of section	Distance to centroid	Second moment of area		
	Thick-ness		Root	Toe			$y–y, z–z$	$u–u$	$v–v$
Size									
$h \times h$	t		r_1	r_2		c	I	I	I
mm	mm	kg/m	mm	mm	cm²	cm	cm⁴	cm⁴	cm⁴
British Standard Notation							$I_{x–x, y–y}$	$I_{u–u}$	$I_{v–v}$
$200 \times 200 \times 24$		71.1	18	9	90.6	5.84	3330	5280	1380
$200 \times 200 \times 20$		59.9	18	9	76.3	5.68	2850	4530	1170
$200 \times 200 \times 18$		54.3	18	9	69.1	5.6	2600	4150	1050
$200 \times 200 \times 16$		48.5	18	9	61.8	5.52	2340	3720	960
$150 \times 150 \times 18$[a]		40.1	16	8	51.2	4.38	1060	1680	440
$150 \times 150 \times 15$		33.8	16	8	43	4.25	898	1430	370
$150 \times 150 \times 12$		27.3	16	8	34.8	4.12	737	1170	303
$150 \times 150 \times 10$		23	16	8	29.3	4.03	624	990	258
$120 \times 120 \times 15$[a]		26.6	13	6.5	34	3.52	448	710	186
$120 \times 120 \times 12$		21.6	13	6.5	27.5	3.4	368	584	152
$120 \times 120 \times 10$		18.2	13	6.5	23.2	3.31	313	497	129
$120 \times 120 \times 8$[a]		14.7	13	6.5	18.8	3.24	259	411	107
$100 \times 100 \times 15$[a]		21.9	12	6	28	3.02	250	395	105
$100 \times 100 \times 12$		17.8	12	6	22.7	2.9	207	328	85.7
$100 \times 100 \times 10$		15	12	6	19.2	2.82	177	280	73
$100 \times 100 \times 8$		12.2	12	6	15.5	2.74	145	230	59.9
$90 \times 90 \times 12$[a]		15.9	11	5.5	20.3	2.66	149	235	62
$90 \times 90 \times 10$		13.4	11	5.5	17.1	2.58	127	201	52.6
$90 \times 90 \times 8$		10.9	11	5.5	13.9	2.5	104	166	43.1
$90 \times 90 \times 7$		9.61	11	5.5	12.2	2.45	92.6	147	38.3
$80 \times 80 \times 10$[b]		11.9	10	5	15.1	2.34	87.5	139	36.4
$80 \times 80 \times 8$[b]		9.63	10	5	12.3	2.26	72.2	115	29.9
$75 \times 75 \times 8$[b]		8.99	9	4.5	11.4	2.14	59.1	93.8	24.5
$75 \times 75 \times 6$[b]		6.85	9	4.5	8.73	2.05	45.8	72.7	18.9
$70 \times 70 \times 7$[b]		7.38	9	4.5	9.40	1.97	42.3	67.1	17.5
$70 \times 70 \times 6$[b]		6.38	9	4.5	8.13	1.93	36.9	58.5	15.3

Radius of gyration			Elastic modulus	Torsional constants	Equivalent slenderness coefficient	Buckling ratio
y–y, z–z	u–u	v–v	y–y, z–z			h/t
i	i	i	W	I_T	ϕ_a	or
cm	cm	cm	cm³	cm⁴		D/T
$r_{x-x,\,y-y}$	r_{u-u}	r_{v-v}	$Z_{x-x,\,y-y}$	J		d/t
6.06	7.64	3.9	235	182	2.50	8
6.11	7.7	3.92	199	107	3.05	10
6.13	7.75	3.9	181	78.9	3.43	11
6.16	7.76	3.94	162	56.1	3.85	13
4.55	5.73	2.93	99.8	58.6	2.48	8
4.57	5.76	2.93	83.5	34.6	3.01	10
4.6	5.8	2.95	67.7	18.2	3.77	13
4.62	5.82	2.97	56.9	10.8	4.51	15
3.63	4.57	2.34	52.8	27	2.37	8
3.65	4.6	2.35	42.7	14.2	2.99	10
3.67	4.63	2.36	36	8.41	3.61	12
3.71	4.67	2.38	29.5	4.44	4.56	15
2.99	3.76	1.94	35.8	22.3	1.92	7
3.02	3.8	1.94	29.1	11.8	2.44	8
3.04	3.83	1.95	24.6	6.97	2.94	10
3.06	3.85	1.96	19.9	3.68	3.70	13
2.71	3.4	1.75	23.5	10.5	2.17	8
2.72	3.42	1.75	19.8	6.2	2.64	9
2.74	3.45	1.76	16.1	3.28	3.33	11
2.75	3.46	1.77	14.1	2.24	3.80	13
2.41	3.03	1.55	15.4	5.45	2.33	8
2.43	3.06	1.56	12.6	2.88	2.94	10
2.27	2.86	1.46	11	2.65	2.76	9
2.29	2.89	1.47	8.41	1.17	3.70	13
2.12	2.67	1.36	8.41	1.69	2.92	10
2.13	2.68	1.37	7.27	1.09	3.41	12

continued

(continued) UK equal angles – dimensions and properties

Section designation		Mass per metre	Radius		Area of section	Distance to centroid	Second moment of area		
Size	Thick-ness		Root	Toe			y–y, z–z	u–u	v–v
$h \times h$	t		r_1	r_2		c	I	I	I
mm	mm	kg/m	mm	mm	cm²	cm	cm⁴	cm⁴	cm⁴
British Standard Notation							$I_{x–x,\ y–y}$	$I_{u–u}$	$I_{v–v}$
$65 \times 65 \times 7^b$		6.83	9	4.5	8.70	1.85	33.4	53.0	13.8
$60 \times 60 \times 8^b$		7.09	8	4	9.03	1.77	29.2	46.1	12.2
$60 \times 60 \times 6^b$		5.42	8	4	6.91	1.69	22.8	36.1	9.44
$60 \times 60 \times 5^b$		4.57	8	4	5.82	1.64	19.4	30.7	8.03
$50 \times 50 \times 6^b$		4.47	7	3.5	5.69	1.45	12.8	20.3	5.34
$50 \times 50 \times 5^b$		3.77	7	3.5	4.80	1.40	11.0	17.4	4.55
$50 \times 50 \times 4^b$		3.06	7	3.5	3.89	1.36	8.97	14.2	3.73
$45 \times 45 \times 4.5^b$		3.06	7	3.5	3.90	1.25	7.14	11.4	2.94
$40 \times 40 \times 5^b$		2.97	6	3	3.79	1.16	5.43	8.60	2.26
$40 \times 40 \times 4^b$		2.42	6	3	3.08	1.12	4.47	7.09	1.86
$35 \times 35 \times 4^b$		2.09	5	2.5	2.67	1.00	2.95	4.68	1.23
$30 \times 30 \times 4^b$		1.78	5	2.5	2.27	0.878	1.80	2.85	0.754
$30 \times 30 \times 3^b$		1.36	5	2.5	1.74	0.835	1.40	2.22	0.585
$25 \times 25 \times 4^b$		1.45	3.5	1.75	1.85	0.762	1.02	1.61	0.43
$25 \times 25 \times 3^b$		1.12	3.5	1.75	1.42	0.723	0.803	1.27	0.334
$20 \times 20 \times 3^b$		0.882	3.5	1.75	1.12	0.598	0.392	0.618	0.165

Notes:
a TATA Sections produced in addition to the range of BS Sections.
b BS Sections not produced by TATÁ.
c Is the distance from the back of the leg to the centre of gravity?

Source: TATA Steel. 2008 – reproduced with the kind permission of TATA.

Radius of gyration			Elastic modulus	Torsional constants	Equivalent slenderness coefficient	Buckling ratio
y–y, z–z	u–u	v–v	y–y, z–z			h/t
i	i	i	W	I_T	ϕ_a	or
cm	cm	cm	cm³	cm⁴		D/T
$r_{x-x,\,y-y}$	r_{u-u}	r_{v-v}	$Z_{x-x,\,y-y}$	J		d/t
1.96	2.47	1.26	7.18	1.58	2.67	9
1.80	2.26	1.16	6.89	2.09	2.14	8
1.82	2.29	1.17	5.29	0.922	2.90	10
1.82	2.3	1.17	4.45	0.550	3.48	12
1.50	1.89	0.968	3.61	0.755	2.38	8
1.51	1.90	0.973	3.05	0.450	2.88	10
1.52	1.91	0.979	2.46	0.240	3.57	13
1.35	1.71	0.870	2.2	0.304	2.84	10
1.20	1.51	0.773	1.91	0.352	2.26	8
1.21	1.52	0.777	1.55	0.188	2.83	10
1.05	1.32	0.678	1.18	0.158	2.50	9
0.892	1.12	0.577	0.85	0.137	2.07	8
0.899	1.13	0.581	0.649	0.0613	2.75	10
0.741	0.931	0.482	0.586	0.1070	1.75	6
0.751	0.945	0.484	0.452	0.0472	2.38	8
0.59	0.742	0.383	0.279	0.0382	1.81	7

UKA unequal angles – dimensions and properties

Section designation		Radius		Dimension		Second moment of area			
Size	Thick-ness	Root	Toe			$y-y$	$z-z$	$u-u$	$v-v$
$h \times b$	t	r_1	r_2	c_y	c_z	I	I	I	I
mm	mm	mm	mm	cm	cm	cm⁴	cm⁴	cm⁴	cm⁴
British Standard Notation						I_{x-x}	I_{y-y}	I_{u-u}	I_{v-v}
$200 \times 150 \times 18$	47.1	15	7.5	6.34	3.86	2390	1160	2920	623
$200 \times 150 \times 15$	39.6	15	7.5	6.21	3.73	2020	979	2480	526
$200 \times 150 \times 12$	32	15	7.5	6.08	3.61	1650	803	2030	430
$200 \times 100 \times 15$	33.8	15	7.5	7.16	2.22	1760	299	1860	193
$200 \times 100 \times 12$	27.3	15	7.5	7.03	2.1	1440	247	1530	159
$200 \times 100 \times 10$	23	15	7.5	6.93	2.01	1220	210	1290	135
$150 \times 90 \times 15$	26.6	12	6	5.21	2.23	761	205	841	126
$150 \times 90 \times 12$	21.6	12	6	5.08	2.12	627	171	694	104
$150 \times 90 \times 10$	18.2	12	6	5	2.04	533	146	591	88.3
$150 \times 75 \times 15$	24.8	12	6	5.52	1.81	713	119	753	78.6
$150 \times 75 \times 12$	20.2	12	6	5.4	1.69	588	99.6	623	64.7
$150 \times 75 \times 10$	17	12	6	5.31	1.61	501	85.6	531	55.1
$125 \times 75 \times 12$	17.8	11	5.5	4.31	1.84	354	95.5	391	58.5
$125 \times 75 \times 10$	15	11	5.5	4.23	1.76	302	82.1	334	49.9
$125 \times 75 \times 8$	12.2	11	5.5	4.14	1.68	247	67.6	274	40.9
$100 \times 75 \times 12$	15.4	10	5	3.27	2.03	189	90.2	230	49.5
$100 \times 75 \times 10$	13	10	5	3.19	1.95	162	77.6	197	42.2
$100 \times 75 \times 8$	10.6	10	5	3.1	1.87	133	64.1	162	34.6
$100 \times 65 \times 10$	12.3	10	5	3.36	1.63	154	51	175	30.1
$100 \times 65 \times 8$	9.94	10	5	3.27	1.55	127	42.2	144	24.8
$100 \times 65 \times 7$	8.77	10	5	3.23	1.51	113	37.6	128	22
$100 \times 50 \times 8$	8.97	8	4	3.6	1.13	116	19.7	123	12.8
$100 \times 50 \times 6$	6.84	8	4	3.51	1.05	89.9	15.4	95.4	9.92
$80 \times 40 \times 8$	7.07	7	4	2.94	0.963	57.6	9.61	60.9	6.34
$80 \times 60 \times 7$	7.36	8	3.5	2.51	1.52	59	28.4	72	15.4
$80 \times 40 \times 6$	5.41	7	3.5	2.85	0.884	44.9	7.59	47.6	4.93
$75 \times 50 \times 8$	7.39	7	3.5	2.52	1.29	52	18.4	59.6	10.8
$75 \times 50 \times 6$	5.65	7	3.5	2.44	1.21	40.5	14.4	46.6	8.36
$70 \times 50 \times 6$	5.41	7	3.5	2.23	1.25	33.4	14.2	39.7	7.92
$65 \times 50 \times 5$	4.35	6	3	1.99	1.25	23.2	11.9	28.8	6.32

Radius of gyration				Elastic modulus		Angle	Torsional constant	Equivalent slenderness coefficient		Mono-symmetry	Area of section	Buckling ratio
y–y	z–z	u–u	v–v	y–y	z–z	y–y to u–u		Min	Max	Index		h/t_f
i	i	i	i	W_{el}	W_{el}	Tan α	I_T	ϕ_a	ϕ_a	Ψ_a		or
cm	cm	cm	cm	cm³	cm³		cm⁴				cm²	D/T
r_{x-x}	r_{y-y}	r_{u-u}	r_{v-v}	Z_{x-x}	Z_{y-y}		J					d/t
6.3	4.38	6.97	3.22	175	104	0.549	64.9	2.93	3.72	4.6	60.1	8
6.33	4.4	7	3.23	147	86.9	0.551	37.9	3.53	4.5	5.55	50.5	10
6.36	4.44	7.04	3.25	119	70.5	0.552	19.6	4.43	5.7	6.97	40.8	13
6.4	2.64	6.59	2.12	137	38.5	0.26	32.3	3.54	5.17	9.19	43	7
6.43	2.67	6.63	2.14	111	31.3	0.262	16.7	4.42	6.57	11.5	34.8	8
6.46	2.68	6.65	2.15	93.2	26.3	0.263	9.73	5.26	7.92	13.9	29.2	10
4.74	2.46	4.98	1.93	77.7	30.4	0.354	25.4	2.58	3.59	5.96	33.9	6
4.77	2.49	5.02	1.94	63.3	24.8	0.358	13.2	3.24	4.58	7.5	27.5	8
4.8	2.51	5.05	1.95	53.3	21	0.36	7.73	3.89	5.56	9.03	23.2	9
4.75	1.94	4.88	1.58	75.2	21	0.253	23.8	2.62	3.74	6.84	31.7	5
4.78	1.97	4.92	1.59	61.3	17.1	0.258	12.3	3.3	4.79	8.6	25.7	6
4.81	1.99	4.95	1.6	51.6	14.5	0.261	7.23	3.95	5.83	10.4	21.7	8
3.95	2.05	4.15	1.61	43.2	16.9	0.354	10.9	2.66	3.73	6.23	22.7	6
3.97	2.07	4.18	1.61	36.5	14.3	0.357	6.37	3.21	4.55	7.5	19.1	8
4	2.09	4.21	1.63	29.6	11.6	0.36	3.31	4	5.75	9.43	15.5	9
3.1	2.14	3.42	1.59	28	16.5	0.54	9.46	2.1	2.64	3.46	19.7	6
3.12	2.16	3.45	1.59	23.8	14	0.544	5.53	2.54	3.22	4.17	16.6	8
3.14	2.18	3.47	1.6	19.3	11.4	0.547	2.88	3.18	4.08	5.24	13.5	9
3.14	1.81	3.35	1.39	23.2	10.5	0.41	5.2	2.52	3.43	5.45	15.6	7
3.16	1.83	3.37	1.4	18.9	8.54	0.413	2.71	3.14	4.35	6.86	12.7	8
3.17	1.83	3.39	1.4	16.6	7.53	0.415	1.83	3.58	5	7.85	11.2	9
18.2	5.08	3.28	1.06	3.19	1.31	0.258	2.61	3.3	4.8	8.61	11.4	6
13.8	3.89	3.31	1.07	3.21	1.33	0.262	1.14	4.38	6.52	11.6	8.71	8
11.4	3.16	2.6	0.838	2.53	1.03	0.253	1.66	2.92	3.72	4.78	9.01	5
10.7	6.34	2.77	1.28	2.51	1.74	0.546	2.05	2.61	3.73	6.85	9.38	9
8.73	2.44	2.63	0.845	2.55	1.05	0.258	0.899	3.48	5.12	9.22	6.89	7
10.4	4.95	2.52	1.07	2.35	1.4	0.43	2.14	2.36	3.18	4.92	9.41	6
8.01	3.81	2.55	1.08	2.37	1.42	0.435	0.935	3.18	4.34	6.6	7.19	8
7.01	3.78	2.4	1.07	2.2	1.43	0.5	0.899	2.96	3.89	5.44	6.89	8
5.14	3.19	2.28	1.07	2.05	1.47	0.577	0.498	3.38	4.26	5.08	5.54	10

continued

(continued) UKA unequal angles – dimensions and properties

Section designation		Radius		Dimension		Second moment of area			
Size	Thick-ness	Root	Toe			y–y	z–z	u–u	v–v
$h \times b$	t	r_1	r_2	c_y	c_z	I	I	I	I
mm	mm	mm	mm	cm	cm	cm^4	cm^4	cm^4	cm^4
British Standard Notation						I_{x-x}	I_{y-y}	I_{u-u}	I_{v-v}
$60 \times 40 \times 6$	4.46	6	3	2	1.01	20.1	7.12	23.1	4.16
$60 \times 30 \times 5$	3.36	5	3	2.17	0.684	15.6	2.63	16.5	1.71
$60 \times 40 \times 5$	3.76	6	2.5	1.96	0.972	17.2	6.11	19.7	3.54
$50 \times 30 \times 5$	2.96	5	2.5	1.73	0.741	9.36	2.51	10.3	1.54
$45 \times 30 \times 4$	2.25	4.5	2.25	1.48	0.74	5.78	2.05	6.65	1.18
$40 \times 20 \times 4$	1.77	4	2	1.47	0.48	3.59	0.6	3.8	0.393
$40 \times 25 \times 4$	1.93	4	2	1.36	0.623	3.89	1.16	4.35	0.7
$30 \times 20 \times 4$	1.46	4	2	1.03	0.541	1.59	0.553	1.81	0.33
$30 \times 20 \times 3$	1.12	4	2	0.99	0.502	1.25	0.437	1.43	0.256

Source: TATA Steel. 2008 – reproduced with the kind permission of TATA.

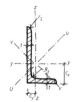

Radius of gyration				Elastic modulus		Angle	Torsional constant	Equivalent slenderness coefficient		Mono-symmetry	Area of section	Buckling ratio
y–y	z–z	u–u	v–v	y–y	z–z	y–y to u–u		Min	Max	Index		h/t_f
i	i	i	i	W_{el}	W_{el}	Tan α	I_T	ϕ_a	ϕ_a	ψ_a		or
cm	cm	cm	cm	cm³	cm³		cm⁴				cm²	D/T
$r_{x\text{-}x}$	$r_{y\text{-}y}$	$r_{u\text{-}u}$	$r_{v\text{-}v}$	$Z_{x\text{-}x}$	$Z_{y\text{-}y}$		J					d/t
5.03	2.38	2.02	0.855	1.88	1.12	0.431	0.735	2.51	3.39	5.26	5.68	7
4.07	1.14	1.97	0.633	1.91	0.784	0.257	0.435	3.02	4.11	6.34	4.28	6
4.25	2.02	2.03	0.86	1.89	1.13	0.434	0.382	3.15	4.56	8.26	4.79	8
2.86	1.11	1.65	0.639	1.57	0.816	0.352	0.34	2.51	3.52	5.99	3.78	6
1.91	0.91	1.52	0.64	1.42	0.85	0.436	0.166	2.85	3.87	5.92	2.87	8
1.42	0.393	1.3	0.417	1.26	0.514	0.252	0.142	2.51	3.48	5.75	2.26	5
1.47	0.619	1.33	0.534	1.26	0.687	0.38	0.131	2.57	3.68	6.86	2.46	6
0.807	0.379	0.988	0.421	0.925	0.546	0.421	0.1096	1.79	2.39	3.95	1.86	5
0.621	0.292	1	0.424	0.935	0.553	0.427	0.0486	2.4	3.28	5.31	1.43	7

Hot finished rectangular hollow sections – dimensions and properties

Section designation		Mass per metre	Area of section	Ratios for local buckling		Second moment of area	
Size	Thickness			−1		y–y	z–z
$h \times b$	t		A	c_w/t	c_f/t	I_y	I_z
mm	mm	kg/m	cm²			cm⁴	cm⁴
British Standard Notation				d/s	–	I_{x-x}	I_{y-y}
$50 \times 30 \times 3.0^a$		3.41	4.34	13.7	7	13.6	5.94
$50 \times 30 \times 3.2$		3.61	4.6	12.6	6.37	14.2	6.2
$50 \times 30 \times 3.6^a$		4.01	5.1	10.9	5.33	15.4	6.67
$50 \times 30 \times 4.0$		4.39	5.59	9.5	4.5	16.5	7.08
$50 \times 30 \times 5.0$		5.28	6.73	7	3	18.7	7.89
$60 \times 40 \times 3.0^a$		4.35	5.54	17	10.3	26.5	13.9
$60 \times 40 \times 3.2$		4.62	5.88	15.7	9.5	27.8	14.6
$60 \times 40 \times 3.6^a$		5.14	6.54	13.7	8.11	30.4	15.9
$60 \times 40 \times 4.0$		5.64	7.19	12	7	32.8	17
$60 \times 40 \times 5.0$		6.85	8.73	9	5	38.1	19.5
$60 \times 40 \times 6.3$		8.31	10.6	6.52	3.35	43.4	21.9
$80 \times 40 \times 3.0^a$		5.29	6.74	23.7	10.3	54.2	18
$80 \times 40 \times 3.2$		5.62	7.16	22	9.5	57.2	18.9
$80 \times 40 \times 3.6^a$		6.27	7.98	19.2	8.11	62.8	20.6
$80 \times 40 \times 4.0$		6.9	8.79	17	7	68.2	22.2
$80 \times 40 \times 5.0$		8.42	10.7	13	5	80.3	25.7
$80 \times 40 \times 6.3$		10.3	13.1	9.7	3.35	93.3	29.2
$80 \times 40 \times 7.1^a$		11.4	14.5	8.27	2.63	99.8	30.7
$80 \times 40 \times 8.0$		12.5	16	7	2	106	32.1
$90 \times 50 \times 3.0^a$		6.24	7.94	27	13.7	84.4	33.5
$90 \times 50 \times 3.2$		6.63	8.44	25.1	12.6	89.1	35.3
$90 \times 50 \times 3.6^a$		7.4	9.42	22	10.9	98.3	38.7
$90 \times 50 \times 4.0$		8.15	10.4	19.5	9.5	107	41.9
$90 \times 50 \times 5.0$		9.99	12.7	15	7	127	49.2
$90 \times 50 \times 6.3$		12.3	15.6	11.3	4.94	150	57
$90 \times 50 \times 7.1^a$		13.6	17.3	9.68	4.04	162	60.9
$90 \times 50 \times 8.0$		15	19.2	8.25	3.25	174	64.6
$100 \times 50 \times 3.0^a$		6.71	8.54	30.3	13.7	110	36.8
$100 \times 50 \times 3.2$		7.13	9.08	28.3	12.6	116	38.8
$100 \times 50 \times 3.6^a$		7.96	10.1	24.8	10.9	128	42.6

Note: TATA Sections produced in addition to the range of BS Sections.

Radius of gyration		Elastic modulus		Plastic modulus		Torsional constants		Buckling ratio
y–y	z–z	y–y	z–z	y–y	z–z			h/t_f
i_y	i_z	W_{ely}	W_{elz}	W_{ply}	W_{plz}	I_T	W_t	or
cm	cm	cm³	cm³	cm³	cm³	cm⁴	cm³	D/T
r_{x-x}	r_{y-y}	Z_{x-x}	Z_{y-y}	S_{x-x}	S_{y-y}	J	C	d/t
1.77	1.17	5.43	3.96	6.88	4.76	13.5	6.51	17
1.76	1.16	5.68	4.13	7.25	5	14.2	6.8	16
1.74	1.14	6.16	4.45	7.94	5.46	15.4	7.31	14
1.72	1.13	6.6	4.72	8.59	5.88	16.6	7.77	13
1.67	1.08	7.49	5.26	10	6.8	19	8.67	10
2.18	1.58	8.82	6.95	10.9	8.19	29.2	11.2	20
2.18	1.57	9.27	7.29	11.5	8.64	30.8	11.7	19
2.16	1.56	10.1	7.93	12.7	9.5	33.8	12.8	17
2.14	1.54	10.9	8.52	13.8	10.3	36.7	13.7	15
2.09	1.5	12.7	9.77	16.4	12.2	43	15.7	12
2.02	1.44	14.5	11	19.2	14.2	49.5	17.6	10
2.84	1.63	13.6	9	17.1	10.4	43.8	15.3	27
2.83	1.63	14.3	9.46	18	11	46.2	16.1	25
2.81	1.61	15.7	10.3	20	12.1	50.8	17.5	22
2.79	1.59	17.1	11.1	21.8	13.2	55.2	18.9	20
2.74	1.55	20.1	12.9	26.1	15.7	65.1	21.9	16
2.67	1.49	23.3	14.6	31.1	18.4	75.6	24.8	13
2.63	1.46	25	15.4	33.8	19.8	80.9	26.2	11
2.58	1.42	26.5	16.1	36.5	21.2	85.8	27.4	10
3.26	2.05	18.8	13.4	23.2	15.3	76.5	22.4	30
3.25	2.04	19.8	14.1	24.6	16.2	80.9	23.6	28
3.23	2.03	21.8	15.5	27.2	18	89.4	25.9	25
3.21	2.01	23.8	16.8	29.8	19.6	97.5	28	23
3.16	1.97	28.3	19.7	36	23.5	116	32.9	18
3.1	1.91	33.3	22.8	43.2	28	138	38.1	14
3.06	1.88	36	24.4	47.2	30.5	149	40.7	13
3.01	1.84	38.6	25.8	51.4	32.9	160	43.2	11
3.58	2.08	21.9	14.7	27.3	16.8	88.4	25	33
3.57	2.07	23.2	15.5	28.9	17.7	93.4	26.4	31
3.55	2.05	25.6	17	32.1	19.6	103	29	28

continued

(continued) Hot finished rectangular hollow sections – dimensions and properties

Section designation		Mass per metre	Area of section	Ratios for local buckling		Second moment of area	
Size	Thickness			−1		y–y	z–z
$h \times b$	t		A	c_w/t	c_f/t	I_y	I_z
mm	mm	kg/m	cm²			cm⁴	cm⁴
British Standard Notation				d/s	–	I_{x-x}	I_{y-y}
$100 \times 50 \times 4.0$		8.78	11.2	22	9.5	140	46.2
$100 \times 50 \times 5.0$		10.8	13.7	17	7	167	54.3
$100 \times 50 \times 6.3$		13.3	16.9	12.9	4.94	197	63
$100 \times 50 \times 7.1^a$		14.7	18.7	11.1	4.04	214	67.5
$100 \times 50 \times 8.0$		16.3	20.8	9.5	3.25	230	71.7
$100 \times 50 \times 8.8^a$		17.6	22.5	8.36	2.68	243	74.8
$100 \times 50 \times 10.0^a$		19.6	24.9	7	2	259	78.4
$100 \times 60 \times 3.0^a$		7.18	9.14	30.3	17	124	55.7
$100 \times 60 \times 3.2$		7.63	9.72	28.3	15.7	131	58.8
$100 \times 60 \times 3.6^a$		8.53	10.9	24.8	13.7	145	64.8
$100 \times 60 \times 4.0$		9.41	12	22	12	158	70.5
$100 \times 60 \times 5.0$		11.6	14.7	17	9	189	83.6
$100 \times 60 \times 6.3$		14.2	18.1	12.9	6.52	225	98.1
$100 \times 60 \times 7.1^a$		15.8	20.2	11.1	5.45	244	106
$100 \times 60 \times 8.0$		17.5	22.4	9.5	4.5	264	113
$100 \times 60 \times 8.8^a$		19	24.2	8.36	3.82	279	119
$100 \times 60 \times 10.0^a$		21.1	26.9	7	3	299	126
$120 \times 60 \times 3.0^a$		8.12	10.3	37	17	194	65.5
$120 \times 60 \times 3.2^a$		8.64	11	34.5	15.7	205	69.2
$120 \times 60 \times 3.6^a$		9.66	12.3	30.3	13.7	227	76.3
$120 \times 60 \times 4.0$		10.7	13.6	27	12	249	83.1
$120 \times 60 \times 5.0$		13.1	16.7	21	9	299	98.8
$120 \times 60 \times 6.3$		16.2	20.7	16	6.52	358	116
$120 \times 60 \times 7.1^a$		18.1	23	13.9	5.45	391	126
$120 \times 60 \times 8.0$		20.1	25.6	12	4.5	425	135
$120 \times 60 \times 8.8^a$		21.8	27.8	10.6	3.82	452	142
$120 \times 60 \times 10.0^a$		24.3	30.9	9	3	488	152
$120 \times 60 \times 12.5^a$		29.1	37.1	6.6	1.8	546	165
$120 \times 80 \times 3.6^a$		10.8	13.7	30.3	19.2	276	147
$120 \times 80 \times 4.0$		11.9	15.2	27	17	303	161
$120 \times 80 \times 5.0$		14.7	18.7	21	13	365	193
$120 \times 80 \times 6.3$		18.2	23.2	16	9.7	440	230

Radius of gyration		Elastic modulus		Plastic modulus		Torsional constants		Buckling ratio
y–y	z–z	y–y	z–z	y–y	z–z			h/t_f
i_y	i_z	W_{ely}	W_{elz}	W_{ply}	W_{plz}	I_T	W_t	or
cm	cm	cm³	cm³	cm³	cm³	cm⁴	cm³	D/T
r_{x-x}	r_{y-y}	Z_{x-x}	Z_{y-y}	S_{x-x}	S_{y-y}	J	C	d/t
3.53	2.03	27.9	18.5	35.2	21.5	113	31.4	25
3.48	1.99	33.3	21.7	42.6	25.8	135	36.9	20
3.42	1.93	39.4	25.2	51.3	30.8	160	42.9	16
3.38	1.9	42.7	27	56.3	33.5	173	46	14
3.33	1.86	46	28.7	61.4	36.3	186	48.9	13
3.29	1.82	48.5	29.9	65.6	38.5	197	51.1	11
3.22	1.77	51.8	31.4	71.2	41.4	209	53.6	10
3.68	2.47	24.7	18.6	30.2	21.2	121	30.7	33
3.67	2.46	26.2	19.6	32	22.4	129	32.4	31
3.65	2.44	28.9	21.6	35.6	24.9	142	35.6	28
3.63	2.43	31.6	23.5	39.1	27.3	156	38.7	25
3.58	2.38	37.8	27.9	47.4	32.9	188	45.9	20
3.52	2.33	45	32.7	57.3	39.5	224	53.8	16
3.48	2.29	48.8	35.3	62.9	43.2	245	58	14
3.44	2.25	52.8	37.8	68.7	47.1	265	62.2	13
3.4	2.22	55.9	39.7	73.6	50.2	282	65.4	11
3.33	2.16	59.9	42.1	80.2	54.4	304	69.3	10
4.33	2.52	32.3	21.8	40	24.6	156	37.2	40
4.32	2.51	34.2	23.1	42.4	26.1	165	39.2	38
4.3	2.49	37.9	25.4	47.2	28.9	183	43.3	33
4.28	2.47	41.5	27.7	51.9	31.7	201	47.1	30
4.23	2.43	49.9	32.9	63.1	38.4	242	56	24
4.16	2.37	59.7	38.8	76.7	46.3	290	65.9	19
4.12	2.34	65.2	41.9	84.4	50.8	317	71.3	17
4.08	2.3	70.8	45	92.7	55.4	344	76.6	15
4.04	2.27	75.3	47.5	99.6	59.2	366	80.8	14
3.97	2.21	81.4	50.5	109	64.4	396	86.1	12
3.84	2.11	91.1	54.9	126	73.1	442	93.8	10
4.48	3.27	46	36.7	55.6	42	301	59.5	33
4.46	3.25	50.4	40.2	61.2	46.1	330	65	30
4.42	3.21	60.9	48.2	74.6	56.1	401	77.9	24
4.36	3.15	73.3	57.6	91	68.2	487	92.9	19

continued

(continued) Hot finished rectangular hollow sections – dimensions and properties

Section designation		Mass per metre	Area of section	Ratios for local buckling		Second moment of area	
Size	Thickness			−1		y–y	z–z
$h \times b$	t		A	c_w/t	c_f/t	I_y	I_z
mm	mm	kg/m	cm²			cm⁴	cm⁴
British Standard Notation				d/s	–	I_{x-x}	I_{y-y}
$120 \times 80 \times 7.1^a$		20.3	25.8	13.9	8.27	482	251
$120 \times 80 \times 8.0$		22.6	28.8	12	7	525	273
$120 \times 80 \times 8.8^a$		24.5	31.3	10.6	6.09	561	290
$120 \times 80 \times 10.0$		27.4	34.9	9	5	609	313
$120 \times 80 \times 12.5^a$		33	42.1	6.6	3.4	692	349
$150 \times 100 \times 4.0^a$		15.1	19.2	34.5	22	607	324
$150 \times 100 \times 5.0$		18.6	23.7	27	17	739	392
$150 \times 100 \times 6.3$		23.1	29.5	20.8	12.9	898	474
$150 \times 100 \times 7.1^a$		25.9	32.9	18.1	11.1	990	520
$150 \times 100 \times 8.0$		28.9	36.8	15.8	9.5	1090	569
$150 \times 100 \times 8.8^a$		31.5	40.1	14	8.36	1170	610
$150 \times 100 \times 10.0$		35.3	44.9	12	7	1280	665
$150 \times 100 \times 12.5$		42.8	54.6	9	5	1490	763
$160 \times 80 \times 4.0^a$		14.4	18.4	37	17	612	207
$160 \times 80 \times 5.0$		17.8	22.7	29	13	744	249
$160 \times 80 \times 6.3$		22.2	28.2	22.4	9.7	903	299
$160 \times 80 \times 7.1^a$		24.7	31.5	19.5	8.27	994	327
$160 \times 80 \times 8.0$		27.6	35.2	17	7	1090	356
$160 \times 80 \times 8.8^a$		30.1	38.3	15.2	6.09	1170	379
$160 \times 80 \times 10.0$		33.7	42.9	13	5	1280	411
$160 \times 80 \times 12.5$		40.9	52.1	9.8	3.4	1490	465
$180 \times 60 \times 4.0^a$		14.4	18.4	42	12	697	121
$180 \times 60 \times 5.0^a$		17.8	22.7	33	9	846	144
$180 \times 60 \times 6.3^a$		22.2	28.2	25.6	6.52	1030	171
$180 \times 60 \times 7.1^a$		24.7	31.5	22.4	5.45	1130	186
$180 \times 60 \times 8.0^a$		27.6	35.2	19.5	4.5	1240	201
$180 \times 60 \times 8.8^a$		30.1	38.3	17.5	3.82	1330	212
$180 \times 60 \times 10.0^a$		33.7	42.9	15	3	1460	228
$180 \times 60 \times 12.5^a$		40.9	52.1	11.4	1.8	1680	251
$180 \times 100 \times 4.0^a$		16.9	21.6	42	22	945	379
$180 \times 100 \times 5.0^a$		21	26.7	33	17	1150	460

Radius of gyration		Elastic modulus		Plastic modulus		Torsional constants		Buckling ratio
y–y	z–z	y–y	z–z	y–y	z–z			h/t_f
i_y	i_z	W_{ely}	W_{elz}	W_{ply}	W_{plz}	I_T	W_t	or
cm	cm	cm³	cm³	cm³	cm³	cm⁴	cm³	D/T
r_{x-x}	r_{y-y}	Z_{x-x}	Z_{y-y}	S_{x-x}	S_{y-y}	J	C	d/t
4.32	3.12	80.3	62.8	100	75.2	535	101	17
4.27	3.08	87.5	68.1	111	82.6	587	110	15
4.24	3.04	93.5	72.4	119	88.7	629	117	14
4.18	2.99	102	78.1	131	97.3	688	126	12
4.05	2.88	115	87.4	153	113	789	141	10
5.63	4.11	81	64.8	97.4	73.6	660	105	38
5.58	4.07	98.5	78.5	119	90.1	807	127	30
5.52	4.01	120	94.8	147	110	986	153	24
5.48	3.97	132	104	163	122	1090	168	21
5.44	3.94	145	114	180	135	1200	183	19
5.4	3.9	156	122	195	146	1300	196	17
5.34	3.85	171	133	216	161	1430	214	15
5.22	3.74	198	153	256	190	1680	246	12
5.77	3.35	76.5	51.7	94.7	58.3	493	88.1	40
5.72	3.31	93	62.3	116	71.1	600	106	32
5.66	3.26	113	74.8	142	86.8	730	127	25
5.62	3.22	124	81.7	158	95.9	804	139	23
5.57	3.18	136	89	175	106	883	151	20
5.53	3.15	147	94.9	189	114	949	161	18
5.47	3.1	161	103	209	125	1040	175	16
5.34	2.99	186	116	247	146	1200	198	13
6.16	2.56	77.4	40.3	99.8	45.2	341	72.2	45
6.1	2.52	94	48.1	122	54.9	411	86.3	36
6.03	2.46	114	57	150	66.6	495	102	29
5.99	2.43	126	61.9	166	73.3	542	111	25
5.94	2.39	138	66.9	184	80.4	590	120	23
5.89	2.35	148	70.8	199	86.2	630	127	20
5.83	2.3	162	75.8	220	94.4	683	137	18
5.68	2.2	187	83.7	260	109	770	151	14
6.61	4.19	105	75.9	128	85.2	852	127	45
6.57	4.15	128	92	157	104	1040	154	36
6.5	4.09	156	111	194	128	1280	186	29

continued

(continued) Hot finished rectangular hollow sections – dimensions and properties

Section designation		Mass per metre	Area of section	Ratios for local buckling		Second moment of area	
Size	Thickness			−1		y–y	z–z
$h \times b$	t		A	c_w/t	c_f/t	I_y	I_z
mm	mm	kg/m	cm²			cm⁴	cm⁴
British Standard Notation				d/s	−	I_{x-x}	I_{y-y}
$180 \times 100 \times 6.3^a$		26.1	33.3	25.6	12.9	1410	557
$180 \times 100 \times 7.1^a$		29.2	37.2	22.4	11.1	1560	613
$180 \times 100 \times 8.0^a$		32.6	41.6	19.5	9.5	1710	671
$180 \times 100 \times 8.8^a$		35.6	45.4	17.5	8.36	1850	720
$180 \times 100 \times 10.0^a$		40	50.9	15	7	2040	787
$180 \times 100 \times 12.5^a$		48.7	62.1	11.4	5	2390	908
$200 \times 100 \times 4.0^a$		18.2	23.2	47	22	1220	416
$200 \times 100 \times 5.0$		22.6	28.7	37	17	1500	505
$200 \times 100 \times 6.3$		28.1	35.8	28.7	12.9	1830	613
$200 \times 100 \times 7.1^a$		31.4	40	25.2	11.1	2020	674
$200 \times 100 \times 8.0$		35.1	44.8	22	9.5	2230	739
$200 \times 100 \times 8.8^a$		38.4	48.9	19.7	8.36	2410	793
$200 \times 100 \times 10.0$		43.1	54.9	17	7	2660	869
$200 \times 100 \times 12.5$		52.7	67.1	13	5	3140	1000
$200 \times 100 \times 14.2^a$		58.9	75	11.1	4.04	3420	1080
$200 \times 100 \times 16.0^{\wedge}r$		65.2	83	9.5	3.25	3680	1150
$200 \times 120 \times 5.0^a$		24.1	30.7	37	21	1690	762
$200 \times 120 \times 6.3$		30.1	38.3	28.7	16	2070	929
$200 \times 120 \times 7.1^a$		33.7	42.9	25.2	13.9	2290	1030
$200 \times 120 \times 8.0$		37.6	48	22	12	2530	1130
$200 \times 120 \times 8.8^a$		41.1	52.4	19.7	10.6	2730	1220
$200 \times 120 \times 10.0$		46.3	58.9	17	9	3030	1340
$200 \times 120 \times 12.5^a$		56.6	72.1	13	6.6	3580	1560
$200 \times 120 \times 14.2^a$		63.3	80.7	11.1	5.45	3910	1690
$200 \times 120 \times 16.0^a$		70.2	89.4	9.5	4.5	4220	1810
$200 \times 150 \times 5.0^a$		26.5	33.7	37	27	1970	1270
$200 \times 150 \times 6.3^a$		33	42.1	28.7	20.8	2420	1550
$200 \times 150 \times 7.1^a$		37	47.1	25.2	18.1	2690	1720
$200 \times 150 \times 8.0$		41.4	52.8	22	15.8	2970	1890
$200 \times 150 \times 8.8^a$		45.3	57.7	19.7	14	3220	2050
$200 \times 150 \times 10.0$		51	64.9	17	12	3570	2260
$200 \times 150 \times 12.5^a$		62.5	79.6	13	9	4240	2670
$200 \times 150 \times 14.2^a$		70	89.2	11.1	7.56	4640	2920

Radius of gyration		Elastic modulus		Plastic modulus		Torsional constants		Buckling ratio
y–y	z–z	y–y	z–z	y–y	z–z			h/t_1
i_y	i_z	W_{ely}	W_{elz}	W_{ply}	W_{plz}	I_T	W_t	or
cm	cm	cm³	cm³	cm³	cm³	cm⁴	cm³	D/T
$r_{x\text{-}x}$	$r_{y\text{-}y}$	$Z_{x\text{-}x}$	$Z_{y\text{-}y}$	$S_{x\text{-}x}$	$S_{y\text{-}y}$	J	C	d/t
6.47	4.06	173	123	215	142	1410	205	25
6.42	4.02	190	134	239	157	1560	224	23
6.38	3.98	205	144	259	170	1690	240	20
6.32	3.93	226	157	288	188	1860	263	18
6.2	3.82	265	182	344	223	2190	303	14
7.26	4.24	122	83.2	150	92.8	983	142	50
7.21	4.19	149	101	185	114	1200	172	40
7.15	4.14	183	123	228	140	1480	208	32
7.11	4.1	202	135	254	155	1630	229	28
7.06	4.06	223	148	282	172	1800	251	25
7.02	4.03	241	159	306	186	1950	270	23
6.96	3.98	266	174	341	206	2160	295	20
6.84	3.87	314	201	408	245	2540	341	16
6.75	3.8	342	216	450	268	2770	368	14
6.66	3.72	368	229	491	290	2980	391	13
7.4	4.98	168	127	205	144	1650	210	40
7.34	4.92	207	155	253	177	2030	255	32
7.3	4.89	229	171	281	197	2250	282	28
7.26	4.85	253	188	313	218	2500	310	25
7.22	4.82	273	203	340	237	2700	334	23
7.17	4.76	303	223	379	263	3000	367	20
7.04	4.66	358	260	455	314	3570	428	16
6.96	4.58	391	282	503	346	3920	464	14
6.87	4.5	422	302	550	377	4250	497	13
7.64	6.12	197	169	234	192	2390	267	40
7.58	6.07	242	207	289	237	2950	326	32
7.55	6.03	268	229	322	264	3280	361	28
7.5	5.99	297	253	359	294	3640	398	25
7.47	5.96	322	273	390	319	3960	430	23
7.41	5.91	357	302	436	356	4410	475	20
7.3	5.8	424	356	525	428	5290	559	16
7.22	5.72	464	389	582	473	5830	610	14
7.13	5.64	504	420	638	518	6370	658	13

continued

(continued) Hot finished rectangular hollow sections – dimensions and properties

Section designation		Mass per metre	Area of section	Ratios for local buckling		Second moment of area	
Size	Thickness			−1		y–y	z–z
$h \times b$	t		A	c_w/t	c_f/t	I_y	I_z
mm	mm	kg/m	cm²			cm⁴	cm⁴
British Standard Notation				d/s	–	I_{x-x}	I_{y-y}
$200 \times 150 \times 16.0^a$		77.7	99	9.5	6.38	5040	3150
$220 \times 120 \times 5.0^a$		25.7	32.7	41	21	2130	829
$220 \times 120 \times 6.3^a$		32	40.8	31.9	16	2610	1010
$220 \times 120 \times 7.1^a$		35.9	45.7	28	13.9	2900	1120
$220 \times 120 \times 8.0^a$		40.2	51.2	24.5	12	3200	1230
$220 \times 120 \times 8.8^a$		43.9	55.9	22	10.6	3470	1320
$220 \times 120 \times 10.0^a$		49.4	62.9	19	9	3840	1460
$220 \times 120 \times 12.5^a$		60.5	77.1	14.6	6.6	4560	1710
$220 \times 120 \times 14.2^a$		67.8	86.3	12.5	5.45	5000	1850
$220 \times 120 \times 16.0^a$		75.2	95.8	10.8	4.5	5410	1990
$250 \times 100 \times 5.0^a$		26.5	33.7	47	17	2610	618
$250 \times 100 \times 6.3^a$		33	42.1	36.7	12.9	3210	751
$250 \times 100 \times 7.1^a$		37	47.1	32.2	11.1	3560	827
$250 \times 100 \times 8.0^a$		41.4	52.8	28.3	9.5	3940	909
$250 \times 100 \times 8.8^a$		45.3	57.7	25.4	8.36	4270	977
$250 \times 100 \times 10.0^a$		51	64.9	22	7	4730	1070
$250 \times 100 \times 12.5^a$		62.5	79.6	17	5	5620	1250
$250 \times 100 \times 14.2^a$		70	89.2	14.6	4.04	6170	1340
$250 \times 100 \times 16.0^a$		77.7	99	12.6	3.25	6690	1430
$250 \times 150 \times 5.0^a$		30.4	38.7	47	27	3360	1530
$250 \times 150 \times 6.3$		38	48.4	36.7	20.8	4140	1870
$250 \times 150 \times 7.1^a$		42.6	54.2	32.2	18.1	4610	2080
$250 \times 150 \times 8.0$		47.7	60.8	28.3	15.8	5110	2300
$250 \times 150 \times 8.8^a$		52.2	66.5	25.4	14	5550	2490
$250 \times 150 \times 10.0$		58.8	74.9	22	12	6170	2760
$250 \times 150 \times 12.5$		72.3	92.1	17	9	7390	3270
$250 \times 150 \times 14.2^a$		81.1	103	14.6	7.56	8140	3580
$250 \times 150 \times 16.0$		90.3	115	12.6	6.38	8880	3870
$260 \times 140 \times 5.0^a$		30.4	38.7	49	25	3530	1350
$260 \times 140 \times 6.3^a$		38	48.4	38.3	19.2	4360	1660
$260 \times 140 \times 7.1^a$		42.6	54.2	33.6	16.7	4840	1840

Radius of gyration		Elastic modulus		Plastic modulus		Torsional constants		Buckling ratio
y–y	z–z	y–y	z–z	y–y	z–z			h/t_f
i_y	i_z	W_{ely}	W_{elz}	W_{ply}	W_{plz}	I_T	W_t	or
cm	cm	cm³	cm³	cm³	cm³	cm⁴	cm³	D/T
r_{x-x}	r_{y-y}	Z_{x-x}	Z_{y-y}	S_{x-x}	S_{y-y}	J	C	d/t
8.06	5.03	193	138	236	155	1880	232	44
8	4.98	237	168	292	191	2320	283	35
7.96	4.94	263	186	326	213	2570	312	31
7.91	4.9	291	205	362	236	2850	343	28
7.87	4.87	315	221	394	256	3090	370	25
7.82	4.81	349	243	440	285	3430	407	22
7.69	4.71	415	285	530	341	4090	476	18
7.61	4.63	454	309	586	376	4490	517	15
7.52	4.55	492	331	643	410	4870	555	14
8.8	4.28	209	124	263	138	1620	217	50
8.73	4.22	257	150	326	169	1980	264	40
8.69	4.19	285	165	363	188	2200	291	35
8.64	4.15	315	182	404	209	2430	319	31
8.6	4.12	341	195	439	226	2630	343	28
8.54	4.06	379	214	491	251	2910	376	25
8.41	3.96	450	249	592	299	3440	438	20
8.31	3.88	493	269	655	329	3750	473	18
8.22	3.8	535	287	719	358	4050	505	16
9.31	6.28	269	204	324	228	3280	337	50
9.25	6.22	331	250	402	283	4050	413	40
9.22	6.19	368	277	449	315	4520	457	35
9.17	6.15	409	306	501	350	5020	506	31
9.13	6.12	444	331	545	381	5460	547	28
9.08	6.06	494	367	611	426	6090	605	25
8.96	5.96	591	435	740	514	7330	717	20
8.87	5.88	651	477	823	570	8100	784	18
8.79	5.8	710	516	906	625	8870	849	16
9.55	5.91	272	193	331	216	3080	326	52
9.49	5.86	335	237	411	267	3800	399	41
9.45	5.82	372	263	459	298	4230	442	37
9.4	5.78	413	290	511	331	4700	488	33

continued

(continued) Hot finished rectangular hollow sections – dimensions and properties

Section designation		Mass per metre	Area of section	Ratios for local buckling		Second moment of area	
Size	Thickness			−1		y–y	z–z
$h \times b$	t		A	c_w/t	c_f/t	I_y	I_z
mm	mm	kg/m	cm²			cm⁴	cm⁴
British Standard Notation				d/s	–	I_{x-x}	I_{y-y}
$260 \times 140 \times 8.0$[a]		47.7	60.8	29.5	14.5	5370	2030
$260 \times 140 \times 8.8$[a]		52.2	66.5	26.5	12.9	5830	2200
$260 \times 140 \times 10.0$[a]		58.8	74.9	23	11	6490	2430
$260 \times 140 \times 12.5$[a]		72.3	92.1	17.8	8.2	7770	2880
$260 \times 140 \times 14.2$[a]		81.1	103	15.3	6.86	8560	3140
$260 \times 140 \times 16.0$[a]		90.3	115	13.3	5.75	9340	3400
$300 \times 100 \times 5.0$[a]		30.4	38.7	57	17	4150	731
$300 \times 100 \times 6.3$[a]		38	48.4	44.6	12.9	5110	890
$300 \times 100 \times 7.1$[a]		42.6	54.2	39.3	11.1	5680	981
$300 \times 100 \times 8.0$		47.7	60.8	34.5	9.5	6310	1080
$300 \times 100 \times 8.8$[a]		52.2	66.5	31.1	8.36	6840	1160
$300 \times 100 \times 10.0$		58.8	74.9	27	7	7610	1280
$300 \times 100 \times 12.5$[a]		72.3	92.1	21	5	9100	1490
$300 \times 100 \times 14.2$[a]		81.1	103	18.1	4.04	10000	1610
$300 \times 100 \times 16.0$[a]		90.3	115	15.8	3.25	10900	1720
$300 \times 150 \times 8.0$[a] rr		54	68.8	34.5	15.8	8010	2700
$300 \times 150 \times 8.8$[a] rr		59.1	75.3	31.1	14	8710	2930
$300 \times 150 \times 10.0$[a] rr		66.7	84.9	27	12	9720	3250
$300 \times 150 \times 12.5$[a] rr		82.1	105	21	9	11700	3860
$300 \times 150 \times 14.2$[a] rr		92.3	118	18.1	7.56	12900	4230
$300 \times 150 \times 16.0$[a] rr		103	131	15.8	6.38	14200	4600
$300 \times 200 \times 5.0$[a]		38.3	48.7	57	37	6320	3400
$300 \times 200 \times 6.3$		47.9	61	44.6	28.7	7830	4190
$300 \times 200 \times 7.1$[a]		53.7	68.4	39.3	25.2	8730	4670
$300 \times 200 \times 8.0$		60.3	76.8	34.5	22	9720	5180
$300 \times 200 \times 8.8$[a]		66	84.1	31.1	19.7	10600	5630
$300 \times 200 \times 10.0$		74.5	94.9	27	17	11800	6280
$300 \times 200 \times 12.5$		91.9	117	21	13	14300	7540
$300 \times 200 \times 14.2$[a]		103	132	18.1	11.1	15800	8330
$300 \times 200 \times 16.0$		115	147	15.8	9.5	17400	9110
$300 \times 250 \times 5.0$[a]		42.2	53.7	57	47	7410	5610

Radius of gyration		Elastic modulus		Plastic modulus		Torsional constants		Buckling ratio
y–y	z–z	y–y	z–z	y–y	z–z			h/t$_f$
i$_y$	i$_z$	W$_{ely}$	W$_{elz}$	W$_{ply}$	W$_{plz}$	I$_T$	W$_t$	or
cm	cm	cm³	cm³	cm³	cm³	cm⁴	cm³	D/T
r$_{x-x}$	r$_{y-y}$	Z$_{x-x}$	Z$_{y-y}$	S$_{x-x}$	S$_{y-y}$	J	C	d/t
9.37	5.75	449	314	557	360	5110	527	30
9.31	5.7	499	347	624	402	5700	584	26
9.18	5.59	597	411	756	485	6840	690	21
9.1	5.52	658	449	840	537	7560	754	18
9.01	5.44	718	486	925	588	8260	815	16
10.3	4.34	276	146	354	161	2040	262	60
10.3	4.29	341	178	439	199	2500	319	48
10.2	4.25	379	196	490	221	2780	352	42
10.2	4.21	420	216	546	245	3070	387	38
10.1	4.18	456	232	594	266	3320	416	34
10.1	4.13	508	255	666	296	3680	458	30
9.94	4.02	607	297	806	354	4350	534	24
9.85	3.94	669	321	896	390	4760	578	21
9.75	3.87	729	344	986	425	5140	619	19
10.8	6.27	534	360	663	407	6450	613	38
10.8	6.23	580	390	723	443	7020	664	34
10.7	6.18	648	433	811	496	7840	736	30
10.6	6.07	779	514	986	600	9450	874	24
10.5	6	862	564	1100	666	10500	959	21
10.4	5.92	944	613	1210	732	11500	1040	19
11.4	8.35	421	340	501	380	6820	552	60
11.3	8.29	522	419	624	472	8480	681	48
11.3	8.26	582	467	698	528	9470	757	42
11.3	8.22	648	518	779	589	10600	840	38
11.2	8.18	705	563	851	643	11500	912	34
11.2	8.13	788	628	956	721	12900	1020	30
11	8.02	952	754	1170	877	15700	1220	24
11	7.95	1060	833	1300	978	17500	1340	21
10.9	7.87	1160	911	1440	1080	19300	1470	19
11.7	10.2	494	449	575	508	9770	697	60
11.7	10.2	613	556	716	633	12200	862	48
11.6	10.1	683	620	802	708	13600	960	42

continued

(continued) Hot finished rectangular hollow sections – dimensions and properties

Section designation		Mass per metre	Area of section	Ratios for local buckling		Second moment of area	
Size	Thickness			−1		y–y	z–z
$h \times b$	t		A	c_w/t	c_f/t	I_y	I_z
mm	mm	kg/m	cm²			cm⁴	cm⁴
British Standard Notation				d/s	−	$I_{x\text{-}x}$	$I_{y\text{-}y}$
$300 \times 250 \times 6.3$[a]		52.8	67.3	44.6	36.7	9190	6950
$300 \times 250 \times 7.1$[a]		59.3	75.5	39.3	32.2	10300	7750
$300 \times 250 \times 8.0$		66.5	84.8	34.5	28.3	11400	8630
$300 \times 250 \times 8.8$[a]		72.9	92.9	31.1	25.4	12400	9390
$300 \times 250 \times 10.0$[a]		82.4	105	27	22	13900	10500
$300 \times 250 \times 12.5$[a]		102	130	21	17	16900	12700
$300 \times 250 \times 14.2$[a]		115	146	18.1	14.6	18700	14100
$300 \times 250 \times 16.0$		141	163	15.8	12.6	20600	15500
$340 \times 100 \times 10.0$		65.1	82.9	31	7	10600	1440
$350 \times 150 \times 5.0$[a]		38.3	48.7	67	27	7660	2050
$350 \times 150 \times 6.3$[a]		47.9	61	52.6	20.8	9480	2530
$350 \times 150 \times 7.1$[a]		53.7	68.4	46.3	18.1	10600	2800
$350 \times 150 \times 8.0$[a]		60.3	76.8	40.8	15.8	11800	3110
$350 \times 150 \times 8.8$[a]		66	84.1	36.8	14	12800	3360
$350 \times 150 \times 10.0$[a]		74.5	94.9	32	12	14300	3740
$350 \times 150 \times 12.5$[a]		91.9	117	25	9	17300	4450
$350 \times 150 \times 14.2$[a]		103	132	21.6	7.56	19200	4890
$350 \times 150 \times 16.0$[a]		115	147	18.9	6.38	21100	5320
$350 \times 250 \times 6.3$[a]		57.8	73.6	52.6	36.7	13200	7890
$350 \times 250 \times 7.1$[a]		64.9	82.6	46.3	32.2	14700	8800
$350 \times 250 \times 8.0$[a]		72.8	92.8	40.8	28.3	16400	9800
$350 \times 250 \times 8.8$[a]		79.8	102	36.8	25.4	17900	10700
$350 \times 250 \times 10.0$[a]		90.2	115	32	22	20100	11900
$350 \times 250 \times 12.5$[a]		112	142	25	17	24400	14400
$350 \times 250 \times 14.2$[a]		126	160	21.6	14.6	27200	16000
$350 \times 250 \times 16.0$[a]		141	179	18.9	12.6	30000	17700
$400 \times 150 \times 6.3$[a]		52.8	67.3	60.5	20.8	13300	2850
$400 \times 150 \times 7.1$[a]		59.3	75.5	53.3	18.1	14800	3170
$400 \times 150 \times 8.0$[a]		66.5	84.8	47	15.8	16500	3510
$400 \times 150 \times 8.8$[a]		72.9	92.9	42.5	14	18000	3800
$400 \times 150 \times 10.0$[a]		82.4	105	37	12	20100	4230
$400 \times 150 \times 12.5$[a]		102	130	29	9	24400	5040
$400 \times 150 \times 14.2$[a]		115	146	25.2	7.56	27100	5550

Radius of gyration		Elastic modulus		Plastic modulus		Torsional constants		Buckling ratio
$y-y$	$z-z$	$y-y$	$z-z$	$y-y$	$z-z$			h/t_f
i_y	i_z	W_{ely}	W_{elz}	W_{ply}	W_{plz}	I_T	W_t	or
cm	cm	cm³	cm³	cm³	cm³	cm⁴	cm³	D/T
r_{x-x}	r_{y-y}	Z_{x-x}	Z_{y-y}	S_{x-x}	S_{y-y}	J	C	d/t
11.6	10.1	761	690	896	791	15200	1070	38
11.6	10.1	829	751	979	864	16600	1160	34
11.5	10	928	840	1100	971	18600	1300	30
11.4	9.89	1120	1010	1350	1190	22700	1560	24
11.3	9.82	1250	1130	1510	1330	25400	1730	21
11.2	9.74	1380	1240	1670	1470	28100	1900	19
11.3	4.16	623	288	823	332	4300	523	34
12.5	6.49	437	274	543	301	5160	477	70
12.5	6.43	542	337	676	373	6390	586	56
12.4	6.4	604	374	756	416	7120	651	49
12.4	6.36	673	414	844	464	7930	721	44
12.3	6.33	732	449	922	506	8620	781	40
12.3	6.27	818	498	1040	566	9630	867	35
12.2	6.17	988	593	1260	686	11600	1030	28
12.1	6.09	1100	652	1410	763	12900	1130	25
12	6.01	1210	709	1560	840	14100	1230	22
13.4	10.4	754	631	892	709	15200	1010	56
13.4	10.3	843	704	999	794	17000	1130	49
13.3	10.3	940	784	1120	888	19000	1250	44
13.3	10.2	1030	853	1220	970	20800	1370	40
13.2	10.2	1150	955	1380	1090	23400	1530	35
13.1	10.1	1400	1160	1690	1330	28500	1840	28
13	10	1550	1280	1890	1490	31900	2040	25
12.9	9.93	1720	1410	2100	1660	35300	2250	22
14	6.51	663	380	836	418	7600	673	63
14	6.47	740	422	936	467	8470	748	56
13.9	6.43	824	468	1050	521	9420	828	50
13.9	6.4	898	507	1140	568	10300	898	45
13.8	6.35	1010	564	1290	636	11500	998	40
13.7	6.24	1220	672	1570	772	13800	1190	32
13.6	6.16	1360	740	1760	859	15300	1310	28
13.5	6.09	1490	805	1950	947	16800	1430	25

continued

(continued) Hot finished rectangular hollow sections – dimensions and properties

Section designation		Mass per metre	Area of section	Ratios for local buckling		Second moment of area	
Size	Thickness			-1		y–y	z–z
$h \times b$	t		A	c_w/t	c_f/t	I_y	I_z
mm	mm	kg/m	cm²			cm⁴	cm⁴
British Standard Notation				d/s	–	$I_{x\text{-}x}$	$I_{y\text{-}y}$
$400 \times 150 \times 16.0$		128	163	22	6.38	29800	6040
$400 \times 200 \times 6.3$[a]		57.8	73.6	60.5	28.7	15700	5380
$400 \times 150 \times 7.1$[a]		64.9	82.6	53.3	25.2	17500	5990
$400 \times 150 \times 8.0$		72.8	92.8	47	22	19600	6660
$400 \times 150 \times 8.8$[a]		79.8	102	42.5	19.7	21300	7240
$400 \times 150 \times 10.0$		90.2	115	37	17	23900	8080
$400 \times 150 \times 12.5$		112	142	29	13	29100	9740
$400 \times 150 \times 14.2$[a]		126	160	25.2	11.1	32400	10800
$400 \times 150 \times 16.0$		141	179	22	9.5	35700	11800
$400 \times 300 \times 8.0$[a]		85.4	109	47	34.5	25700	16500
$400 \times 300 \times 8.8$[a]		93.6	119	42.5	31.1	28100	18000
$400 \times 300 \times 10.0$[a]		106	135	37	27	31500	20200
$400 \times 300 \times 12.5$[a]		131	167	29	21	38500	24600
$400 \times 300 \times 14.2$[a]		148	189	25.2	18.1	43000	27400
$400 \times 300 \times 16.0$[a]		166	211	22	15.8	47500	30300
$450 \times 250 \times 8.0$		85.4	109	53.3	28.3	30100	12100
$450 \times 250 \times 8.8$[a]		93.6	119	48.1	25.4	32800	13200
$450 \times 250 \times 10.0$		106	135	42	22	36900	14800
$450 \times 250 \times 12.5$		131	167	33	17	45000	18000
$450 \times 250 \times 14.2$[a]		148	189	28.7	14.6	50300	20000
$450 \times 250 \times 16.0$		166	211	25.1	12.6	55700	22000
$500 \times 200 \times 8.0$[a]		85.4	109	59.5	22	34000	8140
$500 \times 200 \times 8.8$[a]		93.6	119	53.8	19.7	37200	8850
$500 \times 200 \times 10.0$[a]		106	135	47	17	41800	9890
$500 \times 200 \times 12.5$[a]		131	167	37	13	51000	11900
$500 \times 200 \times 14.2$[a]		148	189	32.2	11.1	56900	13200
$500 \times 200 \times 16.0$[a]		166	211	28.3	9.5	63000	14500
$500 \times 300 \times 8.0$[a]		97.9	125	59.5	34.5	43700	20000
$500 \times 300 \times 8.8$[a]		107	137	53.8	31.1	47800	21800
$500 \times 300 \times 10.0$		122	155	47	27	53800	24400
$500 \times 300 \times 12.5$[a]		151	192	37	21	65800	29800
$500 \times 300 \times 14.2$[a]		170	217	32.2	18.1	73700	33200
$500 \times 300 \times 16.0$		191	243	28.3	15.8	81800	36800
$500 \times 300 \times 20.0$		235	300	22	12	98800	44100

Note:
[a] Special order only. Minimum tonnage applies.
Source: TATA Steel. 2008 – reproduced with the kind permission of TATA.

Radius of gyration		Elastic modulus		Plastic modulus		Torsional constants		Buckling ratio
y–y	z–z	y–y	z–z	y–y	z–z			h/t_f
i_y	i_z	W_{ely}	W_{elz}	W_{ply}	W_{plz}	I_T	W_t	or
cm	cm	cm³	cm³	cm³	cm³	cm⁴	cm³	D/T
r_{x-x}	r_{y-y}	Z_{x-x}	Z_{y-y}	S_{x-x}	S_{y-y}	J	C	d/t
14.6	8.55	785	538	960	594	12600	917	63
14.6	8.51	877	599	1080	665	14100	1020	56
14.5	8.47	978	666	1200	743	15700	1140	50
14.5	8.44	1070	724	1320	811	17200	1230	45
14.4	8.39	1200	808	1480	911	19300	1380	40
14.3	8.28	1450	974	1810	1110	23400	1660	32
14.2	8.21	1620	1080	2030	1240	26100	1830	28
14.1	8.13	1790	1180	2260	1370	28900	2010	25
15.4	12.3	1290	1100	1520	1250	31000	1750	50
15.3	12.3	1400	1200	1660	1360	33900	1910	45
15.3	12.2	1580	1350	1870	1540	38200	2140	40
15.2	12.1	1920	1640	2300	1880	46800	2590	32
15.1	12.1	2150	1830	2580	2110	52500	2890	28
15	12	2380	2020	2870	2350	58300	3180	25
16.6	10.6	1340	971	1620	1080	27100	1630	56
16.6	10.5	1460	1060	1770	1180	29600	1770	51
16.5	10.5	1640	1190	2000	1330	33300	1990	45
16.4	10.4	2000	1440	2460	1630	40700	2410	36
16.3	10.3	2240	1600	2760	1830	45600	2680	32
16.2	10.2	2480	1760	3070	2030	50500	2950	28
17.7	8.65	1360	814	1710	896	21100	1430	63
17.7	8.61	1490	885	1870	979	23000	1560	57
17.6	8.56	1670	989	2110	1100	25900	1740	50
17.5	8.45	2040	1190	2590	1350	31500	2100	40
17.4	8.38	2280	1320	2900	1510	35200	2320	35
17.3	8.3	2520	1450	3230	1670	38900	2550	31
18.7	12.6	1750	1330	2100	1480	42600	2200	63
18.7	12.6	1910	1450	2300	1620	46600	2400	57
18.6	12.6	2150	1630	2600	1830	52500	2700	50
18.5	12.5	2630	1990	3200	2240	64400	3280	40
18.4	12.4	2950	2220	3590	2520	72200	3660	35
18.3	12.3	3270	2450	4010	2800	80300	4040	31
18.2	12.1	3950	2940	4890	3410	97400	4840	25

Hot finished square hollow sections – dimensions and properties

Section desig-nation		Mass per metre	Area of section	Ratio for local buckling	Second moment of area	Radius of gyration	Elastic modulus	Plastic modulus	Torsional constant		Buck-ling ratio
Size	Thick-ness										h/t_f
$h \times h$	t		A	c/t	I	i	W_{el}	W_{pl}	I_T	W_t	or
mm	mm	kg/m	cm²		cm⁴	cm	cm³	cm³	cm⁴	cm³	D/T
British Standard Notation					I	r	Z	S	J	C	d/t
40 × 40 × 3[a]	3.41		4.34	10.3	9.78	1.5	4.89	5.97	15.7	7.1	13
40 × 40 × 3.2	3.61		4.6	9.5	10.2	1.49	5.11	6.28	16.5	7.42	13
40 × 40 × 3.6[a]	4.01		5.1	8.11	11.1	1.47	5.54	6.88	18.1	8.01	11
40 × 40 × 4	4.39		5.59	7	11.8	1.45	5.91	7.44	19.5	8.54	10
40 × 40 × 5	5.28		6.73	5	13.4	1.41	6.68	8.66	22.5	9.6	8
50 × 50 × 3[a]	4.35		5.54	13.7	20.2	1.91	8.08	9.7	32.1	11.8	17
50 × 50 × 3.2	4.62		5.88	12.6	21.2	1.9	8.49	10.2	33.8	12.4	16
50 × 50 × 3.6[a]	5.14		6.54	10.9	23.2	1.88	9.27	11.3	37.2	13.5	14
50 × 50 × 4	5.64		7.19	9.5	25	1.86	9.99	12.3	40.4	14.5	13
50 × 50 × 5	6.85		8.73	7	28.9	1.82	11.6	14.5	47.6	16.7	10
50 × 50 × 6.3	8.31		10.6	4.94	32.8	1.76	13.1	17	55.2	18.8	8
50 × 50 × 7.1[a]	9.14		11.6	4.04	34.5	1.72	13.8	18.3	58.9	19.8	7
50 × 50 × 8[a]	10		12.8	3.25	36	1.68	14.4	19.5	62.3	20.6	6
60 × 60 × 3[a]	5.29		6.74	17	36.2	2.32	12.1	14.3	56.9	17.7	20
60 × 60 × 3.2	5.62		7.16	15.7	38.2	2.31	12.7	15.2	60.2	18.6	19
60 × 60 × 3.6[a]	6.27		7.98	13.7	41.9	2.29	14	16.8	66.5	20.4	17
60 × 60 × 4	6.9		8.79	12	45.4	2.27	15.1	18.3	72.5	22	15
60 × 60 × 5	8.42		10.7	9	53.3	2.23	17.8	21.9	86.4	25.7	12
60 × 60 × 6.3	10.3		13.1	6.52	61.6	2.17	20.5	26	102	29.6	10
60 × 60 × 7.1[a]	11.4		14.5	5.45	65.8	2.13	21.9	28.2	110	31.6	8
60 × 60 × 8	12.5		16	4.5	69.7	2.09	23.2	30.4	118	33.4	8
70 × 70 × 3[a]	6.24		7.94	20.3	59	2.73	16.9	19.9	92.2	24.8	23
70 × 70 × 3.2	6.63		8.44	18.9	62.3	2.72	17.8	21	97.6	26.1	22
70 × 70 × 3.6[a]	7.4		9.42	16.4	68.6	2.7	19.6	23.3	108	28.7	19
70 × 70 × 4	8.15		10.4	14.5	74.7	2.68	21.3	25.5	118	31.2	18
70 × 70 × 5	9.99		12.7	11	88.5	2.64	25.3	30.8	142	36.8	14
70 × 70 × 6.3	12.3		15.6	8.11	104	2.58	29.7	36.9	169	42.9	11
70 × 70 × 7.1[a]	13.6		17.3	6.86	112	2.54	32	40.3	185	46.1	10

(continued) Hot finished square hollow sections – dimensions and properties

Section desig-nation		Mass per metre	Area of section	Ratio for local buckling	Second moment of area	Radius of gyration	Elastic modulus	Plastic modulus	Torsional constant		Buck-ling ratio
Size	Thick-ness										h/t_f
$h \times h$	t	A	c/t	I	i	W_{el}	W_{pl}	I_T	W_t	or
mm	mm	kg/m	cm²		cm⁴	cm	cm³	cm³	cm⁴	cm³	D/T
British Standard Notation					I	r	z	S	J	C	d/t
$70 \times 70 \times 8$		15	19.2	5.75	120	2.5	34.2	43.8	200	49.2	9
$70 \times 70 \times 8.8$[a]		16.3	20.7	4.95	126	2.46	35.9	46.6	212	51.6	8
$80 \times 80 \times 3$[a]		7.18	9.14	23.7	89.8	3.13	22.5	26.3	140	33	27
$80 \times 80 \times 3.2$		7.63	9.72	22	95	3.13	23.7	27.9	148	34.9	25
$80 \times 80 \times 3.6$[a]		8.53	10.9	19.2	105	3.11	26.2	31	164	38.5	22
$80 \times 80 \times 4$		9.41	12	17	114	3.09	28.6	34	180	41.9	20
$80 \times 80 \times 5$		11.6	14.7	13	137	3.05	34.2	41.1	217	49.8	16
$80 \times 80 \times 6.3$		14.2	18.1	9.7	162	2.99	40.5	49.7	262	58.7	13
$80 \times 80 \times 7.1$[a]		15.8	20.2	8.27	176	2.95	43.9	54.5	286	63.5	11
$80 \times 80 \times 8$		17.5	22.4	7	189	2.91	47.3	59.5	312	68.3	10
$80 \times 80 \times 8.8$[a]		19	24.2	6.09	200	2.87	50	63.7	332	72	9
$80 \times 80 \times 10$[a]		21.1	26.9	5	214	2.82	53.5	69.3	360	76.8	8
$80 \times 80 \times 12.5$[a]		25.2	32.1	3.4	234	2.7	58.6	78.9	404	83.8	6
$90 \times 90 \times 3.6$[a]		9.66	12.3	22	152	3.52	33.8	39.7	237	49.7	25
$90 \times 90 \times 4$		10.7	13.6	19.5	166	3.5	37	43.6	260	54.2	23
$90 \times 90 \times 5$		13.1	16.7	15	200	3.45	44.4	53	316	64.8	18
$90 \times 90 \times 6.3$		16.2	20.7	11.3	238	3.4	53	64.3	382	77	14
$90 \times 90 \times 7.1$[a]		18.1	23	9.68	260	3.36	57.7	70.8	419	83.7	13
$90 \times 90 \times 8$		20.1	25.6	8.25	281	3.32	62.6	77.6	459	90.5	11
$90 \times 90 \times 8.8$[a]		21.8	27.8	7.23	299	3.28	66.5	83.4	492	96	10
$90 \times 90 \times 10$[a]		24.3	30.9	6	322	3.23	71.6	91.3	536	103	9
$90 \times 90 \times 12.5$[a]		29.1	37.1	4.2	359	3.11	79.8	105	612	114	7
$100 \times 100 \times 3.6$		10.8	13.7	24.8	212	3.92	42.3	49.5	328	62.3	28
$100 \times 100 \times 4$		11.9	15.2	22	232	3.91	46.4	54.4	361	68.2	25
$100 \times 100 \times 5$		14.7	18.7	17	279	3.86	55.9	66.4	439	81.8	20
$100 \times 100 \times 6.3$		18.2	23.2	12.9	336	3.8	67.1	80.9	534	97.8	16
$100 \times 100 \times 7.1$[a]		20.3	25.8	11.1	367	3.77	73.4	89.2	589	107	14
$100 \times 100 \times 8$		22.6	28.8	9.5	400	3.73	79.9	98.2	646	116	13

continued

(continued) Hot finished square hollow sections – dimensions and properties

Section designation		Mass per metre	Area of section	Ratio for local buckling	Second moment of area	Radius of gyration	Elastic modulus	Plastic modulus	Torsional constant		Buckling ratio
Size	Thickness										h/t_f
$h \times h$	t		A	c/t	I	i	W_{el}	W_{pl}	I_T	W_t	or
mm	mm	kg/m	cm²		cm⁴	cm	cm³	cm³	cm⁴	cm³	D/T
British Standard Notation					I	r	z	S	J	C	d/t
$100 \times 100 \times 8.8^a$		24.5	31.3	8.36	426	3.69	85.2	106	694	123	11
$100 \times 100 \times 10$		27.4	34.9	7	462	3.64	92.4	116	761	133	10
$100 \times 100 \times 12.5^a$		33	42.1	5	522	3.52	104	135	879	150	8
$120 \times 120 \times 4^a$		14.4	18.4	27	410	4.72	68.4	79.7	635	101	30
$120 \times 120 \times 5$		17.8	22.7	21	498	4.68	83	97.6	777	122	24
$120 \times 120 \times 6.3$		22.2	28.2	16	603	4.62	100	120	950	147	19
$120 \times 120 \times 7.1^a$		24.7	31.5	13.9	663	4.59	110	133	1050	161	17
$120 \times 120 \times 8$		27.6	35.2	12	726	4.55	121	146	1160	176	15
$120 \times 120 \times 8.8^a$		30.1	38.3	10.6	779	4.51	130	158	1250	189	14
$120 \times 120 \times 10$		33.7	42.9	9	852	4.46	142	175	1380	206	12
$120 \times 120 \times 12.5$		40.9	52.1	6.6	982	4.34	164	207	1620	236	10
$140 \times 140 \times 5$		21	26.7	25	807	5.5	115	135	1250	170	28
$140 \times 140 \times 6.3$		26.1	33.3	19.2	984	5.44	141	166	1540	206	22
$140 \times 140 \times 7.1^a$		29.2	37.2	16.7	1090	5.4	155	184	1710	227	20
$140 \times 140 \times 8$		32.6	41.6	14.5	1200	5.36	171	204	1890	249	18
$140 \times 140 \times 8.8^a$		35.6	45.4	12.9	1290	5.33	184	221	2050	268	16
$140 \times 140 \times 10$		40	50.9	11	1420	5.27	202	246	2270	294	14
$140 \times 140 \times 12.5$		48.7	62.1	8.2	1650	5.16	236	293	2700	342	11
$150 \times 150 \times 5$		22.6	28.7	27	1000	5.9	134	156	1550	197	30
$150 \times 150 \times 6.3$		28.1	35.8	20.8	1220	5.85	163	192	1910	240	24
$150 \times 150 \times 7.1^a$		31.4	40	18.1	1350	5.81	180	213	2120	264	21
$150 \times 150 \times 8$		35.1	44.8	15.8	1490	5.77	199	237	2350	291	19
$150 \times 150 \times 8.8^a$		38.4	48.9	14	1610	5.74	214	257	2550	313	17
$150 \times 150 \times 10$		43.1	54.9	12	1770	5.68	236	286	2830	344	15
$150 \times 150 \times 12.5$		52.7	67.1	9	2080	5.57	277	342	3380	402	12
$150 \times 150 \times 14.2^a$		58.9	75	7.56	2260	5.49	302	377	3710	436	11
$150 \times 150 \times 16^a r$		65.2	83	6.38	2430	5.41	324	411	4030	467	9

(continued) Hot finished square hollow sections – dimensions and properties

Section designation		Mass per metre	Area of section	Ratio for local buckling	Second moment of area	Radius of gyration	Elastic modulus	Plastic modulus	Torsional constant		Buckling ratio
Size	Thickness										h/t_f
$h \times h$	t		A	c/t	I	i	W_{el}	W_{pl}	I_T	W_t	or
mm	mm	kg/m	cm²		cm⁴	cm	cm³	cm³	cm⁴	cm³	D/T
British Standard Notation					I	r	Z	S	J	C	d/t
$160 \times 160 \times 5$[a]	24.1	30.7	29	1230	6.31	153	178	1890	226	32	
$160 \times 160 \times 6.3$	30.1	38.3	22.4	1500	6.26	187	220	2330	275	25	
$160 \times 160 \times 7.1$[a]	33.7	42.9	19.5	1660	6.22	207	245	2600	304	23	
$160 \times 160 \times 8$	37.6	48	17	1830	6.18	229	272	2880	335	20	
$160 \times 160 \times 8.8$[a]	41.1	52.4	15.2	1980	6.14	247	295	3130	361	18	
$160 \times 160 \times 10$	46.3	58.9	13	2190	6.09	273	329	3480	398	16	
$160 \times 160 \times 12.5$	56.6	72.1	9.8	2580	5.98	322	395	4160	467	13	
$160 \times 160 \times 14.2$[a]	63.3	80.7	8.27	2810	5.9	351	436	4580	508	11	
$160 \times 160 \times 16$[a]	70.2	89.4	7	3030	5.82	379	476	4990	546	10	
$180 \times 180 \times 5$[a]	27.3	34.7	33	1770	7.13	196	227	2720	290	36	
$180 \times 180 \times 6.3$	34	43.3	25.6	2170	7.07	241	281	3360	355	29	
$180 \times 180 \times 7.1$[a]	38.1	48.6	22.4	2400	7.04	267	314	3740	393	25	
$180 \times 180 \times 8$	42.7	54.4	19.5	2660	7	296	349	4160	434	23	
$180 \times 180 \times 8.8$[a]	46.7	59.4	17.5	2880	6.96	320	379	4520	469	20	
$180 \times 180 \times 10$	52.5	66.9	15	3190	6.91	355	424	5050	518	18	
$180 \times 180 \times 12.5$	64.4	82.1	11.4	3790	6.8	421	511	6070	613	14	
$180 \times 180 \times 14.2$[a]	72.2	92	9.68	4150	6.72	462	566	6710	670	13	
$180 \times 180 \times 16$	80.2	102	8.25	4500	6.64	500	621	7340	724	11	
$200 \times 200 \times 5$	30.4	38.7	37	2450	7.95	245	283	3760	362	40	
$200 \times 200 \times 6.3$	38	48.4	28.7	3010	7.89	301	350	4650	444	32	
$200 \times 200 \times 7.1$[a]	42.6	54.2	25.2	3350	7.85	335	391	5190	493	28	
$200 \times 200 \times 8$	47.7	60.8	22	3710	7.81	371	436	5780	545	25	
$200 \times 200 \times 8.8$[a]	52.2	66.5	19.7	4020	7.78	402	474	6290	590	23	
$200 \times 200 \times 10$	58.8	74.9	17	4470	7.72	447	531	7030	655	20	
$200 \times 200 \times 12.5$	72.3	92.1	13	5340	7.61	534	643	8490	778	16	
$200 \times 200 \times 14.2$[a]	81.1	103	11.1	5870	7.54	587	714	9420	854	14	
$200 \times 200 \times 16$	90.3	115	9.5	6390	7.46	639	785	10300	927	13	

continued

(continued) Hot finished square hollow sections – dimensions and properties

Section designation		Mass per metre	Area of section	Ratio for local buckling	Second moment of area	Radius of gyration	Elastic modulus	Plastic modulus	Torsional constant		Buckling ratio
Size	Thickness										h/t_f
$h \times h$	t		A	c/t	I	i	W_{el}	W_{pl}	I_T	W_t	or
mm	mm	kg/m	cm²		cm⁴	cm	cm³	cm³	cm⁴	cm³	D/T
British Standard Notation					I	r	Z	S	J	C	d/t
250 × 250 × 5[a]	38.3	48.7	47	4860	9.99	389	447	7430	577	50
250 × 250 × 6.3	47.9	61	36.7	6010	9.93	481	556	9240	712	40
250 × 250 × 7.1[a]	53.7	68.4	32.2	6700	9.9	536	622	10300	792	35
250 × 250 × 8	60.3	76.8	28.3	7460	9.86	596	694	11500	880	31
250 × 250 × 8.8[a]	66	84.1	25.4	8110	9.82	649	758	12600	955	28
250 × 250 × 10	74.5	94.9	22	9060	9.77	724	851	14100	1070	25
250 × 250 × 12.5	91.9	117	17	10900	9.66	873	1040	17200	1280	20
250 × 250 × 14.2[a]	103	132	14.6	12100	9.58	967	1160	19100	1410	18
250 × 250 × 16	115	147	12.6	13300	9.5	1060	1280	21100	1550	16
260 × 260 × 6.3[a]	49.9	63.5	38.3	6790	10.3	522	603	10400	773	41
260 × 260 × 7.1[a]	56	71.3	33.6	7570	10.3	582	674	11600	861	37
260 × 260 × 8[a]	62.8	80	29.5	8420	10.3	648	753	13000	956	33
260 × 260 × 8.8[a]	68.8	87.6	26.5	9160	10.2	705	822	14200	1040	30
260 × 260 × 10[a]	77.7	98.9	23	10200	10.2	788	924	15900	1160	26
260 × 260 × 12.5[a]	95.8	122	17.8	12400	10.1	951	1130	19400	1390	21
260 × 260 × 14.2[a]	108	137	15.3	13700	9.99	1060	1260	21700	1540	18
260 × 260 × 16[a]	120	153	13.3	15100	9.91	1160	1390	23900	1690	16
300 × 300 × 6.3[a]	57.8	73.6	44.6	10500	12	703	809	16100	1040	48
300 × 300 × 7.1[a]	64.9	82.6	39.3	11800	11.9	785	906	18100	1160	42
300 × 300 × 8	72.8	92.8	34.5	13100	11.9	875	1010	20200	1290	38
300 × 300 × 8.8[a]	79.8	102	31.1	14300	11.9	954	1110	22100	1410	34
300 × 300 × 10	90.2	115	27	16000	11.8	1070	1250	24800	1580	30
300 × 300 × 12.5	112	142	21	19400	11.7	1300	1530	30300	1900	24
300 × 300 × 14.2[a]	126	160	18.1	21600	11.6	1440	1710	33900	2110	21
300 × 300 × 16	141	179	15.8	23900	11.5	1590	1900	37600	2330	19
350 × 350 × 8	85.4	109	40.8	21100	13.9	1210	1390	32400	1790	44
350 × 350 × 8.8[a]	93.6	119	36.8	23100	13.9	1320	1520	35400	1950	40
350 × 350 × 10	106	135	32	25900	13.9	1480	1720	39900	2190	35
350 × 350 × 12.5	131	167	25	31500	13.7	1800	2110	48900	2650	28
350 × 350 × 14.2[a]	148	189	21.6	35200	13.7	2010	2360	54900	2960	25
350 × 350 × 16	166	211	18.9	38900	13.6	2230	2630	61000	3260	22

(continued) Hot finished square hollow sections – dimensions and properties

Section desig-nation		Mass per metre	Area of section	Ratio for local buckling	Second moment of area	Radius of gyration	Elastic modulus	Plastic modulus	Torsional constant		Buck-ling ratio
Size	Thick-ness										h/t_f
$h \times h$	t		A	c/t	I	i	W_{el}	W_{pl}	I_T	W_t	or
mm	mm	kg/m	cm²		cm⁴	cm	cm³	cm³	cm⁴	cm³	D/T
British Standard Notation					I	r	z	S	J	C	d/t
$400 \times 400 \times 8$[a]		97.9	125	47	31900	16	1590	1830	48700	2360	50
$400 \times 400 \times 8.8$[a]		107	137	42.5	34800	15.9	1740	2000	53300	2580	45
$400 \times 400 \times 10$		122	155	37	39100	15.9	1960	2260	60100	2900	40
$400 \times 400 \times 12.5$		151	192	29	47800	15.8	2390	2780	73900	3530	32
$400 \times 400 \times 14.2$[a]		170	217	25.2	53500	15.7	2680	3130	83000	3940	28
$400 \times 400 \times 16$		191	243	22	59300	15.6	2970	3480	92400	4360	25
$400 \times 400 \times 20$[a]		235	300	17	71500	15.4	3580	4250	112000	5240	20

Note:
[a] SAW process (single longitudinal seam weld, slightly proud).

Source: TATA Steel. 2008 – reproduced with the kind permission of TATA.

Hot finished circular hollow sections – dimensions and properties

Section designation		Mass per metre	Area of section	Ratio for local buckling	Second moment of area	Radius of gyration	Elastic modulus	Plastic modulus	Torsional constant		Buck-ling ratio
Outside diameter	Thick-ness										h/t_f
d	t		A	d/t	I	i	W_{el}	W_{pl}	I_T	W_t	or
mm	mm	kg/m	cm²		cm⁴	cm	cm³	cm³	cm⁴	cm³	D/T
British Standard Notation				D/t	I	r	Z	S	J	C	d/t
21.3 × 2.6[a]		1.2	1.53	8.19	0.681	0.668	0.639	0.915	1.36	1.28	8
21.3 × 2.9[a]		1.32	1.68	7.34	0.727	0.659	0.683	0.99	1.45	1.37	7
21.3 × 3.2		1.43	1.82	6.66	0.768	0.65	0.722	1.06	1.54	1.44	7
26.9 × 2.6[a]		1.56	1.98	10.3	1.48	0.864	1.1	1.54	2.96	2.2	10
26.9 × 2.9[a]		1.72	2.19	9.28	1.6	0.855	1.19	1.68	3.19	2.38	9
26.9 × 3.2		1.87	2.38	8.41	1.7	0.846	1.27	1.81	3.41	2.53	8
26.9 × 3.6[a]		2.07	2.64	7.47	1.83	0.834	1.36	1.97	3.66	2.72	7
33.7 × 2.6[a]		1.99	2.54	13	3.09	1.1	1.84	2.52	6.19	3.67	13
33.7 × 2.9[a]		2.2	2.81	11.6	3.36	1.09	1.99	2.76	6.71	3.98	12
33.7 × 3.2		2.41	3.07	10.5	3.6	1.08	2.14	2.99	7.21	4.28	11
33.7 × 3.6[a]		2.67	3.4	9.36	3.91	1.07	2.32	3.28	7.82	4.64	9
33.7 × 4		2.93	3.73	8.43	4.19	1.06	2.49	3.55	8.38	4.97	8
33.7 × 4.5[a]		3.24	4.13	7.49	4.5	1.04	2.67	3.87	9.01	5.35	7
33.7 × 5[a]		3.54	4.51	6.74	4.78	1.03	2.84	4.16	9.57	5.68	7
42.4 × 2.6[a]		2.55	3.25	16.3	6.46	1.41	3.05	4.12	12.9	6.1	16
42.4 × 2.9[a]		2.82	3.6	14.6	7.06	1.4	3.33	4.53	14.1	6.66	15
42.4 × 3.2		3.09	3.94	13.2	7.62	1.39	3.59	4.93	15.2	7.19	13
42.4 × 3.6[a]		3.44	4.39	11.8	8.33	1.38	3.93	5.44	16.7	7.86	12
42.4 × 4		3.79	4.83	10.6	8.99	1.36	4.24	5.92	18	8.48	11
42.4 × 4.5[a]		4.21	5.36	9.42	9.76	1.35	4.6	6.49	19.5	9.2	9
42.4 × 5[a]		4.61	5.87	8.48	10.5	1.33	4.93	7.04	20.9	9.86	8
48.3 × 2.6[a]		2.93	3.73	18.6	9.78	1.62	4.05	5.44	19.6	8.1	19
48.3 × 2.9[a]		3.25	4.14	16.7	10.7	1.61	4.43	5.99	21.4	8.86	17
48.3 × 3.2		3.56	4.53	15.1	11.6	1.6	4.8	6.52	23.2	9.59	15
48.3 × 3.6[a]		3.97	5.06	13.4	12.7	1.59	5.26	7.21	25.4	10.5	13
48.3 × 4		4.37	5.57	12.1	13.8	1.57	5.7	7.87	27.5	11.4	12
48.3 × 4.5[a]		4.86	6.19	10.7	15	1.56	6.21	8.66	30	12.4	11
48.3 × 5		5.34	6.8	9.66	16.2	1.54	6.69	9.42	32.3	13.4	10
48.3 × 5.6[a]		5.9	7.51	8.63	17.4	1.52	7.21	10.3	34.8	14.4	9
48.3 × 6.3		6.53	8.31	7.67	18.7	1.5	7.76	11.2	37.5	15.5	8

(continued) Hot finished circular hollow sections – dimensions and properties

Section designation		Mass per metre	Area of section	Ratio for local buckling	Second moment of area	Radius of gyration	Elastic modulus	Plastic modulus	Torsional constant		Buckling ratio
Outside diameter	Thickness										h/t_f
d	t		A	d/t	I	i	W_{el}	W_{pl}	I_T	W_t	or
mm	mm	kg/m	cm²		cm⁴	cm	cm³	cm³	cm⁴	cm³	D/T
British Standard Notation				D/t	I	r	z	S	J	C	d/t
60.3×2.6^a		3.7	4.71	23.2	19.7	2.04	6.52	8.66	39.3	13	23
60.3×2.9^a		4.11	5.23	20.8	21.6	2.03	7.16	9.56	43.2	14.3	21
60.3×3.2		4.51	5.74	18.8	23.5	2.02	7.78	10.4	46.9	15.6	19
60.3×3.6^a		5.03	6.41	16.8	25.9	2.01	8.58	11.6	51.7	17.2	17
60.3×4		5.55	7.07	15.1	28.2	2	9.34	12.7	56.3	18.7	15
60.3×4.5^a		6.19	7.89	13.4	30.9	1.98	10.2	14	61.8	20.5	13
60.3×5		6.82	8.69	12.1	33.5	1.96	11.1	15.3	67	22.2	12
60.3×5.6^a		7.55	9.62	10.8	36.4	1.94	12.1	16.8	72.7	24.1	11
60.3×6.3		8.39	10.7	9.57	39.5	1.92	13.1	18.5	79	26.2	10
60.3×8^a		10.3	13.1	7.54	46	1.87	15.3	22.1	92	30.5	8
76.1×2.9		5.24	6.67	26.2	44.7	2.59	11.8	15.5	89.5	23.5	26
76.1×3.2		5.75	7.33	23.8	48.8	2.58	12.8	17	97.6	25.6	24
76.1×3.6^a		6.44	8.2	21.1	54	2.57	14.2	18.9	108	28.4	21
76.1×4		7.11	9.06	19	59.1	2.55	15.5	20.8	118	31	19
76.1×4.5^a		7.95	10.1	16.9	65.1	2.54	17.1	23.1	130	34.2	17
76.1×5		8.77	11.2	15.2	70.9	2.52	18.6	25.3	142	37.3	15
76.1×5.6^a		9.74	12.4	13.6	77.5	2.5	20.4	27.9	155	40.8	14
76.1×6.3		10.8	13.8	12.1	84.8	2.48	22.3	30.8	170	44.6	12
76.1×8		13.4	17.1	9.51	101	2.42	26.4	37.3	201	52.9	10
88.9×2.9^a		6.15	7.84	30.7	72.5	3.04	16.3	21.5	145	32.6	31
88.9×3.2^a		6.76	8.62	27.8	79.2	3.03	17.8	23.5	158	35.6	28
88.9×3.6^a		7.57	9.65	24.7	87.9	3.02	19.8	26.2	176	39.5	25
88.9×4		8.38	10.7	22.2	96.3	3	21.7	28.9	193	43.3	22
88.9×4.5^a		9.37	11.9	19.8	107	2.99	24	32.1	213	47.9	20
88.9×5		10.3	13.2	17.8	116	2.97	26.2	35.2	233	52.4	18
88.9×5.6^a		11.5	14.7	15.9	128	2.95	28.7	38.9	255	57.5	16
88.9×6.3		12.8	16.3	14.1	140	2.93	31.5	43.1	280	63.1	14
88.9×8		16	20.3	11.1	168	2.87	37.8	52.5	336	75.6	11
88.9×10^a		19.5	24.8	8.89	196	2.81	44.1	62.6	392	88.2	9

continued

(continued) Hot finished circular hollow sections – dimensions and properties

Section designation		Mass per metre	Area of section	Ratio for local buckling	Second moment of area	Radius of gyration	Elastic modulus	Plastic modulus	Torsional constant		Buckling ratio
Outside diameter	Thickness										h/t_f
d	t		A	d/t	I	i	W_{el}	W_{pl}	I_T	W_t	or
mm	mm	kg/m	cm²		cm⁴	cm	cm³	cm³	cm⁴	cm³	D/T
British Standard Notation				D/t	I	r	Z	S	J	C	d/t
101.6 × 3.2[a]		7.77	9.89	31.7	120	3.48	23.6	31	240	47.2	32
101.6 × 3.6[a]		8.7	11.1	28.2	133	3.47	26.2	34.6	266	52.5	28
101.6 × 4[a]		9.63	12.3	25.4	146	3.45	28.8	38.1	293	57.6	25
101.6 × 4.5[a]		10.8	13.7	22.6	162	3.44	31.9	42.5	324	63.8	23
101.6 × 5[a]		11.9	15.2	20.3	177	3.42	34.9	46.7	355	69.9	20
101.6 × 5.6[a]		13.3	16.9	18.1	195	3.4	38.4	51.7	390	76.9	18
101.6 × 6.3[a]		14.8	18.9	16.1	215	3.38	42.3	57.3	430	84.7	16
101.6 × 8[a]		18.5	23.5	12.7	260	3.32	51.1	70.3	519	102	13
101.6 × 10[a]		22.6	28.8	10.2	305	3.26	60.1	84.2	611	120	10
114.3 × 3.2[a]		8.77	11.2	35.7	172	3.93	30.2	39.5	345	60.4	36
114.3 × 3.6		9.83	12.5	31.8	192	3.92	33.6	44.1	384	67.2	32
114.3 × 4		10.9	13.9	28.6	211	3.9	36.9	48.7	422	73.9	29
114.3 × 4.5[a]		12.2	15.5	25.4	234	3.89	41	54.3	469	82	25
114.3 × 5		13.5	17.2	22.9	257	3.87	45	59.8	514	89.9	23
114.3 × 5.6[a]		15	19.1	20.4	283	3.85	49.6	66.2	566	99.1	20
114.3 × 6.3		16.8	21.4	18.1	313	3.82	54.7	73.6	625	109	18
114.3 × 8		21	26.7	14.3	379	3.77	66.4	90.6	759	133	14
114.3 × 10[a]		25.7	32.8	11.4	450	3.7	78.7	109	899	157	11
139.7 × 3.2[a]		10.8	13.7	43.7	320	4.83	45.8	59.6	640	91.6	44
139.7 × 3.6[a]		12.1	15.4	38.8	357	4.81	51.1	66.7	713	102	39
139.7 × 4[a]		13.4	17.1	34.9	393	4.8	56.2	73.7	786	112	35
139.7 × 4.5[a]		15	19.1	31	437	4.78	62.6	82.3	874	125	31
139.7 × 5		16.6	21.2	27.9	481	4.77	68.8	90.8	961	138	28
139.7 × 5.6[a]		18.5	23.6	24.9	531	4.75	76.1	101	1060	152	25
139.7 × 6.3		20.7	26.4	22.2	589	4.72	84.3	112	1180	169	22
139.7 × 8		26	33.1	17.5	720	4.66	103	139	1440	206	17
139.7 × 10		32	40.7	14	862	4.6	123	169	1720	247	14
139.7 × 12.5[a]		39.2	50	11.2	1020	4.52	146	203	2040	292	11

(continued) Hot finished circular hollow sections – dimensions and properties

Section designation		Mass per metre	Area of section	Ratio for local buckling	Second moment of area	Radius of gyration	Elastic modulus	Plastic modulus	Torsional constant		Buckling ratio
Outside diameter	Thickness										h/t_f
d	t		A	d/t	I	i	W_{el}	W_{pl}	I_T	W_t	or
mm	mm	kg/m	cm²		cm⁴	cm	cm³	cm³	cm⁴	cm³	D/T
British Standard Notation				D/t	I	r	Z	S	J	C	d/t
168.3 × 5		20.1	25.7	33.7	856	5.78	102	133	1710	203	34
168.3 × 5.6[a]		22.5	28.6	30.1	948	5.76	113	148	1900	225	30
168.3 × 6.3		25.2	32.1	26.7	1050	5.73	125	165	2110	250	27
168.3 × 8		31.6	40.3	21	1300	5.67	154	206	2600	308	21
168.3 × 10		39	49.7	16.8	1560	5.61	186	251	3130	372	17
168.3 × 12.5		48	61.2	13.5	1870	5.53	222	304	3740	444	13
193.7 × 5		23.3	29.6	38.7	1320	6.67	136	178	2640	273	39
193.7 × 5.6[a]		26	33.1	34.6	1470	6.65	151	198	2930	303	35
193.7 × 6.3		29.1	37.1	30.7	1630	6.63	168	221	3260	337	31
193.7 × 8		36.6	46.7	24.2	2020	6.57	208	276	4030	416	24
193.7 × 10		45.3	57.7	19.4	2440	6.5	252	338	4880	504	19
193.7 × 12.5		55.9	71.2	15.5	2930	6.42	303	411	5870	606	15
193.7 × 16[a]		70.1	89.3	12.1	3550	6.31	367	507	7110	734	12
219.1 × 4.5[a]		23.8	30.3	48.7	1750	7.59	159	207	3490	319	49
219.1 × 5[a]		26.4	33.6	43.8	1930	7.57	176	229	3860	352	44
219.1 × 5.6[a]		29.5	37.6	39.1	2140	7.55	195	255	4280	391	39
219.1 × 6.3		33.1	42.1	34.8	2390	7.53	218	285	4770	436	35
219.1 × 8		41.6	53.1	27.4	2960	7.47	270	357	5920	540	27
219.1 × 10		51.6	65.7	21.9	3600	7.4	328	438	7200	657	22
219.1 × 12.5		63.7	81.1	17.5	4350	7.32	397	534	8690	793	18
219.1 × 14.2[a]		71.8	91.4	15.4	4820	7.26	440	597	9640	880	15
219.1 × 16		80.1	102	13.7	5300	7.2	483	661	10600	967	14
244.5 × 5[a]		29.5	37.6	48.9	2700	8.47	221	287	5400	441	49
244.5 × 5.6[a]		33	42	43.7	3000	8.45	245	320	6000	491	44
244.5 × 6.3[a]		37	47.1	38.8	3350	8.42	274	358	6690	547	39
244.5 × 8[a]		46.7	59.4	30.6	4160	8.37	340	448	8320	681	31
244.5 × 10[a]		57.8	73.7	24.5	5070	8.3	415	550	10100	830	24
244.5 × 12.5		71.5	91.1	19.6	6150	8.21	503	673	12300	1010	20
244.5 × 14.2[a]		80.6	103	17.2	6840	8.16	559	754	13700	1120	17
244.5 × 16		90.2	115	15.3	7530	8.1	616	837	15100	1230	15

(continued) Hot finished circular hollow sections – dimensions and properties

Section designation		Mass per metre	Area of section	Ratio for local buckling	Second moment of area	Radius of gyration	Elastic modulus	Plastic modulus	Torsional constant		Buck-ling ratio
Outside diameter	Thick-ness										h/t_f
d	t		A	d/t	I	i	W_{el}	W_{pl}	I_T	W_t	or
mm	mm	kg/m	cm²		cm⁴	cm	cm³	cm³	cm⁴	cm³	D/T
British Standard Notation				D/t	I	r	Z	S	J	C	d/t

continued

273 × 5[a]		33	42.1	54.6	3780	9.48	277	359	7560	554	55
273 × 5.6[a]		36.9	47	48.8	4210	9.46	308	400	8410	616	49
273 × 6.3[a]		41.4	52.8	43.3	4700	9.43	344	448	9390	688	43
273 × 8[a]		52.3	66.6	34.1	5850	9.37	429	562	11700	857	34
273 × 10		64.9	82.6	27.3	7150	9.31	524	692	14300	1050	27
273 × 12.5		80.3	102	21.8	8700	9.22	637	849	17400	1270	22
273 × 14.2[a]		90.6	115	19.2	9700	9.16	710	952	19400	1420	19
273 × 16		101	129	17.1	10700	9.1	784	1060	21400	1570	17
323.9 × 5[a]		39.3	50.1	64.8	6370	11.3	393	509	12700	787	65
323.9 × 5.6[a]		44	56	57.8	7090	11.3	438	567	14200	876	58
323.9 × 6.3[a]		49.3	62.9	51.4	7930	11.2	490	636	15900	979	51
323.9 × 8[a]		62.3	79.4	40.5	9910	11.2	612	799	19800	1220	40
323.9 × 10		77.4	98.6	32.4	12200	11.1	751	986	24300	1500	32
323.9 × 12.5		96	122	25.9	14800	11	917	1210	29700	1830	26
323.9 × 14.2[a]		108	138	22.8	16600	11	1030	1360	33200	2050	23
323.9 × 16		121	155	20.2	18400	10.9	1140	1520	36800	2270	20
355.6 × 6.3[a]		54.3	69.1	56.4	10500	12.4	593	769	21100	1190	56
355.6 × 8[a]		68.6	87.4	44.5	13200	12.3	742	967	26400	1490	44
355.6 × 10[a]		85.2	109	35.6	16200	12.2	912	1200	32400	1830	36
355.6 × 12.5[a]		106	135	28.4	19900	12.1	1120	1470	39700	2230	28
355.6 × 14.2[a]		120	152	25	22200	12.1	1250	1660	44500	2500	25
355.6 × 16		134	171	22.2	24700	12	1390	1850	49300	2770	22
406.4 × 6.3[a]		62.2	79.2	64.5	15800	14.1	780	1010	31700	1560	65
406.4 × 8[a]		78.6	100	50.8	19900	14.1	978	1270	39700	1960	51
406.4 × 10		97.8	125	40.6	24500	14	1210	1570	49000	2410	41
406.4 × 12.5[a]		121	155	32.5	30000	13.9	1480	1940	60100	2960	33

(continued) Hot finished circular hollow sections –
dimensions and properties

Section designation		Mass per metre	Area of section	Ratio for local buckling	Second moment of area	Radius of gyration	Elastic modulus	Plastic modulus	Torsional constant		Buck-ling ratio
Outside diameter	Thick-ness										h/t_f
d	t	A	d/t	I	i	W_{el}	W_{pl}	I_T	W_t	or
mm	mm	kg/m	cm²		cm⁴	cm	cm³	cm³	cm⁴	cm³	D/T
British Standard Notation			D/t	I	r	Z	S	J	C	d/t
406.4 × 14.2[a]		137	175	28.6	33700	13.9	1660	2190	67400	3320	29
406.4 × 16		154	196	25.4	37400	13.8	1840	2440	74900	3690	25
457 × 6.3[a]		70	89.2	72.5	22700	15.9	991	1280	45300	1980	73
457 × 8[a]		88.6	113	57.1	28400	15.9	1250	1610	56900	2490	57
457 × 10		110	140	45.7	35100	15.8	1540	2000	70200	3070	46
457 × 12.5[a]		137	175	36.6	43100	15.7	1890	2470	86300	3780	37
457 × 14.2[a]		155	198	32.2	48500	15.7	2120	2790	96900	4240	32
457 × 16		174	222	28.6	54000	15.6	2360	3110	108000	4720	29
508 × 6.3[a]		77.9	99.3	80.6	31200	17.7	1230	1590	62500	2460	81
508 × 8[a]		98.6	126	63.5	39300	17.7	1550	2000	78600	3090	64
508 × 10[a]		123	156	50.8	48500	17.6	1910	2480	97000	3820	51
508 × 12.5		153	195	40.6	59800	17.5	2350	3070	120000	4710	41
508 × 14.2[a]		173	220	35.8	67200	17.5	2650	3460	134000	5290	36
508 × 16		194	247	31.8	74900	17.4	2950	3870	150000	5900	32

Note:
[a] TATA Sections produced in addition to the range of BS Sections.

Source: TATA Steel. 2008 – reproduced with the kind permission of TATA.

Hot finished elliptical hollow sections – dimensions and properties

Section designation		Mass per metre	Area of section	Second moment of area		Radius of gyration		Elastic modulus		Plastic modulus		Torsional constants		Buckling ratio
				y–y	z–z	y–y	z–z	y–y	z–z	y–y	z–z			h/t_f
Size	Thickness													
$h \times b$	t		A	I_y	I_z	i_y	i_z	W_{ely}	W_{elz}	W_{ply}	W_{plz}	I_T	W_t	or
mm	mm	kg/m	cm²	cm⁴	cm⁴	cm	cm	cm³	cm³	cm³	cm³	cm⁴	cm³	D/T
British Standard Notation				I_{x-x}	I_{y-y}	r_{x-x}	r_{y-y}	Z_{x-x}	Z_{y-y}	S_{x-x}	S_{y-y}	J	C	d/t
150 × 75 × 4ª	10.7	13.6	301	101	4.7	2.72	40.1	26.9	56.1	34.4	303	60.1	38	
150 × 75 × 5ª	13.3	16.9	367	122	4.66	2.69	48.9	32.5	68.9	42	367	72.2	30	
150 × 75 × 6.3ª	16.5	21	448	147	4.62	2.64	59.7	39.1	84.9	51.5	443	86.3	24	
200 × 100 × 5ª	17.9	22.8	897	302	6.27	3.64	89.7	60.4	125	76.8	905	135	40	
200 × 100 × 6.3ª	22.3	28.4	1100	368	6.23	3.6	110	73.5	155	94.7	1110	163	32	
200 × 100 × 8ª	28	35.7	1360	446	6.17	3.54	136	89.3	193	117	1350	197	25	
200 × 100 × 10ª	34.5	44	1640	529	6.1	3.47	164	106	235	141	1610	232	20	
200 × 100 × 12.5ª	42.4	54	1950	619	6.02	3.39	195	124	284	169	1890	269	16	
250 × 125 × 6.3ª	28.2	35.9	2210	742	7.84	4.55	176	119	246	151	2220	265	40	
250 × 125 × 8ª	35.4	45.1	2730	909	7.78	4.49	219	145	307	188	2730	323	31	
250 × 125 × 10ª	43.8	55.8	3320	1090	7.71	4.42	265	174	376	228	3290	385	25	
250 × 125 × 12.5ª	53.9	68.7	4000	1290	7.63	4.34	320	207	458	276	3920	453	20	
300 × 150 × 8ª	42.8	54.5	4810	1620	9.39	5.44	321	215	449	275	4850	481	38	
300 × 150 × 10ª	53	67.5	5870	1950	9.32	5.37	391	260	551	336	5870	577	30	
300 × 150 × 12.5	65.5	83.4	7120	2330	9.24	5.29	475	311	674	409	7050	686	24	
300 × 150 × 16ª	82.5	105	8730	2810	9.12	5.17	582	374	837	503	8530	818	19	
400 × 200 × 8ª	57.6	73.4	11700	3970	12.6	7.35	584	397	811	500	11900	890	50	
400 × 200 × 10ª	71.5	91.1	14300	4830	12.5	7.28	717	483	1000	615	14500	1080	40	
400 × 200 × 12.5	88.6	113	17500	5840	12.5	7.19	877	584	1230	753	17600	1300	32	
400 × 200 × 16	112	143	21700	7140	12.3	7.07	1090	714	1540	936	21600	1580	25	
500 × 250 × 10ª	90	115	28500	9680	15.8	9.19	1140	775	1590	976	29000	1740	50	
500 × 250 × 12.5ª	112	142	35000	11800	15.7	9.1	1400	943	1960	1200	35300	2110	40	
500 × 250 × 16ª	142	180	43700	14500	15.6	8.98	1750	1160	2460	1500	43700	2590	31	

Source: TATA Steel. 2008 – reproduced with the kind permission of TATA .

Mild steel rounds typically available

Bar diameter (mm)	Weight (kg/m)	Bar diameter (mm)	Weight (kg/m)	Bar diameter (mm)	Weight (kg/m)	Bar diameter (mm)	Weight (kg/m)
6	0.22	16	1.58	40	9.86	65	26.0
8	0.39	20	2.47	45	12.5	75	34.7
10	0.62	25	3.85	50	15.4	90	49.9
12	0.89	32	6.31	60	22.2	100	61.6

Mild steel square bars typically available

Bar size (mm)	Weight (kg/m)	Bar size (mm)	Weight (kg/m)	Bar size (mm)	Weight (kg/m)
8	0.50	25	4.91	50	19.60
10	0.79	30	7.07	60	28.30
12.5	1.22	32	8.04	75	44.20
16	2.01	40	12.60	90	63.60
20	3.14	45	15.90	100	78.50

Mild steel flats typically available

Bar size (mm)	Weight (kg/m)	Bar size (mm)	Weight (kg/m)	Bar size (mm)	Weight (kg/m)
13 × 3	0.307	45 × 25	8.830	75 × 20	11.78
13 × 6	0.611	50 × 3	1.180	75 × 25	14.72
16 × 3	0.378	50 × 5	1.960	75 × 30	17.68
20 × 3	0.466	50 × 6	2.360	80 × 6	3.77
20 × 5	0.785	50 × 8	3.140	80 × 8	5.02
20 × 6	0.940	50 × 10	3.93	80 × 10	6.28
20 × 10	1.570	50 × 12	4.71	80 × 12	7.54
25 × 3	0.589	50 × 15	5.89	80 × 15	9.42
25 × 5	0.981	50 × 20	7.85	80 × 20	12.60
25 × 6	1.18	50 × 25	9.81	80 × 25	15.70
25 × 8	1.570	50 × 30	11.80	80 × 30	18.80
25 × 10	1.960	50 × 40	15.70	80 × 40	25.10
25 × 12	2.360	55 × 10	4.56	80 × 50	31.40
30 × 3	0.707	60 × 8	3.77	90 × 6	4.24
30 × 5	1.180	60 × 8	3.77	90 × 10	7.07
30 × 6	1.410	60 × 10	4.71	90 × 12	8.48
30 × 8	1.880	60 × 12	5.65	90 × 15	10.60
30 × 10	2.360	60 × 15	7.07	90 × 20	14.10
30 × 12	2.830	60 × 20	9.42	90 × 25	17.70
30 × 20	4.710	60 × 25	11.80	100 × 5	3.93
35 × 6	1.650	60 × 30	14.14	100 × 6	4.71
35 × 10	2.750	65 × 5	2.55	100 × 8	6.28
35 × 12	3.300	65 × 6	3.06	100 × 10	7.85
35 × 20	5.500	65 × 8	4.05	100 × 12	9.42
40 × 3	0.942	65 × 10	5.10	100 × 15	11.80
40 × 5	1.570	65 × 12	6.12	100 × 20	15.70
40 × 6	1.880	65 × 15	7.65	100 × 25	19.60
40 × 8	2.510	65 × 20	10.20	100 × 30	23.60
40 × 10	3.140	65 × 25	12.80	100 × 40	31.40
40 × 12	3.770	65 × 30	15.30	100 × 50	39.30
40 × 15	4.710	65 × 40	20.40	110 × 6	5.18
40 × 20	6.280	70 × 8	4.40	110 × 10	8.64
40 × 25	7.850	70 × 10	5.50	110 × 12	10.40
40 × 30	9.420	70 × 12	6.59	110 × 20	17.30
45 × 3	1.060	70 × 20	11.0	110 × 50	43.20
45 × 6	2.120	70 × 25	13.70	120 × 6	5.65
45 × 8	2.830	75 × 6	3.54	120 × 10	9.42
45 × 10	3.530	75 × 8	4.71	120 × 12	11.30
45 × 12	4.240	75 × 10	5.90	120 × 15	14.10
45 × 15	5.295	75 × 12	7.07	120 × 20	18.80
45 × 20	7.070	75 × 15	8.84	120 × 25	23.60

(continued) Mild steel flats typically available

Bar size (mm)	Weight (kg/m)	Bar size (mm)	Weight (kg/m)	Bar size (mm)	Weight (kg/m)
130 × 6	6.10	160 × 10	12.60	220 × 15	25.87
130 × 8	8.16	160 × 12	15.10	220 × 20	34.50
130 × 10	10.20	160 × 15	18.80	220 × 25	43.20
130 × 12	12.20	160 × 20	25.20	250 × 10	19.60
130 × 15	15.30	180 × 6	8.50	250 × 12	23.60
130 × 20	20.40	180 × 10	14.14	250 × 15	29.40
130 × 25	25.50	180 × 12	17.00	250 × 20	39.20
140 × 6	6.60	180 × 15	21.20	250 × 25	49.10
140 × 10	11.00	180 × 20	28.30	250 × 40	78.40
140 × 12	13.20	180 × 25	35.30	250 × 50	98.10
140 × 20	22.00	200 × 6	9.90	280 × 12.5	27.48
150 × 6	7.06	200 × 10	15.70	300 × 10	23.55
150 × 8	9.42	200 × 12	18.80	300 × 12	28.30
150 × 10	11.80	200 × 15	23.60	300 × 15	35.30
150 × 12	14.10	200 × 20	31.40	300 × 20	47.10
150 × 15	17.70	200 × 25	39.20	300 × 25	58.80
150 × 20	23.60	200 × 30	47.20	300 × 40	94.20
150 × 25	29.40	220 × 10	17.25		

Hot rolled mild steel plates typically available

Thickness (mm)	Weight (kg/m²)	Thickness (mm)	Weight (kg/m²)	Thickness (mm)	Weight (kg/m²)
3	23.55	22.5	176.62	65	510.25
3.2	25.12	25	196.25	70	549.50
4	31.40	30	235.50	75	588.75
5	39.25	32	251.20	80	628.00
6	47.10	35	274.75	90	706.50
8	62.80	40	314.00	100	785.00
10	78.50	45	353.25	110	863.50
12.5	98.12	50	392.50	120	942.00
15	117.75	55	431.75	130	1050.00
20	157.00	60	471.00	150	1177.50

Durbar mild steel floor plates typically available

Basic size (mm)	Weight (kg/m²)
2500 × 1250 × 3 3000 × 1500 × 3	26.19
2000 × 1000 × 4.5 2500 × 1250 × 4.5 3000 × 1250 × 4.5 3700 × 1830 × 4.5 4000 × 1750 × 4.5	37.97
2000 × 1000 × 6 2500 × 1250 × 6 3000 × 1500 × 6 3700 × 1830 × 6 4000 × 1750 × 6	49.74
2000 × 1000 × 8 2500 × 1250 × 8 3000 × 1500 × 8	65.44
3700 × 1830 × 8 4000 × 1750 × 8 6100 × 1830 × 8	65.44
2000 × 1000 × 10 2500 × 1250 × 10 3000 × 1500 × 10 3700 × 1830 × 10	81.14
2000 × 1000 × 12.5 2500 × 1250 × 12.5 3000 × 1500 × 12.5 3700 × 1830 × 12.5 4000 × 1750 × 12.5	100.77

The depth of pattern ranges from 1.9 to 2.4 mm.

Member axes and dimension notation

BS EN 1993 has made some changes to the notation used for member axes that will probably take some time for most UK engineers to get used to. The X–X axis is no longer the principal cross-section axis, but instead is the axis along the member. Axis Y–Y is now the principal axis of bending and Z–Z the minor axis for bending. Although the EC3 axes have long been used in computer analysis packages, and this change will bring steel into line with other Eurocodes, this change will undoubtedly cause considerable confusion during transition.

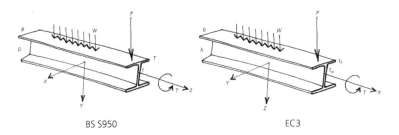

BS S950   EC3

Slenderness

Slenderness and elastic buckling

The slenderness (λ) of a structural element indicates how much load the element can carry in compression. Short stocky elements have low values of slenderness and are likely to fail by crushing, while elements with high slenderness values will fail by elastic (reversible) buckling. Slender columns will buckle when the axial compression reaches the critical load. Slender beams will buckle when the compressive stress causes the compression flange to buckle and twist sideways. This is called 'lateral torsional buckling' and it can be avoided (and the load capacity of the beam increased) by restraining the compression flange at intervals or over its full length. Full lateral restraint can be assumed if the construction fixed to the compression flange is capable of resisting a force of not less than 2.5% of the maximum force in that flange distributed uniformly along its length.

Slenderness limits

Slenderness $\lambda = L_e/r$, where L_e is the effective length and r is the radius of gyration – generally about the weaker axis.

For robustness, members should be selected so that their slenderness does not exceed the following limits:

Members resisting load other than wind	$\lambda \leq 180$
Members resisting self-weight and wind only	$\lambda \leq 250$
Members normally acting as ties but subject to load reversal due to wind	$\lambda \leq 350$

Effective length

BS EN 1993 refers to effective length as L_{CR} instead of L_E and provides little or no guidance on how to estimate values. In the absence of other information, BS 5950 guidance is still relevant.

Effective length of beams – end restraint

			Effective length	
Conditions of restraint at the ends of the beams			**Normal loading**	**Destabilising loading**
Compression flange laterally restrained; beam fully restrained against torsion (rotation about the longitudinal axis)	Both flanges fully restrained against rotation on plan		0.70L	0.85L
	Compression flange fully restrained against rotation on plan		0.75L	0.90L
	Both flanges partially restrained against rotation on plan		0.80L	0.95L
	Compression flange partially restrained against rotation on plan		0.85L	1.00L
	Both flanges free to rotate on plan		1.00L	1.20L
Compression flange laterally unrestrained; both flanges free to rotate on plan	Partial torsional restraint against rotation about the longitudinal axis provided by connection of bottom flange to supports		1.0L + 2D	1.2L + 2D
	Partial torsional restraint against rotation about the longitudinal axis provided only by the pressure of the bottom flange bearing onto the supports		1.2L + 2D	1.4L + 2D

Note: The illustrated connections are not the only methods of providing the restraints noted in the table.

Source: BS 5950 Part 1:2000.

Effective length of cantilevers

Conditions of restraint		Effective length[a]	
Support	Cantilever tip	Normal loading	Destabilising loading
Continuous with lateral restraint to top flange	Free	3.0L	7.5L
	Top flange laterally restrained	2.7L	7.5L
	Torsional restraint	2.4L	4.5L
	Lateral and torsional restraint	2.1L	3.6L
Continuous with partial torsional restraint	Free	2.0L	5.0L
	Top flange laterally restrained	1.8L	5.0L
	Torsional restraint	1.6L	3.0L
	Lateral and torsional restraint	1.4L	2.4L
Continuous with lateral and torsional restraint	Free	1.0L	2.5L
	Top flange laterally restrained	0.9L	2.5L
	Torsional restraint	0.8L	1.5L
	Lateral and torsional restraint	0.7L	1.2L
Restrained laterally, torsionally and against rotation on plan	Free	0.8L	1.4L
	Top flange laterally restrained	0.7L	1.4L
	Torsional restraint	0.6L	0.6L
	Lateral and torsional restraint	0.5L	0.5L

Cantilever tip restraint conditions

Free	Top flange laterally restrained	Torsional restraint	Lateral and torsional restraint
'Not braced on plan'	'Braced on plan in at least one bay'	'Not braced on plan'	'Braced on plan in at least one bay'

Note:
[a] Increases length effectively by 30% for moments applied at cantilever tip.

Effective length of braced columns – restraint provided by cross bracing or shear wall

Conditions of restraint at the ends of the columns		Effective length
Effectively held in position at both ends	Effectively restrained in direction at both ends	0.70L
	Partially restrained in direction at both ends	0.85L
	Restrained in direction at one end	0.85L
	Not restrained in direction at either end	1.00L

Effective length of unbraced columns – restraint provided by sway of columns

Conditions of restraint at the ends of the columns		Effective length
Effectively held in position and restrained in direction at one end	Other end effectively restrained in direction	1.20L
	Other end partially restrained in direction	1.50L
	Other end not restrained in direction	2.00L

Source: BS 5950 Part 1:2000.

Durability and fire resistance

Corrosion mechanism and protection

$$4Fe + 3O_2 + 2H_2O = 2Fe_2O_3 \cdot H_2O$$

Iron/Steel + Oxygen + Water = Rust

For corrosion of steel to take place, oxygen and water must both be present. The corrosion rate is affected by the atmospheric pollution and the length of time the steelwork remains wet. Sulphates (typically from industrial pollution) and chlorides (typically in marine environments – coastal is considered to be a 2-km strip around the coast in the United Kingdom) can accelerate the corrosion rate. All corrosion occurs at the anode (negative where electrons are lost) and the products of corrosion are deposited at the cathode (positive where the electrons are gained). Both anodic and cathodic areas can be present on a steel surface.

The following factors should be considered in relation to the durability of a structure: the environment, degree of exposure, shape of the members, structural detailing, protective measures and whether inspection and maintenance are possible. Bi-metallic corrosion should also be considered in damp conditions.

Durability exposure conditions

Corrosive environments are classified by BS EN ISO 12944: Part 2 and ISO 9223, and the corrosivity of the environment must be assessed for each project.

Corrosivity category and risk	Examples of typical environments in a temperate climate[a]	
	Exterior	Interior
C1 – Very low	–	Heated buildings with clean atmospheres, for example, offices, shops, schools and hotels (theoretically no protection is needed)
C2 – Low	Atmospheres with low levels of pollution. Mostly rural areas	Unheated buildings where condensation may occur, for example, depots and sports halls
C3 – Medium	Urban and industrial atmospheres with moderate sulphur dioxide pollution. Coastal areas with low salinity	Production rooms with high humidity and some air pollution, for example, food processing plants, laundries, breweries and dairies
C4 – High	Industrial areas and coastal areas with moderate salinity	Chemical plants, swimming pools, coastal ship and boatyards
C51 – Very high (industrial)	Industrial areas with high humidity and aggressive atmosphere	Buildings or areas with almost permanent condensation and high pollution
C5M – Very high (marine)	Coastal and offshore areas with high salinity	Buildings or areas with almost permanent condensation and high pollution

Note:
[a] A hot and humid climate increases the corrosion rate and steel will require additional protection than in a temperate climate.

BS EN ISO 12944: Part 3 gives advice on steelwork detailing to avoid crevices where moisture and dirt can be caught and accelerate corrosion. Some acidic timbers should be isolated from steelwork. Get advice for each project: Corus can give advice on all steelwork coatings. The Galvanizers' Association, Thermal Spraying and Surface Engineering Association and paint manufacturers also give advice on system specifications.

Methods of corrosion protection

A corrosion protection system should consist of good surface preparation and application of a suitable coating with the required durability and minimum cost.

Mild steel surface preparation to BS EN ISO 8501

Hot-rolled structural steelwork (in mild steel) leaves the last rolling process at about 1000°C. As it cools, its surface reacts with the air to form a blue-grey coating called mill scale, which is unstable, will allow rusting of the steel and will cause problems with the adhesion of protective coatings. The steel must be degreased to ensure that any contaminants which might affect the coatings are removed. The mill scale can then be removed by abrasive blast cleaning. Typical blast cleaning surface grades are:

Sa 1   Light blast cleaning

Sa 2   Thorough blast cleaning

Sa 2½   Very thorough blast cleaning

Sa 3   Blast cleaning to visually clean steel

Sa 2½ is used for most structural steel. Sa 3 is often used for surface preparation for metal spray coatings.

Metallic and non-metallic particles can be used to blast clean the steel surface. Chilled angular metallic grit (usually grade G24) provides a rougher surface than round metallic shot, so that the coatings have better adhesion to the steel surface. Acid pickling is often used after blast cleaning to Sa 2½ to remove final traces of mill scale before galvanising. Coatings must be applied very quickly after the surface preparation to avoid rust reforming and the requirement for reblasting.

Paint coatings for structural steel

Paint provides a barrier coating to prevent corrosion and is made up of pigment (for colour and protection), binder (for formation of the coating film) and solvent (to allow application of the paint before it evaporates and the paint hardens). When first applied, the paint forms a wet film thickness which can be measured and the dry film thickness (DFT – which is normally the specified element) can be predicted when the percentage volume of solids in the paint is known. Primers are normally classified on their protective pigment (e.g. zinc phosphate primer). Intermediate (which build the coating thickness) and finish coats are usually classified on their binders (e.g. epoxies, vinyls and urethanes). Shop primers (with a DFT of 15–25 μm) can be applied before fabrication but these only provide a couple of weeks' worth of protection. Zinc-rich primers generally perform best. Application of paint can be by brush, roller, air spray and airless spray – the latter is the most common in the United Kingdom. Application can be done on site or in the shop and where the steel is to be exposed; the method of application should be chosen for practicality and the surface finish. Shop-applied coatings tend to need touching up on site if they are damaged in transit.

Metallic coatings for structural steel

Hot dip galvanising: Degreased, blast cleaned (generally Sa 2(1/2)) and then acid pickled steel is dipped into a flux agent and then into a bath of molten zinc. The zinc reacts with the surface of the steel, forming alloys and as the steel is lifted out, a layer of pure zinc is deposited on the outer surface of the alloys. The zinc coating is chemically bonded to the steel and is sacrificial. The Galvanizers' Association can provide details of galvanising baths around the country, but the average bath size is about 10 m long × 1.2 m wide × 2 m deep. The largest baths available in 2002 in the United Kingdom are 21 m × 1.5 m × 2.4 m and 7.6 m × 2.1 m × 3 m. The heat can cause distortions in fabricated, asymmetric or welded elements. Galvanising is typically 85–140 μm thick and should be carried out to BS EN ISO 1461 and 14713. Paint coatings can be applied on top of the galvanising for aesthetic or durability reasons and an etch primer is normally required to ensure that the paint properly adheres to the galvanising.

Thermal spray: Degreased and blast cleaned (generally Sa 3) steel is sprayed with molten particles of aluminium or zinc. The coating is particulate and the pores normally need to be sealed with an organic sealant in order to prevent rust staining. Metal-sprayed coatings are mechanically bonded to the steel and work partly by anodic protection and partly by barrier protection. There are no limits on the size of elements which can be coated and there are no distortion problems. Thermal spray is typically 150–200 μm thick in aluminium, 100–150 μm thick in zinc and should be carried out to BS EN 22063 and BS EN ISO 14713. Paint coatings can be applied for aesthetic or durability reasons. Bi-metallic corrosion issues should be considered when selecting fixings for aluminium-sprayed elements in damp or external environments.

Weathering steel

Weathering steels are high-strength, low-alloy, weldable structural steels which form a protective rust coating in air that reaches a critical level within 2–5 years and prevents further corrosion. Corten is the Corus proprietary brand of weathering steel, which has material properties comparable to S355, but the relevant material standard is BS EN 10155. To optimise the use of weathering steel, avoid contact with absorbent surfaces (e.g. concrete), prolonged wetting (e.g. north faces of buildings in the United Kingdom), burial in soils, contact with dissimilar metals and exposure to aggressive environments. Even if these conditions are met, rust staining can still affect adjacent materials during the first few years. Weathering bolts (ASTM A325, Type 3 or Cor-ten X) must be used for bolted connections. Standard black bolts should not be used as the zinc coating will be quickly consumed and the fastener corroded. Normal welding techniques can be used.

Stainless steel

Stainless steel is the most corrosion resistant of all the steels due to the presence of chromium in its alloys. The surface of the steel forms a self-healing invisible oxide layer which prevents ongoing corrosion and so the surface must be kept clean and exposed to provide the oxygen required to maintain the corrosion resistance. Stainless steel is resistant to most things, but special precautions should be taken in chlorinated environments. Alloying elements are added in different percentages to alter the durability properties:

SS 304	18% Cr, 10% Ni	Used for general cladding, brick support angles and so on
SS 409	11% Cr	Sometimes used for lintels
SS 316	17% Cr, 12% Ni, 2.5% Mo	Used in medium marine/aggressive environments
SS Duplex 2205	22% Cr, 5.5% Ni, 3% Mo	Used in extreme marine and industrial environments

Summary of methods of fire protection

System	Typical thickness[b] for 60-min protection	Advantages	Disadvantages
Boards Up to 4 h protection. Most popular system in the United Kingdom	25–30 mm	Clean 'boxed in' appearance; dry application; factory quality boards; needs no steel surface preparation	High cost; complex fitting around details; slow to apply
Vermiculite concrete spray Up to 4 h protection. Second most popular system in the United Kingdom	20 mm	Cheap; easy on complex junctions; needs no steel surface preparation; often boards used on columns, with spray on the beams	Poor appearance; messy application needs screening; the wet trade will affect following trades; compatibility with corrosion protection needs to be checked
Intumescent paint Maximum 2 h protection. Charring starts at 200–250°C	1–4 mm[a]	Good aesthetic; shows off form of steel; easy to cover complex details; can be applied in shop or on site	High cost; not suited to all environments; short periods of resistance; soft, thick, easily damaged coatings; difficult to get a really high-quality finish; compatibility with corrosion protection needs to be checked
Flexible blanket Cheap alternative to sprays	20–30 mm	Low cost; dry fixing	Not good aesthetics
Concrete encasement Generally only used when durability is a requirement	25–50 mm	Provides resistance to abrasion, impact, corrosion and weather exposure	Expensive; time consuming; heavy; large thickness required
Concrete-filled columns Used for up to 2 h protection or to reduce intumescent paint thickness on hollow sections	–	Takes up less plan area; acts as permanent shutter; good durability	No data for CHS posts; minimum section size which can be protected 140 × 140SHS; expensive
Water-filled columns Columns interconnected to allow convection cooling. Only used if no other option	–	Long periods of fire resistance	Expensive; lots of maintenance required to control water purity and chemical content
Block-filled column Webs up to 30 min protection	–	Reduced cost; less plan area; good durability	Limited protection times; not advised for steel in partition walls

Note:

[a] Coating thickness specified on the basis of the sections' dimensions and the number of sides that will be exposed to fire.

[b] Castellated beams need about 20% more fire protection than is calculated for the basic parent material.

Preliminary sizing of steel elements

Typical span/depth ratios

Element	Typical span (L) (m)	Beam depth
Primary beams/trusses (heavy point loads)	4–12	$L/10$–15
Secondary beams/trusses (distributed loads)	4–20	$L/15$–25
Transfer beams/trusses carrying floors	16–30	$L/10$
Castellated beams	4–12	$L/10$–15
Plate girders	10–30	$L/10$–12
Vierendeel girders	6–18	$L/8$–10
Parallel chord roof trusses	10–100	$L/12$–20
Pitched roof trusses	8–20	$L/5$–10
Light roof beams	6–60	$L/18$–30
Conventional lattice roof girders	5–20	$L/12$–15
Space frames (allow for $l/250$ pre-camber)	10–100	$L/15$–30
Hot rolled universal column	Single storey 2–8	$L/20$–25
	Multi-storey 2–4	$L/7$–18
Hollow section column	Single storey 2–8	$L/20$–35
	Multi-storey 2–4	$L/7$–28
Lattice column	4–10	$L/20$–25
Portal leg and rafter (haunch depth <0.11)	9–60	$L/35$–40

Preliminary sizing

Beams – There are no shortcuts. Deflection will tend to govern long spans, while shear will govern short spans with heavy loading. Plate girders or trusses are used when the loading is beyond the capacity of rolled sections.

Columns – typical maximum column section size for braced frames

203 UC   Buildings 2 to 3 storeys high and spans up to 7 m

254 UC   Buildings up to 5 storeys high

305 UC   Buildings up to 8 storeys high or supports for low-rise buildings with long spans

354 UC   Buildings from 8 to 12 storeys high

Columns – enhanced loads for preliminary axial design: An enhanced axial load for columns subject to out-of-balance loads can be used for preliminary design:

Top storey:	Total axial load $+ 4Z - Z + 2Y - Y$ (using EC3 axes)
Intermediate storey:	Total axial load $+ Z - Z + Y - Y$ (using EC3 axes)

where X–X and Y–Y are the net axial load differences in each direction.

Trusses with parallel chord – Axial force in chord, $F = M_{applied}/d$ where d is the distance between the chord centroids. $I_{truss} = \sum (A_c d^2/4)$ where A_c is the area of each chord.

For equal chords, this can be simplified to $I_{truss} = A_c d^2/2$.

Portal frames

The *Institution of Structural Engineers' Grey Book* for steel design gives the following preliminary method for sizing plastic portal frames with the following assumptions:

- Plastic hinges are formed at the eaves (in the stanchion) and near the apex; therefore, class
  1 sections as defined in BS 5950 and EC3 should be used.
- Moment at the end of the haunch is 0.87 M_p.
- Wind loading does not control the design.
- Stability of the frame should be checked separately.
- Load, W = vertical rafter load per metre run.

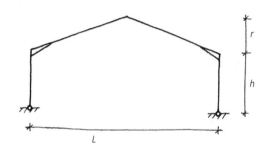

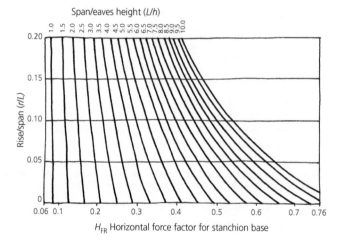

H_{FR} Horizontal force factor for stanchion base

Design moment for rafter, $M_{p\,rafter} = M_{PR}WL^2$

Also consider the high axial force which will be in the rafter and design for combined axial and
bending!

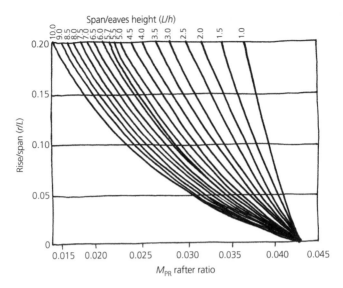

Design moment for stanchion, $M_{\text{p stanchion}} = M_{PR}WL^2$

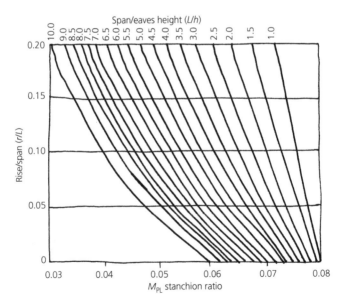

Source: IStructE. 2002.

Structural steel design to BS EN 1993

Section classification and local buckling

Sections are classified by BS EN 1993 in the same way as BS 5950, depending on how their cross section behaves under compressive load. Structural sections in thinner plate will tend to buckle locally and this reduces the overall compressive strength of the section and means that the section cannot achieve its full plastic moment capacity. Sections with tall webs tend to be slender under axial compression, while cross sections with wide out-stand flanges tend to be slender in bending. Combined bending and compression can change the classification of a cross section to slender, when that cross section might not be slender under either bending or compression when applied independently.

For plastic design, the designer must therefore establish the classification of a section (for the given loading conditions) in order to select the appropriate design method from those available. For calculations without capacity tables or computer packages, this can mean many design iterations.

BS EN 1993 has four types of section classification:

Class 1:   Plastic   Cross sections with plastic hinge rotation capacity
Class 2:   Compact   Cross sections with plastic moment capacity
Class 3:   Semi-compact   Cross sections in which the stress at the extreme compression fibre can reach the design strength, but the plastic moment capacity cannot be developed
Class 4:   Slender   Cross sections in which it is necessary to make explicit allowance for the effects of local buckling

Table 5.2 in BS EN 1993 classifies different hot-rolled and fabricated sections based on the limiting width to thickness ratios for each section class. None of the UB, UC, RSJ or PFC sections is slender in pure bending. Under pure axial compression, none of the UC, RSJ or PFC sections is slender, but some UB and hollow sections can be:

UB   Slender if $d/t > 40\varepsilon$
SHS and RHS (hot rolled)   Slender if $d/t > 40\varepsilon$
CHS   Slender if $D/t > 80\varepsilon^2$
Tee stem   Slender if $d/t > 18\varepsilon$

where D = overall depth, t = plate thickness, d = web depth, design, f_y = yield strength and $\varepsilon = \sqrt{235/f_y}$.

For simplicity, only design methods for class 1 and 2 sections are covered in this book.

Source: BS EN 1993-1-1: 2005.

Material properties for design and specification

Selected nominal mild steel design strengths

Steel grade[a]	Steel thickness less than or equal to (mm)	Minimum yield strength, f_y (N/mm^2)	Ultimate tensile strength, f_u (N/mm^2)
S275	16	275	430
	40	265	410
	63	255	410
S355	16	355	490
	40	345	470
	63	335	470

Note:
[a] These values are from BS EN 10025 as referred to by BS EN 1993 even though higher values are used in the main code text.

Generally, it is more economic to use S275 where it is required in small quantities (less than 40 tonnes), where deflection instead of strength limits design, or for members such as nominal ties where the extra strength is not required. In other cases, it is more economical to consider S355.

Ductility and steel grading

In addition to the strength of the material, steel must be specified for a suitable ductility to avoid brittle fracture, which is controlled by the minimum service temperature, the thickness of steel, the steel grade, the type of detail and the stress and strain levels. Ductility is measured by the Charpy V notch test. In the United Kingdom, the minimum service temperature expected to occur over the design life of the structure should be taken as −5°C for internal steelwork or −15°C for external steelwork. For steelwork in cold stores or cold climates, appropriate lower temperatures should be selected. Steel grading has become more important now that the UK construction industry is using more imported steel of varying quality.

Current grading references BS EN 10025-2: 2004 and BS 5950: Part 1: 2000				Superseded grading references[a] BS 5950: Part 1: 1990 and BS 4360: 1990				
Grade	Charpy test temperature (°C)	Steel use	Maximum steel thickness (mm)	Grade	Charpy test temperature (°C)	Steel use	Maximum steel thickness (mm)	
							<100 (N/mm^2)	>100 (N/mm^2)
S275	Untested	Internal only	25	43A	Untested	Internal	50	25
						External	30	15
S275 JR	Room temperature (20°C)	Internal only	30	43B	Room temperature (20°C)	Internal	50	25
						External	30	15
S275 J0	0°C	Internal	65	43 C	0°C	Internal	n/a	60
		External	54			External	80	40
S275 J2	− 20°C	Internal	94	43D	− 20°C	Internal	n/a	n/a
		External	78			External	n/a	90

Note:
[a] Where the superseded equivalent for grades S355 and S460 are Grades 50 and 55, respectively.

Source: BS EN 1993-1-1:2005. Table 3.1 and clause 3.2.5; BS EN 10025-2:2004. Table 7.

Partial safety factors for section resistance in buildings

BS EN 1993 applies γ_m to design resistance rather than provide a design strength. Therefore, unlike BS 5950, γ_m is applied to accommodate both material and section modelling/geometric uncertainties:

Resistance of cross-section whatever class $\gamma_{m0} = 1.0$

Resistance of members to instability assessed by member checks $\gamma_{m1} = 1.0$

Resistance of cross-section in tension to fracture $\gamma_{m2} = 1.10$

Resistance of joints $\gamma_{mz} = 1.25$

Axial load – tension

The design resistance $N_{t,Rd}$ of an axially loaded tension member is the lesser of:

$N_{pl,Rd} = \dfrac{Af_y}{\gamma_{mo}} \equiv Af_y$ – the yield resistance of the gross cross section

$N_{u,Rd} = \dfrac{0.9A_{net}f_u}{\gamma_{m2}} \equiv 0.81A_{net}f_u$ – the ultimate (or fracture) resistance of the cross-section containing holes

Bending – shear

Design shear resistance, $V_{c,Rd}$

$V_{cRd} = A_v \dfrac{f_y}{\sqrt{3}} \equiv 0.577A_v f_y$ (c.f. 0.6 A_v f_y from BS 5950) for a stocky web.

where A_v is the shear area, which should be taken as

tD	For rolled I sections (loaded parallel to the web) and rolled T sections
$AD/(D + B)$	For rectangular hollow sections
$t(D–T)$	For welded T sections
$0.6A$	For circular hollow sections
$0.9A$	Solid bars and plates

t = web thickness, A = cross-sectional area, D = overall depth, B = overall breadth, T = flange thickness.

If $(d_w/t_w) > 72\sqrt{235/f_y}$ for a rolled section, shear buckling must be allowed for (see BS EN 1993-1-1: 2005; clause 6.2.6).

All webs of UBs and UCs in S275 steel satisfy $(d_w/t_w) \leq 72\sqrt{235/f_y}$.

Source: NA to BS EN 1993-1-1:2005; NAZ.5
BS EN 1993-1-1:2005; clause 6.2.6

Bending – flexion

Moment resistance $M_{c,Rd}$ – The basic moment capacity ($M_{c,Rd}$) depends on the provision of full lateral restraint and the interaction of shear and bending stresses. M_c is limited to $1.2f_yW_{el}$ to avoid irreversible deformation under serviceability loads. Full lateral restraint can be assumed if the construction fixed to the compression flange is capable of resisting not less than 2.5% of the maximum compression force in the flange, distributed uniformly along the length of the flange. Moment capacity is generally the controlling capacity for class 1 and 2 sections in the following cases:

- Bending about the minor axis
- CHS, SHS or small solid circular or square bars
- RHS in some cases given in clause 4.3.6.1 of BS 5950
- UB, UC, RSJ, PFC, SHS or RHS if $\lambda < 34$ for S275 steel and $\lambda < 30$ for S355 steel in class 1 and 2 sections where $\lambda = L_E/r$

For low shear ($V_{c,Ed} < 0.5\,V_{c,Rd}$) $M_{c,Rd}$ is typically

$$M_{pl,Rd} = \frac{f_yW_{pl}}{\gamma_{mo}} \equiv f_yW_{pl}$$

for gross, plastic class 1 or 2 sections bending about one axis, without holes in the tension flange.

For high shear ($V_{c,Ed} \geq 0.5\,V_{c,Rd}$) $M_{c,Rd}$ is typically

$$M_{y,V,Rd} = \frac{1}{\gamma_{mo}}\left[W_{pl} - \rho\frac{A_w^2}{4t_w}\right]f_y$$

for I-sections with equal flange bending about the major axis, where $A_w = h_wt_w$

where $\rho = ((2V_{Ed}/V_{pl,Ed}) - 1)^2$ and W_{pl} is the plastic modulus of the shear area used to calculate $V_{pl,Ed}$.

Source: BS EN 1993-1-1:2005; clauses 6.2.6 & 6.2.8.

Lateral torsional buckling resistance, $M_{b,Rd}$

Lateral torsional buckling resistance occurs in tall sections or long beams in bending if insufficient restraint is provided to the compression flange. Instability of the compression flange results in buckling of the beam, preventing the section from developing its full plastic capacity, $M_{c,Rd}$. The reduced bending moment capacity, $M_{b,Rd}$, depends on the slenderness of the section and the engineer's ability to calculate the elastic buckling moment for a perfectly beam, M_{ce}, and the buckling factor, x_{LT}. These calculations render hand calculation virtually impossible.

$$M_{b,LT} = x_{LT,mod} W_{pl,y} \frac{f_y}{\gamma_{m1}} \text{ for class 1 and 2 sections}$$

The modified building factor, $x_{LT,mod} = (x_{LT}/f)$ where enhancement factor (f) is:

$$f = 1 - 0.5(1 - k_c)[1 - 2(\bar{\lambda}_{LT} - 0.8)^2] \not> 1.0 \text{ (if } C_1 = 1.0 \text{ then } f = 1.0)$$

where $k_c = (1/\sqrt{C_1})$ and non-dimensional slenderness, $\bar{\lambda}_{LT} = \sqrt{(f_y W_{pl,y}/M_{CR})}$.

$C_1 = (M_{CR}$ for actual bending moment diagram/M_{CR} for uniform bending moment diagram) and typical tabulated values are given overleaf with the critical buckling moment (M_{CR}) formulae

Buckling factor, $x_{LT} = (1/[\bar{\Phi}_{LT} + \sqrt{\bar{\Phi}_{LT}^2 - \beta\bar{\lambda}_{LT}^2}]) \geq 1.0$ and $(1/\bar{\lambda}_{LT}^2)$

and buckling parameter, $\bar{\Phi}_{LT} = 0.5(1 + \alpha_{LT}(\bar{\lambda}_{LT} - \bar{\lambda}_{LT,0}) + \beta\bar{\lambda}_{LT}^2)$

where

limiting slenderness, $\bar{\lambda}_{LT,0} = 0.4$ and correction factor, $\beta = 0.75$ for rolled sections and imperfections factor:
$\alpha_{LT} = 0.34$ for rolled I sections where $h/b \leq 2.0$,
$\alpha_{LT} = 0.49$ for rolled I sections where $2.0 < h/b \leq 3.1$
$\alpha_{LT} = 0.76$ for rolled I sections where $h/b > 3.1$ as other cross sections.
For slenderness, $\lambda < \lambda_{LT,0}$ buckling effect can be ignored and only cross-sectional checks apply.

Source: BS EN 1993-1-1:2005; Table 6.3, 6.3.2.2; NA to BS EN 1993-1-1:2005; NA 2.17 + NA 2.18.

Critical buckling moment, M_{CR}

BS EN 1993 gives no guidance on the calculation of M_{CR}, so NCCI (Non-Contradictory Complementary Information) must be relied on. Free software can be downloaded from www. cficm.com, or Access Steel provides a shortcut method for calculation of M_{CR} for simply loaded symmetric beams not subject to destabilising loads as follows:

$$M_{CR} = \frac{C_1 x E I_z}{L^2} \left[\sqrt{\frac{I_w}{I_z} + \frac{L^2 G I_w}{\pi^2 E I_z} + (C_2 Z_g)^2} - C_2 Z_g \right] \quad \text{where } k = k_w = 1.0$$

where E is Young's modulus of elasticity (210 kN/mm²), G is the shear modulus (81 kN/mm²), I_z is the minor axis second moment of area, I_t is the torsional constant, I_w is the warping constant, L is the beam span and C_1 and C_2 are factors which account for (section properties, support conditions, applied bending moments and destabilising loads). For a beam without destabilising loads (i.e. loads which can move with the beam as it buckles) $C_2 Z_g = 0$. For simplicity, the values of C_1 below assume no destabilising loads:

Shape of bending moment diagram	C_1				
	$\psi = 1.0$ $C_1 = 1.0$				
	$\psi = 0.8$ $C_1 = 1.11$	$\psi = 0.6$ $C_1 = 1.24$	$\psi = 0.4$ $C_1 = 1.39$	$\psi = 0.2$ $C_1 = 1.57$	$\psi = 0.0$ $C_1 = 1.77$
	$\psi = -0.2$ $C_1 = 1.99$	$\psi = -0.4$ $C_1 = 2.22$	$\psi = -0.6$ $C_1 = 2.43$	$\psi = -0.8$ $C_1 = 2.51$	$\psi = -1.0$ $C_1 = 2.55$
	$C_1 = 1.13$				
	$C_1 = 2.88$				
	$C_1 = 2.58$				
	$C_1 = 1.35$				
	$C_1 = 1.38$				
	$C_1 = 1.68$				

Refer to the full Access Steel document for design methods with applied end moments and/or destabilising loads.

The actual capacity of the beam is less than both M_{CR} and $M_{c,Rd}$, so further reductions must be made for slenderness and tolerances.

Source: Access Steel SN003b. 2010.

Axial load-compression

The design compression resistance of class 1 and 2 sections can be calculated as $N_{Rd} = x f_y A$ using the simplified method in clause 6.3.1.2 of BS EN 1993, where A is the gross area of the section and x can be estimated depending on the expected buckling axis and the section type for steel of $\leq$40 mm thickness, in S275 or S355.

Type of section	Struct curve for value of p_c	
	Axis of buckling	
	y–y	z–z
Hot finished structural hollow section	Any axis: a	
Rolled I section ($h/6 > 1.2$)	a	b
Rolled H section ($h/6 \leq 1.2$)	b	c
Angle section	Any axis: b	
Rolled, channel or T section/paired rolled sections/ compound rolled and solid sections	Any axis: c	

Ultimate compression stresses for rolled sections, p_c.

Buckling factor, x

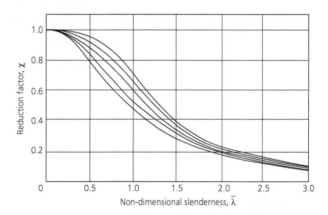

where the non-dimensional slenderness $\lambda = \sqrt{(A f_y / N_{CR})}$ and the elastic critical axial buckling load $N_{CR} = (\pi^2 EI/L^2)$ for class 1 and 2 sections.

Source: BS EN 1993-1-1:2005. Table 6.2 and Figure 6.4, clause 6.3.

Combined bending and compression

Although each section should have its classification checked for combined bending and axial compression, the capacities from the previous tables can be checked against the following simplified relationship for section classes 1 and 2:

$$\frac{N_{Ed}}{N_{Rd}} + \frac{M_{y,Ed}}{M_{y,b,Rd}} + 1.5\frac{M_{z,Ed}}{f_y W_{pL,z}} \cdot \gamma\mu_1 \leq 1.0$$

Section 6.3.3 in BS EN 1993 should be referred to in detail for all the relevant checks.

Source: Access Steel SN048b. 2011.

Connections

In BS EN 1993 connection design is dealt with in Part 8 of the code and its associated national Annex.

Welded connections–simplified method

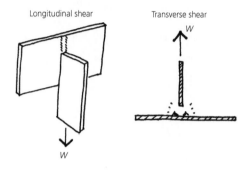

Longitudinal shear   Transverse shear

W

W

The simplified design method in BS EN 1993 states that the design resistance of a fillet held may be assumed to be adequate it, at every point along its length, the resultant of all the forces per unit length transmitted by the weld satisfy the following:

$$F_{w,Ed} \leq F_{w,Rd}$$

Ultimate fillet weld capacities for s275 elements joined at 90°

Leg length s (mm)	Throat thickness $a = 0.7s$ (mm)	Longitudinal capacity[a] $F_{w,L},R_d$ (kN/mm)	Transverse capacity[a] $F_{w,T},R_d$ (kN/mm)
4	2.8	0.62	0.76
6	4.2	0.94	1.15
8	5.6	1.25	1.53
12	8.4	1.87	2.29

Note:

[a] Based on values for S275, $f_{w,d} = \dfrac{f_u/\sqrt{3}}{\beta_w \gamma_{m2}} = 222$ N/mm^2, where $\beta_w = 0.85, \gamma_{m2} = 1.25$ and f_u

$= 410$ N/mm^2 and $a \geq 0.5\, t_p$, where t_p is the thickness of the parent metal being welded.

Source: BS EN 1993-1-8: 2005. Eurocode 3. Table 4.1. NA to BS EN 1993-1-8: 2005; Table NA.1. BS EN 10025-2: 2004; Table 7.

Bolted connections

Limiting bolt spacings

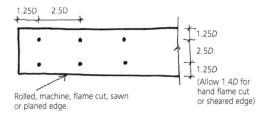

Rolled, machine, flame cut, sawn or planed edge.

(Allow 1.4D for hand flame cut or sheared edge)

Direct shear

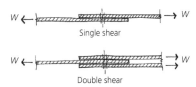

Simple moment connection bolt groups

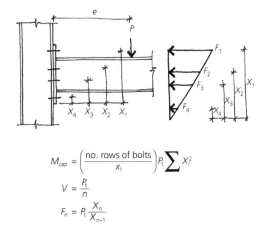

$$M_{cap} = \left(\frac{\text{no. rows of bolts}}{X_1} \right) P_t \sum X_i^2$$

$$V = \frac{P_t}{n}$$

$$F_n = P_t \frac{X_n}{X_{n-1}}$$

where $x_1 = \max x_i$, and $x_i =$ depth from point of rotation to centre of bolt being considered, P_t is the tension capacity of the bolts, n is the number of bolts, V is the shear on each bolt and F is the tension in each bolt. This is a simplified analysis which assumes that the bolt furthest from the point of rotation carries the most load. As the connection elements are likely to be flexible, this is unlikely to be the case; however, more complicated analysis requires a computer or standard tables.

Bolt capacity checks: For bolts in shear or tension, see the following tabulated values. For bolts in shear and tension check: $(F_s/P_s) + (F_t/P_t) \leq 1.4$ where F indicates the factored design load and P indicates the ultimate bolt capacity.

Selected ultimate bolt capacities for non-preloaded ordinary hexagonal head bolts in S275 Steel to BS EN 1993

| Diameter of bolt d (mm) | Tensile stress area A_s (mm²) | Tension resistance $F_{t,Rd}$ (kN) | Min plate thickness for punching shear t_{min} (mm) | Shear resistance | | Bearing resistance (kN) Assuming minimum 1.5d edge distance, 2.0d end distance and 3d for pitch/gauge — Thickness of ply, t (mm) | | | | | | | | | | |
				Single shear $F_{v,Rd}$ (kN)	Double shear $2 \times F_{v,Rd}$ (kN)	5	6	7	8	9	10	12	15	20	25	30
Grade 4.6																
12	84.3	24.3	2.1	13.8	27.5	25.9	31	36.2	41.4	46.6	51.7	62.1	77.6	103	129	155
16	157	45.2	3.2	30.1	60.3	34	40.8	47.7	54.5	61.3	68.1	81.7	102	136	170	204
20	245	70.6	3.9	47	94.1	42.1	50.5	58.9	67.4	75.8	84.2	101	126	168	211	253
24	353	102	4.7	67.8	136	50.1	60.1	70.2	80.2	90.2	100	120	150	200	251	301
30	561	162	5.8	108	215	63.2	75.8	88.4	101	114	126	152	189	253	316	379
Grade 8.8																
12	84.3	48.6	4.3	27.5	55	25.9	31	36.2	41.4	46.6	51.7	62.1	77.6	103	129	155
16	157	90.4	6.3	60.3	121	34	40.8	47.7	**54.5**	61.3	68.1	81.7	102	136	170	204
20	245	141	7.8	94.1	188	**42.1**	50.5	58.9	67.4	75.8	**84.2**	101	126	168	211	253
24	353	203	9.4	136	271	**50.1**	60.1	70.2	**80.2**	**90.2**	**100**	**120**	150	200	251	301
30	561	323	11.6	215	431	**63.2**	**75.8**	**88.4**	**101**	**114**	**126**	**152**	**189**	253	316	379

Notes:
- For M12 bolts, the design shear resistance $F_{v,Rd}$ has been calculated as 0.85 times the value given in BS EN 1993-1-8, Table 3.4 (§3.6.1(5)).
- See clause 3.7(1) of BS EN 1993-1-8: 2005 for calculation of the design resistance of a group of fasteners.
- For bolts with cut threads that do not comply with EN 1090, the given values for tension and shear should be multiplied by 0.85.
- Values of bearing resistance in **bold** are greater than the single shear resistance of the bolt.
- Values of bearing resistance in *italic* are greater than the double shear resistance of the bolt.
- Bearing values assume standard clearance holes (2 mm clearance for bolts $d < 24$ mm or 3 mm clearance holes for $d \geq 24$ mm).
- If oversize or short slotted holes are used, bearing values should be multiplied by 0.8.
- If long slotted or kidney-shaped holes are used, bearing values should be multiplied by 0.6.
- In single lap joints with only one bolt row, the design bearing resistance for each bolt should be limited to $1.5 f_u \, d \, t / \gamma_{M2}$.
- Shear capacity should be reduced for large packing ($>4d/3$), grip lengths or long joints.

Selected ultimate bolt capacities for non-preloaded countersunk head bolts in S275 steel to BS EN 1993

Diameter of bolt d (mm)	Tensile stress area A_s (mm²)	Tension resistance $F_{t,Rd}$ (kN)	Min plate thickness to fit depth of bolt head t_{min} (mm)	Shear resistance		Bearing resistance (kN) Assuming minimum 1.5d edge distance, 2.0d end distance and 3d for pitch/gauge										
				Single shear $F_{v,Rd}$ (kN)	Double shear $2 \times F_{v,Rd}$ (kN)	Thickness of ply, t (mm)										
						5	6	7	8	9	10	12	15	20	25	30
Grade 4.6																
12	84.3	17	7	13.8	27.5	10.3	15.5	20.7	25.9	31	36.2	46.6	62.1	87.9	114	140
16	157	31.7	8	30.1	60.3	–	–	–	27.2	34	40.8	54.5	74.9	109	143	177
20	245	49.4	9	47.0	94.1	–	–	–	–	–	42.1	58.9	84.2	126	168	211
24	353	71.2	14	67.8	136	–	–	–	–	–	–	–	90.2	140	190	241
30	561	113	20	108	215	–	–	–	–	–	–	–	–	158	221	284
Grade 8.8																
12	84.3	34	27	27.5	55	**10.3**	**15.5**	**20.7**	**25.9**	31	36.2	46.6	62.1	87.9	114	140
16	157	63.3	34	60.3	121	–	–	–	**27.2**	**34**	**40.8**	**54.5**	74.9	109	143	177
20	245	98.8	41	94.1	188	–	–	–	–	–	**42.1**	**58.9**	**84.2**	126	168	211
24	353	142	41	136	271	–	–	–	–	–	–	–	**90.2**	140	190	241
30	561	226	50	215	431	–	–	–	–	–	–	–	–	**158**	221	284

Notes:

• For M12 bolts the design shear resistance $F_{v,Rd}$ has been calculated as 0.85 times the value given in BS EN 1993-1-8, Table 3.4 (§3.6.1(5)).
• See clause 3.7(1) of BS EN 1993-1-8: 2005 for calculation of the design resistance of a group of fasteners.
• For bolts with cut threads that do not comply with EN 1090, the given values for tension and shear should be multiplied by 0.85.
• Values of bearing resistance in **bold** are less than the single shear resistance of the bolt.
• Values of bearing resistance in *italic* are greater than the double shear resistance of the bolt.
• Bearing values assume standard clearance holes.
• If oversize or short slotted holes are used, bearing values should be multiplied by 0.8.
• If long slotted or kidney shaped holes are used, bearing values should be multiplied by 0.6.
• In single lap joints with only one bolt row, the design bearing resistance for each bolt should be limited to $1.5 f_u \, d \, t / \gamma_{M2}$.
• Shear capacity should be reduced for large packing (> 4d/3), grip lengths or long joints.

Summary of differences with BS 5950: Structural steel

	BS EN 1993	BS 5950
Layout	The code of practice is set out in sections based on load phenomenon, for example, bending, shear and deflection. Designers will have to refer to a number of standards. For example, beam design will need reference to BS EN 1993 Part 1 for bending and shear, Part 5 for web and stiffener design and Part 8 for connection design	The code of practice is set out in sections based on building element
Member axes	The Eurocode adopts the sign convention traditionally used in computer analysis packages, but this one 'simple' change is likely to be troublesome, not least because the use of the BS convention persists in many supplementary documents!	–
Symbols	Section modulus: W_{el} Plastic modulus: W_{pl}	Section modulus: Z Plastic modulus: S
Materials factor	γ_M is applied to the cross-sectional resistance and so takes into account both material and modelling/geometric uncertainties	γ_m is a basic material properties factor
Steel properties	Refers to BS EN 10025-2 which proposes lower values than the core standard document. Use values quoted in main code with care!	Refers to BS EN 10025-2
Section classification	BS EN 1993 is slightly less onerous for web width to thickness ratios, while being slightly more onerous for out-stand flanges	–
Buckling factor	$\varepsilon = \sqrt{(235/f_y)}$	$\varepsilon = \sqrt{(275/p_y)}$
Lateral torsional buckling	Requires calculation of the critical elastic buckling moment of a perfectly straight beam, M_{CR}. Designers must refer to Non-Contradictory Complementary Information (NCCI) published by Access Steel	Methodology provides buckling curves and formulae
Buckling lengths	Method uses critical buckling length, L_{CR}, but provides no guidance on its calculation. $L_{CR} \approx L_E$ from BS 5950 in the absence of alternative guidance	Method uses effective length, L_E
Combined bending and compression	Very cumbersome methods if doing hand calculations in full. Access Steel publishes some useful shortcut methods	$F_c/P_c + M_x/(M_{cx}$ or $M_b) + M_y/M_{cy} \leq 1.0$
Connections	Similar methodology to BS 5950, but results for both bolting and welding appearing to be slightly more conservative	–

Stainless steel design to BS EN 1993

Stainless steels are a family of corrosion- and heat-resistant steels containing a minimum of 10.5% chromium which results in the formation of a very thin self-healing transparent skin of chromium oxide – which is described as a passive layer. Alloy proportions can be varied to produce different grades of material with differing strength and corrosion properties. The stability of the passive layer depends on the alloy composition. There are five basic groups: austenitic, ferritic, duplex, martensitic and precipitation hardened. Of these, only austenitic and duplex are really suitable for structural use.

Austenitic stainless steels provide a good combination of corrosion resistance, forming and fabrication properties. Duplex stainless steel has high strength and wear resistance, with very good resistance to stress corrosion checking.

Austenitic

Austenitic is the most widely used for structural applications and contains 17–18% chromium, 8–11% nickel and sometimes molybdenum. Austenitic stainless steel has good corrosion resistance, high ductility and can be readily cold formed or welded. Commonly used alloys are 304 (European grade 1.4301) and 316 (European grade 1.4401). Low-carbon versions are available: 304 L (European grade 1.4307) and 316 L (European grade 1.4404).

Duplex

Duplex stainless steels are so named because they share the strength and corrosion resistance properties of both the austenitic and ferritic grades. They typically contain 21–26% chromium, 4–8% nickel and 0.1–4.5% molybdenum. These steels are readily weldable but are not so easily cold rolled. Duplex stainless steel is normally used where an element is under high stress in a severely corrosive environment. A commonly used alloy is Duplex 2205 (European grade 1.44062) and a new 'lean' grade for load-bearing applications is European grade 1.4162.

Material properties

The material properties vary between cast, hot-rolled and cold-rolled elements.

Density	78–80 (kN/m³)
Tensile strength	200–500 N/mm² 0.2% proof stress depending on grade
Poisson's ratio	0.3
Modulus of elasticity	E varies with the stress in the section and the direction of the stresses. As the stress increases, the stiffness decreases and therefore deflection calculations must be done on the basis of the secant modulus.
Shear modulus	76.9 kN/mm²
Linear coefficient of thermal expansion	$17 \times 10^{-6}/°C$ for 304 L (1.4301)
	$16.5 \times 10^{-6}/°C$ for 316 L (1.4401)
	$13 \times 10^{-6}/°C$ for Duplex 2205 (1.4462)
Ductility	Stainless steel is much tougher than mild steel and so BS EN 1993 does not apply any limit on the thickness of stainless steel sections as it does for mild steel

Elastic properties of stainless steel alloys for design

The secant modulus, $E_S = ((E_s1 + E_s2)/2)$, where

$$E_{si} = \frac{E}{(1 + k(f_{1 \text{ or } 2}/P_y)^m)}$$

where $i = 1$ or 2, $k = 0.002E/P_y$ and m is a constant.

Values of the secant modulus are calculated below for different stress ratios (f_i/P_y).

Values of secant modulus for selected stainless steel alloys for structural design

Stress ratio[a] $\frac{f_i}{P_y}$	Secant modulus (kN/mm²)					
	304 L		316 L		Duplex 2205	
	Longitudinal	Transverse	Longitudinal	Transverse	Longitudinal	Transverse
0.0	200	200	190	195	200	205
0.2	200	200	190	195	200	205
0.3	199	200	190	195	199	204
0.4	197	200	188	195	196	200
0.5	191	198	184	193	189	194
0.6	176	191	174	189	179	183
0.7	152	173	154	174	165	168

Note:
[a] $i = 1$ or 2 for the applied stress in the tension and compression flanges, respectively.

Typical stock stainless steel sections

There is no UK-based manufacturer of stainless steel and so all stainless steel sections are imported. Two importers who will send out information on the sections they produce are Valbruna and IMS Group. The sections available are limited. IMS has a larger range including hot-rolled equal angles (from $20 \times 20 \times 3$ up to $100 \times 100 \times 10$), unequal angles ($20 \times 10 \times 3$ up to $200 \times 100 \times 13$), I beams (80×46 up to 400×180), H beams (50×50 up to 300×300), channels (20×10 up to 400×110) and tees ($20 \times 20 \times 3$ up to $120 \times 120 \times 13$) in 1.4301 and 1.4571. Valbruna has a smaller selection of plate, bars and angles in 1.4301 and 1.4404. Perchcourt are stainless steel section stockists based in the Midlands who can supply fabricators. Check their website for availability.

Source: Nickel Development Institute. 1994.

Durability and fire resistances

Suggested grades of stainless steel for different atmospheric conditions

Stainless steel grade	Location											
	Rural			Urban			Industrial			Marine		
	Low	Medium	High	Low	Medium	High	Low	Medium	High	Low	Medium	High
304 L (1.4301)	✓	✓	✓	✓	✓	(✓)	(✓)	(✓)	x	✓	(✓)	x
316 L (1.4401)	○	○	○	○	✓	✓	✓	✓	(✓)	✓	✓	(✓)
Duplex 2205 (1.4462)	○	○	○	○	○	○	○	○	✓	○	○	✓

Note: ✓ = Optimum specification, (✓) = may require additional protection, x = unsuitable, ○ = overspecified.

Note that this table does not apply to chlorinated environments which are very corrosive to stainless steel. Grade 304 L (1.4301) can tarnish and is generally only used where aesthetics are not important; however, marine Grade 316 L (1.4401) will maintain a shiny surface finish.

Corrosion mechanisms

Durability can be reduced by heat treatment and welding. The surface of the steel forms a self-healing invisible oxide layer which prevents ongoing corrosion and so the surface must be kept clean and exposed to provide the oxygen required to maintain the corrosion resistance.

Pitting: Mostly results in the staining of architectural components and is not normally a structural problem. However, chloride attack can cause pitting which can cause cracking and eventual failure. Alloys rich in molybdenum should be used to resist chloride attack.

Crevice corrosion: Chloride attack and lack of oxygen in small crevices, for example, between nuts and washers.

Bi-metallic effects: The larger the cathode, the greater the rate of attack. Mild steel bolts in a stainless steel assembly would be subject to very aggressive attack. Austenitic grades typically only react with copper to produce an unsightly white powder, with little structural effect. Prevent bi-metallic contact by using paint or tape to exclude water as well as using isolation gaskets, nylon/Teflon bushes and washers.

Fire resistance

Stainless steels retain more of their strength and stiffness than mild steels in fire conditions, but typically as stainless steel structure is normally exposed, its fire resistance generally needs to be calculated as part of a fire-engineered scheme.

Source: Nickel Development Institute. 1994.

Preliminary sizing

Assume a reduced Young's modulus depending on how heavily stressed the section will be and assume an approximate value of maximum bending stress for working loads of 130 N/mm². A section size can then be selected for checking to BS EN 1993.

Stainless steel design to BS EN 1993-1

The design is based on ultimate loads calculated on the same partial safety factors as for mild steel.

Ultimate mechanical properties for stainless steel design to BS EN 1993-1

Alloy type	Steel designation	European grade (UK grade)	Minimum 0.2% proof stress (N/mm²)	Ultimate tensile strength (N/mm²)	Minimum elongation after fracture (%)
Basic austenitic[a]	X5CrNi 18-9	304 (1.4301)	210	520–720	45
Molybdenum austenitic[b]	X2CrNiMo 17-12-2	316 (1.4401)	220	520–670	40
Duplex	X2CrNi MoN 22-5-3	Duplex 2205 (1.4462)	460	640–840	20

Notes:
[a] Most commonly used for structural purposes.
[b] Widely used in more corrosive situations.

The alloys listed in the table above are low-carbon alloys which provide good corrosion resistance after welding and fabrication.

As for mild steel, the element cross section must be classified to BS EN 1993 in order to establish the appropriate design method. Generally, this method is as given for mild steels; however, as there are few standard section shapes, the classification and design methods can be laborious. Free software is available at www.steel-stainless.org.

Source: Nickel Development Institute. 1994.

Connections

Bolted and welded connections can be used. Design data for fillet and butt welds require detailed information about which particular welding method is to be used. The information about bolted connections is more general.

Bolted connections

Requirements for stainless steel fasteners are set out in BS EN ISO 3506 which split fixings into three groups: A = austenitic, F = ferritic and C = martensitic. Grade A fasteners are normally used for structural applications. Grade A2 is equivalent to Grade 304 L (1.4301) with a 0.2% proof stress of 210 N/mm^2 and Grade A4 is equivalent to Grade 316 L (1.4401) with a 0.2% proof stress of 450 N/mm^2. There are three further property classes within Grade A: 50, 70 and 80 to BS EN ISO 3506. An approximate ultimate bearing strength for connected parts can be taken as 460 N/mm^2 for preliminary sizing.

Ultimate stress values for bolted connection design

Grade A property class	Shear strength[a] (N/mm^2)	Bearing strength[a] (N/mm^2)	Tensile strength[a] (N/mm^2)
50	140	510	210
70 (most common)	310	820	450
80	380	1000	560

Note:
[a] These values are appropriate with bolt diameters less than M24 and bolts less than 8 diameters long.

Source: Nickel Development Institute. 1994.

Steel design to BS 449

BS 449: Part 2 is the 'old' steel design code issued in 1969 but it is (with amendments) used by older engineers. The code is based on elastic bending and working stresses and is very simple to use. It is therefore invaluable for preliminary design, for simple steel elements and for checking existing structures. It is normal to compare the applied and allowable stresses. BS 449 refers to the old steel grades where Grade 43 is S275, Grade 50 is S355 and Grade 55 is S460.

Notation for BS 449: Part 2

Symbols		Stress subscripts	
f	Applied stress	c or bc	Compression or bending compression
P	Permissible stress	t or bt	Tension or bending tension
l/r	Slenderness ratio	q	Shear
D	Overall section depth	b	Bearing
T	Flange thickness	e	Equivalent

Allowable stresses

The allowable stresses may be exceeded by 25% where the member has to resist an increase in stress which is solely due to wind forces – provided that the stresses in the section before considering wind are within the basic allowable limits.

Applied stresses are calculated using the gross elastic properties of the section, Z or A, where appropriate.

Allowable stress in axial tension P_t

Form	Steel grade	Thickness of steel (mm)	P_t (N/mm²)
Sections, bars, plates, wide flats and hollow sections	43 (S275)	$t \leq 40$	170
		$40 < t \leq 100$	155

Source: BS 449: Part 2: 1969.

Maximum allowable bending stresses P_{bc} or P_{bt} to BS 449

Form	Steel grade	Thickness of steel (mm)	P_{bc} or P_{bt} (N/mm²)
Sections, bars, plates, wide flats and hollow sections	43 (S275)	$t = 40$	180
Compound beams – hot-rolled sections with additional plates		$40 < t \leq 100$	165
Double channel sections acting as an I beam			
Plate girders	43 (S275)	$t \leq 40$	170
		$40 < t \leq 100$	155
Slab bases	All steels		185

Upstand webs or flanges in compression have a reduced capacity and need to be checked in accordance with clause 20, BS 449. These tabulated values of P_{bc} can be used only where full lateral restraint is provided, where bending is about the minor axis or for hollow sections in bending.

Source: BS 449: Part 2: Table 2: 1969.

Allowable compressive bending stresses to BS 449

The maximum allowable bending stress is reduced as the slenderness increases, to allow for the effects of buckling. The reduced allowable bending stress, P_{bc}, can be obtained from the following graph from the ratio of depth of section to thickness of flange (D/T) and the slenderness ($\lambda = L/r$).

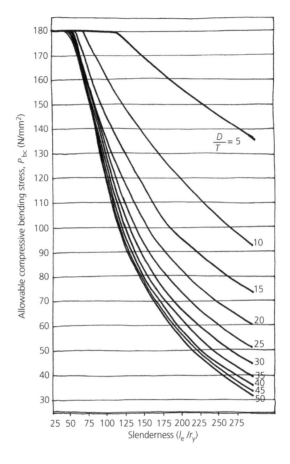

Allowable compressive stresses to BS 449

For uncased compression members, allowable compressive stresses must be reduced by 10% for thick steel sections: if $t > 40$ mm for Grade 43 (S275), $t > 63$ mm for Grade 50 (S355) and $t > 25$ mm for Grade 55 (S460). The allowable axial stress, P_c, reduces as the slenderness of the element increases as shown in the following chart:

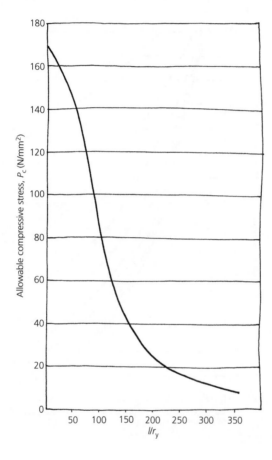

Allowable average shear stress P_v in unstiffened webs to BS 449

Form	Steel grade	Thickness (mm)	P_v[a] (N/mm²)
Sections, bars, plates, wide flats and hollow sections	43 (S275)	$d \leq 40$	110
		$40 < d \leq 100$	100
	50 (S355)	$d \leq 63$	140
		$63 < d \leq 100$	130
	55 (S460)	$d \leq 25$	170

[a] See Table 12 in BS 449: Part 2 for allowable average shear stress in stiffened webs.

Section capacity checks to BS 449

Combined bending and axial load

Compression: $\dfrac{f_c}{p_c} + \dfrac{f_{bcx}}{p_{bcx}} + \dfrac{f_{bcy}}{p_{bcy}} \leq 1.0$

Tension: $\dfrac{f_t}{p_t} + \dfrac{f_{bt}}{p_{bt}} \leq 1.0$ and $\dfrac{f_{bcx}}{p_{bcx}} + \dfrac{f_{bcy}}{p_{bcy}} \leq 1.0$

Combined bending and shear

$$f_e = \sqrt{\left(f_{bt}^2 + 3f_q^2\right)} \quad \text{or} \quad f_e = \sqrt{(f_{bc}^2 + 3f_q^2)} \text{ and } f_e < P_e \quad \text{and} \quad (f_{bc}/p_o)^2 + (f_q'/p_q')^2 \leq 1.25$$

where f_e is the equivalent stress, f'_q is the average shear stress in the web, P_0 is defined in BS 449 subclause 20 item 2b iii and P'_q is defined in clause 23. From BS 449: Table 1, the allowable equivalent stress $P_e = 250$ N/mm² for Grade 43 (S275) steel <40 mm thick.

Combined bending, shear and bearing

$$f_e = \sqrt{(f_{bt}^2 + f_b^2 + f_{bt}f_b + 3f_q^2)} \quad \text{or} \quad f_e = \sqrt{(f_{bc}^2 + f_b^2 + f_{bc}f_b + 3f_q^2)} \quad \text{and} \quad f_e < P_e \quad \text{and}$$
$$(f_{bc}/p_o)^2 + (f_q'/p_q')^2 + (f_{cw}/p_{cw}) \leq 1.25$$

Source: BS 449: Part 2: 1969.

Connections to BS 449

Selected fillet weld working capacities for grade 43 (S275) steel

Leg length s (mm)	Throat thickness $a = 0.7s$ (mm)	Weld capacity[a] (kN/mm)
4	2.8	0.32
6	4.2	0.48
8	5.6	0.64
12	8.4	0.97

Note:
[a] When a weld is subject to a combination of stresses, the combined effect should be checked using the same checks as used for combined loads on sections to BS 449.

Selected full penetration butt weld working capacities for grade 43 (S275) steel

Thickness (mm)	Shear capacity (kN/mm)	Tension or compression capacity[a] (kN/mm)
6	0.60	0.93
15	1.50	2.33
20	2.00	3.10
30	3.00	4.65

Note:
[a] When a weld is subject to a combination of stresses, the combined effect should be checked using the same checks as used for combined loads on sections to BS 449.

Source: BS 449: Part 2:1969.

Allowable stresses in non-preloaded bolts to BS 449

Description	Bolt grade	Axial tension (N/mm²)	Shear (N/mm²)	Bearing (N/mm²)
Close tolerance and turned bolts	4.6	120	100	300
	8.8	280	230	350
Bolts in clearance holes	4.6	120	80	250
	8.8	280	187	350

Allowable stresses on connected parts of bolted connections to BS 449

Description	Allowable stresses on connected parts for different steel grades (N/mm²)		
	43 (S275)	50 (S355)	55 (S460)
Close tolerance and turned bolts	300	420	480
Bolts in clearance holes	250	350	400

Bolted connection capacity check for combined tension and shear to BS 449

$$\frac{f_t}{P_t} + \frac{f_s}{P_s} \leq 1.4$$

Source: BS 449: Part 2: 1969.

Selected working load bolt capacities for non-preloaded ordinary bolts in grade 43 (S275) steel to BS 449

Diameter of molt φ (mm)	Tensile stress area (mm²)	Tension capacity (kN)	Shear capacity		Bearing capacity for end distance = 2φ (kN)						
			Single (kN)	Double (kN)	Thickness of steel passed through (mm)						
					5	6	8	10	12	15	20
Grade 4.6											
6	20.1	1.9	1.6	3.2	7.5	9.0	12.0	15.0	18.0	22.5	30.0
8	36.6	3.5	2.9	5.9	10.0	12.0	16.0	20.0	24.0	30.0	40.0
10	58	5.6	4.6	9.3	12.5	15.0	20.0	25.0	30.0	37.5	50.0
12	84.3	8.1	6.7	13.5	15.0	18.0	24.0	30.0	36.0	45.0	60.0
16	157	15.1	12.6	25.1	*20.0*	*24.0*	32.0	40.0	48.0	60.0	80.0
20	245	23.5	19.6	39.2	*25.0*	*30.0*	40.0	50.0	60.0	75.0	100.0
24	353	33.9	28.2	50.5	*30.0*	*36.6*	*40.0*	60.0	72.0	90.0	120.0
30	561	53.9	44.9	89.8	*37.5*	*45.0*	*60.0*	*75.0*	90.0	112.5	150.0
Grade 8.8											
6	20.1	4.5	3.8	7.5	7.5	9.0	12.0	15.0	18.0	22.5	30.0
8	36.6	8.2	6.8	13.7	*10.0*	*12.0*	16.0	20.0	24.0	30.0	40.0
10	58	13.0	10.8	21.7	*12.5*	*15.0*	20.0	25.0	30.0	37.5	50.0
12	84.3	18.9	15.8	31.5	**15.0**	**18.0**	*24.0*	*30.0*	36.0	45.0	60.0
16	157	35.2	29.4	58.7	**20.0**	**24.0**	*32.0*	*40.0*	*48.0*	60.0	80.0
20	245	54.9	45.8	91.6	**25.0**	**30.0**	**40.0**	*50.0*	*60.0*	75.0	100.0
24	353	79.1	66.0	132.0	**30.0**	**36.0**	**48.0**	**60.0**	*72.0*	90.0	120.0
30	561	125.7	104.9	209.8	**37.5**	**45.0**	**60.0**	**75.0**	**90.0**	*112.5*	*150.0*

Notes:
- 2 mm clearance holes for φ < 24 or 3 mm clearance holes for φ ≥ 24.
- Bearing values shown in **bold** are less than the single shear capacity of the bolt.
- Bearing values shown in *italic* are less than the double shear capacity of the bolt.
- Multiply tabulated bearing values by 0.7 if oversized or short-slotted holes are used.
- Multiply tabulated bearing values by 0.5 if kidney-shaped or long-slotted hole are used.
- Shear capacity should be reduced for large packing, grip lengths or long joints.
- Tabulated tension capacities are nominal tension capacity $= 0.8A_t p_t$ that accounts for prying forces.

Selected working load bolt capacities for non-preloaded countersunk ordinary bolts in grade 43 (S275) to BS 449

Diameter of bolt (ϕ mm)	Tensile stress area (mm²)	Tension capacity (kN)	Shear capacity		Bearing capacity for end distance = 2ϕ (kN)						
			Single (kN)	Double (kN)	Thickness of steel passed through (mm)						
					5	6	8	10	12	15	20
Grade 4.6											
6	20.1	1.9	1.6	3.2	4.7	6.2	9.2	12.2	15.2	19.7	27.2
8	36.6	3.5	2.9	5.9	–	7.0	11.0	15.0	19.0	25.0	35.0
10	58	5.6	4.6	9.3	–	–	11.9	16.9	21.9	29.4	41.9
12	84.3	8.1	6.7	13.5	–	–	–	18.8	24.8	33.8	48.8
16	157	15.1	12.6	25.1	–	–	–	–	30.0	42.0	62.0
20	245	23.5	19.6	39.2	–	–	–	–	33.8	48.8	73.8
24	353	33.9	28.2	56.5	–	–	–	–	–	46.5	76.5
Grade 8.8											
6	20.1	4.5	3.8	7.5	*4.7*	*6.2*	*9.2*	*12.2*	*15.2*	*19.7*	*27.2*
8	36.6	8.2	6.8	13.7	–	**7.0**	*11.0*	15.0	19.0	25.0	35.0
10	58	13.0	10.8	21.7	–	–	*11.9*	*16.9*	21.9	29.4	41.9
12	84.3	18.9	15.8	31.5	–	–	–	**18.8**	*24.8*	33.8	48.8
16	157	35.2	29.4	58.7	–	–	–	–	**30.0**	*42.0*	62.0
20	245	54.9	45.8	91.6	–	–	–	–	**33.8**	**48.8**	*73.8*
24	353	79.1	66.0	132.0	–	–	–	–	–	**46.5**	*76.5*

Notes:
- Values are omitted from the table where the bolt head is too deep to be countersunk into the thickness of the plate.
- 2 mm clearance holes for $\phi < 24$ or 3 mm clearance holes for $\phi \geq 24$.
- Tabulated tension capacities are nominal tension capacity = $0.8A_t p_t$ which accounts for prying forces.
- Bearing values shown in **bold** are less than the single shear capacity of the bolt.
- Bearing values shown in *italic* are less than the double shear capacity of the bolt.
- Multiply tabulated bearing values by 0.7 if oversized or short-slotted holes are used.
- Multiply tabulated bearing values by 0.5 if kidney-shaped or long-slotted holes are used.
- Shear capacity should be reduced for large packing, grip lengths or long joints.
- Grade 4.6 $p_s = 160$ N/mm², $p_t = 240$ N/mm².
- Grade 8.8 $p_s = 375$ N/mm², $p_t = 560$ N/mm².
- Total packing at a shear plane should not exceed ($4\phi/3$).
- Table based on Unbrako machine screw dimensions.

9
Composite Steel and Concrete

Composite steel and concrete flooring, as used today, was developed in the 1960s to economically increase the spans of steel-framed floors while minimising the required structural depths.

Composite flooring elements

Concrete slab: There are various types of slab: solid *in situ, in situ* on profiled metal deck and *in situ* on precast concrete units. Solid slabs are typically 125–150 mm thick and require formwork. The precast and metal deck systems both act as permanent formwork, which may need propping to control deflections. The profiled metal deck sheets have a 50–60 mm depth to create a 115–175 mm slab, which can span 2.5–3.6 m. At present, there are no EN standards for the wide range of profiled sheets available. Precast concrete units 75–100 mm thick with 50–200 mm topping can span 3–8 m.

Steelwork: Generally the steel section is sized to support the wet concrete and construction loads with limited deflection, followed by the full design loading on the composite member. Secondary beams carry the deck and are in turn supported on primary beams which are supported on the columns. The steel beams can be designed as simply supported or continuous. Long span beams can be adversely affected by vibration, and should be used with caution in dynamic loading situations.

Shear studs: Typically 19 mm diameter with typical nominal heights of 95 or 120 mm. Other heights for deep profiled decks are available with longer lead times. Larger-diameter studs are available but not many subcontractors have the automatic welding guns to fix them. Welded studs will carry about twice the load of proprietary 'shot fired' studs.

Economic arrangement: Secondary beam spacing is limited to about arrangement 2.5–3 m in order to keep the slab thickness down and its fire resistance up. The most economic geometrical arrangement is for the primary span to be about 3/4 of the secondary span.

Summary of material properties

The basic properties of steel and concrete are as set out in their separate sections.

Concrete grade	Normal-weight concrete C25/30–C40/50 and lightweight concrete C20/25–C32/40
Density	Normal-weight concrete 24 kN/m³ and lightweight concrete 17 kN/m³
Modular ratio $(\alpha_E = E_s/E_c)$	Normal concrete $\alpha_E = 6$ short term and $\alpha_E = 18$ long term
	Lightweight concrete $\alpha_E = 10$ short term $\alpha_E = 25$ long term
Steel grade	S275 is used where it is required in small quantities (less than 40 tonnes) or where deflection, not strength, limits the design. Otherwise, S355 is more economical, but will increase the minimum number of shear studs which are required by the code

Durability and fire resistance

- The basic durability requirements of steel and concrete are as set out in their separate sections.
- Concrete slabs have an inherent fire resistance. The slab thickness may be controlled by the minimum thickness required for fire separation between floors, rather than by deflection or strength.
- Reinforcing mesh is generally added to the top face of the slab to control surface cracking. The minimum required is 0.1% of the concrete area, but more may be required for continuous spans or in some fire conditions.
- Additional bars are often suspended in the troughs of profiled metal decks to ensure adequate stability under fire conditions. Deck manufacturers provide guidance on bar areas and spacing for different slab spans, loading and thickness for different periods of fire resistance.
- Precast concrete composite planks have a maximum fire resistance of about 2 h.
- The steel frame has to be fire and corrosion protected as set out in the section on structural steelwork.

Preliminary sizing of composite elements

Typical span/depth ratios

Element	Typical spans (m)		Total structural depth (including slab and beams) for simply supported beams	
	Primary	Secondary	Primary	Secondary
Universal beam sections	6–10	8–18	$L/19$	$L/23$
Universal column sections	6–10	8–18	$L/22$	$L/29$
Fabricated sections	>12	>12	$L/15$	$L/25$
Fixed end/haunched beams (Haunch length $L/10$ with maximum depth 2D)	>12	>12	$L/25$ (midspan)	$L/32$ (midspan)
Castellated beams (circular holes $\phi = 2D/3$ at about 1.5ϕ c/c, D is the beam depth)	N/A	6–16	$L/17$	$L/20$
Proprietary composite trusses	>12	>12	$L/12$	$L/16$

Preliminary sizing – Estimate the unfactored moment which will be applied to the beam in its final (rather than construction) condition. Use an allowable working stress of 160 N/mm^2 for S275, or 210 N/mm^2 for S355, to estimate the required section modulus (Z) for a non-composite beam. A preliminary estimate of a composite beam size can be made by selecting a steel beam with 60–70% of the non-composite Z. Commercial office buildings normally have about 1.8–2.2 shear studs (19 mm diameter) per square metre of floor area. Deflections, response to vibration and service holes should be checked for each case.

Approximate limits on holes in rolled steel beams – Reduced section capacity due to holes through the webs of steel beams must be considered for both initial and detailed calculations.

Where D is the depth of the steel beam, limit the size of openings to 0.6D depth and 1.5D length in unstiffened webs, and to 0.7D and 2D, respectively, where stiffeners are provided above and below the opening. Holes should be a minimum of 1.5D apart and be positioned centrally in the depth of the web, in the middle third of the span for uniformly loaded beams. Holes should be a minimum of D from any concentrated loads and 2D from a support position. Should the position of the holes be moved off centre of depth of the beam, the remaining portions of web above and below the hole should not differ by a factor of 1.5–2.

Preliminary composite beam sizing tables for S275 and normal weight concrete 4 kN/m² live loading + 1 kN/m² for partitions

Primary span (m)	Secondary span (m)	No. of secondary beam per grid	Secondary beam spacing (m)	Beam sizes for minimum steel weight			Beam sizes for minimum steel weight		
				Primary beam	Secondary beam	Steel weight (kN/m³)	Primary beam	Secondary beam	Steel weight (kN/m³)
6	8	2	3.00	457 × 152 UB 60	356 × 171 UB 57	0.27	254 × 254 UC 89	254 × 254 UC 89	0.41
	12	2	3.00	533 × 210 UB 82	533 × 210 UB 109	0.43	254 × 254 UC 132	305 × 305 UC 240	0.91
	15	2	3.00	533 × 210 UB 92	762 × 267 UB 147	0.55	305 × 305 UC 158	356 × 406 UC 467	1.66
8	8	3	2.67	457 × 152 UB 82	356 × 171 UB 51	0.29	305 × 305 UC 158	254 × 254 UC 89	0.53
	12	3	2.67	610 × 229 UB 101	533 × 210 UB 101	0.46	305 × 305 UC 240	305 × 305 UC 240	1.10
	15	3	2.67	762 × 267 UB 125	762 × 267 UB 134	0.59	305 × 305 UC 283	356 × 406 UC 393	1.66
9	8	3	3.00	610 × 229 UB 101	356 × 171 UB 57	0.32	305 × 305 UC 198	254 × 254 UC 89	0.54
	12	3	3.00	610 × 229 UB 125	533 × 210 UB 109	0.47	305 × 305 UC 283	305 × 305 UC 240	1.04
	15	3	3.00	762 × 267 UB 147	762 × 267 UB 147	0.59	356 × 406 UC 287	356 × 406 UC 467	1.75
10	8	4	2.50	610 × 229 UB 125	356 × 171 UB 51	0.36	305 × 305 UC 240	254 × 254 UC 89	0.66
	12	4	2.50	762 × 267 UB 147	533 × 210 UB 101	0.53	356 × 406 UC 287	305 × 305 UC 240	1.20
	15	4	2.50	838 × 292 UB 176	762 × 267 UB 134	0.65	356 × 406 UC 340	356 × 406 UC 393	1.80
12	8	4	3.00	762 × 267 UB 147	356 × 171 UB 57	0.37	356 × 406 UC 393	254 × 254 UC 89	0.79
	12	4	3.00	838 × 292 UB 194	533 × 210 UB 109	0.53	356 × 406 UC 551	305 × 305 UC 240	1.26
	15	4	3.00	914 × 305 UB 224	762 × 267 UB 147	0.64	356 × 406 UC 634	356 × 406 UC 467	1.98

Note: Check floor natural frequency <4.5 Hz. Construction deflections limited to span/360 or 25 mm.

Preliminary composite beam sizing tables for S275 and lightweight concrete 4 kN/m2 live loading + 1 kN/m2 for partitions

Primary span (m)	Secondary span (m)	No. of secondary beams per grid	Secondary beam spacing (m)	Minimum steel weight			Minimum floor depth		
				Primary beam	Secondary beam	Steel weight (kN/m³)	Primary beam	Secondary beam	Steel weight (kN/m³)
6	8	2	3.00	457 × 152 UB 60	356 × 171 UB 57	0.27	254 × 254 UC 89	254 × 254 UC 89	0.41
	12	2	3.00	533 × 210 UB 82	533 × 210 UB 109	0.43	254 × 254 UC 132	305 × UC 240	0.91
	15	2	3.00	533 × 210 UB 92	762 × 267 UB 147	0.55	305 × 305 UC 158	356 × 406 UC 467	1.66
8	8	3	2.67	457 × 152 UB 82	356 × 171 UB 51	0.29	305 × 305 UC 158	254 × 254 UC 89	0.53
	12	3	2.67	457 × 229 UB 101	533 × 210 UB 101	0.46	305 × 05 UC 240	305 × 305 UC 240	1.10
	15	3	2.67	610 × 229 UB 125	762 × 267 UB 134	0.59	305 × 305 UC 240	356 × 406 UC 393	1.66
9	8	3	3.00	610 × 229 UB 101	356 × 171 UB 57	0.32	305 × 305 UC 198	254 × 254 UC 89	0.54
	12	3	3.00	610 × 229 UB 125	533 × 210 UB 109	0.47	305 × 305 UC 240	305 × 305 UC 240	1.04
	15	3	3.00	762 × 267 UB 147	762 × 267 UB 147	0.59	356 × 406 UC 467	356 × 406 UC 467	1.75
10	8	4	2.30	610 × 229 UB 125	356 × 171 UB 51	0.36	305 × 305 UC 240	254 × 254 UC 89	0.66
	12	4	2.30	762 × 267 UB 147	533 × 210 UB 101	0.53	356 × 406 UC 287	305 × 305 UC 240	1.20
	15	4	2.30	838 × 292 UB 176	762 × 267 UB 134	0.65	356 × 406 UC 340	356 × 406 UC 393	1.80
12	8	4	3.00	762 × 267 UB 147	356 × 171 UB 57	0.37	356 × 406 UC 393	254 × 254 UC 89	0.76
	12	4	3.00	838 × 292 UB 194	533 × 210 UB 109	0.53	356 × 406 UC 551	305 × 305 UC 240	1.26
	15	4	3.00	914 × 305 UB 224	762 × 267 UB 147	0.64	356 × 406 UC 634	356 × 406 UC 467	1.98

Note: Check floor natural frequency <4.5 Hz.

Composite design to BS EN 1994

BS EN 1994 is based on ultimate loads and plastic design of sections, so the partial safety factors; axes and design strengths of materials from BS EN 1990, 1991, 1992 and 1993 apply as appropriate. Steel sections should be classified for local buckling to determine whether their design should be plastic or elastic.

The composite moment capacity depends on the position of the neutral axis – whether in the concrete slab, the steel flange or the steel web. This depends on the relative strength of the concrete and steel sections. The steel beam should be designed for the non-composite temporary construction situation as well as for composite action in the permanent condition.

The code design methods have been significantly summarised for this book, which only deals with basic moment capacity in relation to uniform loads. The code also provides guidance on how concentrated loads and holes in beams should be designed for in detail. Serviceability and vibration checks are also required.

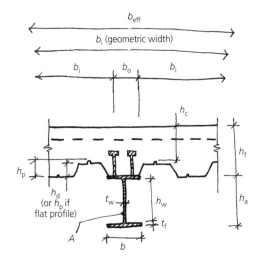

Effective slab width

Where elastic global analysis is *not* used, BS EN 1994 values of effective slab width are based on effective lengths (i.e. distance between points of contraflexure). Simplified values are given in clause 5.4.1.2 for effective lengths in continuous beams as follows:

Midspan end bay	$L_e = 0.85\, L$
First internal support	$L_e = 0.25(L_{span1} + L_{span2})$
Midspan internal bay	$L_e = 0.7\, L$
Cantilever	$L_e = 2.0\, L$

Therefore, different effective slab widths should be calculated for cross-section checks at mid-spans and supports using the following:

Midspan or internal support	$b_{eff} = b_0 + \Sigma b_{ei}$
End support	$b_{eff} = b_0 + \Sigma \beta_i\, b_{ei}$

where
   b_o is the width between outstand shear connectors
   b_{ei} is the width of the concrete flange on either side of the supporting steel, taken as the lesser of $L_e/8$ or the geometrical width, b_i
   b_i is the geometrical width taken from the outstand shear connector to a point midway between adjacent webs, or to a free edge
   $\beta_i = (0.55 + 0.025 L_e/b_{ei}) \leq 1.0$

In the vast majority of cases where elastic analysis is used to calculate moments and shear forces, a constant effective width may be assumed over the whole of each span as follows:

Internal beams	b_{eff} = the lesser of $L_e/4$ or the geometrical spacing
Edge beams	b_{eff} = half of the internal beam valve

However, for sections other than class 1, a more detailed check may be required.

Composite plastic moment capacity for simply supported beams

Assuming that the steel section is compact and uniformly loaded, check that the applied moment is less than the plastic moment of the composite section. These equations are for a profiled metal deck slab where the shear is low ($F_v < 0.5 P_v$):

Compression capacity of concrete slab, $N_{c,f} = 0.85 f_{cd}\, b_{eff}\, x_c$

Tensile capacity of steel element, $N_{pl,a} = f_{yd}\, A_{\text{area of steel section}}$

Tensile capacity of steel web (thickness t and clear depth d), $N_w = 0.95\, f_{yd}\, t_w\, h_w$

Moment capacity of the fully restrained steel section, $M_{pl,a,Rd} = W_{pl}\, f_{yd}/\gamma_{mo}$

Plastic moment capacity of the composite section, $M_{pl,Rd}$

Applied shear stress, F_v

Shear capacity, $V_{pl,Rd}$

Plastic moment capacity for low shear: Neutral axis in concrete slab $N_{c,f} \geq N_{pl,Rd}$

$$M_{pl,Rd} = N_{pl,a}\left(\frac{h_a}{2} + h_s - \frac{h_c N_{pla}}{2 N_{c,f}} \right)$$

Plastic moment capacity for low shear: Neutral axis in steel flange $N_{pl,a} > N_{c,f}$

$$M_{pl,Rd} = N_{pl,a}\left(\frac{h_a}{2} \right) + N_{c,f}\left(\frac{h_c}{2} + h_d \right)$$

Plastic moment capacity for low shear: Neutral axis in steel web

$$M_{pl,Rd} = M_{pl,a,Rd} + N_{c,f}\left(\frac{h_c + 2h_d + h_a}{2} \right) - \left(\frac{N_{c,f}^2}{N_w} \cdot \frac{h_a}{4} \right)$$

Reduced plastic moment capacity for combined high shear and moment

It is accounted for by reducing the design steel strength as follows:

$$(1 - p)f_{yd}, \quad \text{where } p = \left(\frac{2V_{Ed}}{V_{Rd}} - 1 \right)^2 \text{ for class 1 and 2 sections only}$$

Shear capacity

The shear capacity of the beam

$$V_{pl,Rd} = \frac{A_v(f_y/\sqrt{3})}{\gamma_{mo}}, \quad \text{where shear area } A_v = A - 2bt_f + (t_w + 2_r)t_f \geq h_w t_w$$

Shear is considered low if the applied vertical shear load $V_{Ed} = 0.5\ V_{pl,Rd}$. However, in simply supported beams, high shear and moment forces normally only coexist at the positions of heavy point loads. Therefore, generally where there are no point loads and if $M_{applied} < M_{pl,Ed}$ and $V_{Ed} < V_{pl,Rd}$ no further checks are required.

For high shear, $V_{Ed} > 0.5\ V_{pl,Rd}$ a reduced moment capacity, $M_{pl,Rd\ reduced}$ is calculated using the reduced design steel strength as previously indicated.

Longitudinal shear

Shear stud strengths for normal-weight concrete are given in Clause 6.6.4 of BS EN 1994. Studs of 19 mm diameter have characteristic resistances of about 80–100 kN for normal-weight concrete, depending on the height and the concrete strength. The strength of the studs in lightweight concrete can be taken as 90% of the normal concrete weight values. Allowances and reductions must be made for groups of studs as well as the deck shape and its contact area (due to the profiled soffit) with the steel beam. The horizontal shear force on the interface between the steel and concrete should be estimated and an arrangement of shear studs selected to resist that force. The minimum centre-to-centre spacing of studs is 6ϕ longitudinally and 5ϕ transversely, where ϕ is the stud diameter and the maximum longitudinal spacing of the studs is the lesser of $6h_f$ or 800 mm.

Mesh reinforcement is generally required in the top face of the slab to spread the stud shear forces across the effective breadth of the slab (therefore increasing the longitudinal shear resistance) and to minimise cracking in the top of the slab.

Serviceability

For simply supported beams, the second moment of area can be calculated on the basis of the midspan breadth of the concrete flange, which is assumed to be uncracked.

The natural frequency of the structure can be estimated by $f = 18/\sqrt{\delta}$, where δ is the elastic deflection (in mm) under dead and permanent loads. In most cases, problems due to vibrations can be avoided as the natural frequency of the floor is kept greater than 4–4.5 Hz.

10
Timber and Plywood

Commercial timbers are defined as hardwoods and softwoods according to their botanical classification rather than their physical strength. Hardwoods are from broad-leaved trees, which are deciduous in temperate climates. Softwoods are from conifers, which are typically evergreen with needle-shaped leaves.

Structural timber is specified by a strength class that combines the timber species and strength grade. Strength grading is the measurement or estimation of the strength of individual timbers, to allow each piece to be used to its maximum efficiency. This can be done visually or by machine. The strength classes referred to in BS EN 1995 are C14 to C40 for softwoods (C is for coniferous) and D30 to D70 for hardwoods (D is for deciduous). The number refers to the ultimate bending strength in N/mm^2 before application of safety factors for use in design. C16 is the most commonly available softwood, followed by the slightly stronger C24. Specification of C24 should generally be accompanied by checks to confirm that it has actually been used on site in preference to the more readily available C16.

Timber products

Wood-based sheet materials are the main structural timber products, containing substantial amounts of wood in the form of strips, veneers, chips, flakes or fibres. These products are normally classified as:

Laminated panel products: Plywood, laminated veneered lumber (LVL) and glue-laminated timber (glulam) for structural use. Made out of laminations 2–43 mm thick depending on the product.

Particleboard: Chipboard, orientated strandboard (OSB) and wood wool. Developed to use forest thinnings and sawmill waste to create cheap panelling for building applications. Limited structural uses.

Fibreboard: Such as hardboard and medium-density fibreboard. Fine particles bonded together with adhesive to form general, non-structural, utility boards.

Summary of material properties

Density: 1.2–10.7 kN/m^3. Softwood is normally assumed to be between 4 and 6 kN/m^3.

Moisture content: After felling, timber will lose moisture to align itself with atmospheric conditions and becomes harder and stronger as it loses water. In the United Kingdom, the atmospheric humidity is normally about 14%. Seasoning is the name of the controlled process where moisture content is reduced to a level appropriate for the timber's proposed use. Air seasoning within the United Kingdom can achieve a moisture content of 17–23% in several months for softwood, and over a period of years for hardwoods. Kiln drying can be used to achieve the similar moisture contents over several days for softwoods or two to three weeks for hardwoods.

Moisture content should be lower than 20% to stop fungal attack.

Shrinkage: Shrinkage occurs as a result of moisture loss. Typical new structural softwood will reduce in depth across the grain by as much as 3–4% once it is installed in a heated environment. Shrinkage should be allowed for in structural details.

BS EN 1995-1 sets out service classes 1, 2 and 3 which define timber as having moisture contents of <12%, ≤20% and >20%, respectively.

Sizes and processing of timber

Sawn: Most basic cut of timber (rough or fine – although rough is most common) for use where tolerances of ±3 mm are not significant.

Regularised: Two parallel faces planed where there is a dimensional requirement regarding depth or width. Also known as 'Surfaced 2 sides' (S2S).

Planed all round: Used for exposed or dimensional accuracy on all four sides. Also known as 'Surfaced 4 sides' (S4S).

Planing and processing reduces the sawn or 'work size' (normally quoted at 20% moisture content) to a 'target size' (for use in calculations) ignoring permitted tolerances and deviations.

Timber section sizes

Selected timber section sizes and section properties

Timber over the standard maximum length, of about 5.5 m, is more expensive and must be pre-ordered.

Basic size[a]		Area	Z_{yy}	Z_{zz}	I_{yy}	I_{zz}		
D (mm)	B (mm)	10^2 mm²	10^3 mm³	10^3 mm³	10^6 mm⁴	10^6 mm⁴	r_{yy}, mm	r_{zz} mm
100	38	38	63.3	24.1	3.17	0.46	28.9	11.0
100	50	50	83.3	41.7	4.17	1.04	28.9	14.4
100	63	63	105.0	66.2	5.25	2.08	28.9	18.2
100	75	75	125.0	93.8	6.25	3.52	28.9	21.7
100	100	100	166.7	166.7	8.33	8.33	28.9	28.9
150	38	57	142.5	36.1	10.69	0.69	43.3	11.0
150	50	75	187.5	62.5	14.06	1.56	43.3	14.4
150	63	94	236.3	99.2	17.72	3.13	43.3	18.2
150	75	112	281.3	140.6	21.09	5.27	43.3	21.7
150	100	150	375.0	250.0	28.13	12.50	43.3	28.9
150	150	225	562.5	562.5	42.19	42.19	43.3	43.3
175	38	66	194.0	42.1	16.97	0.80	50.5	11.0
175	50	87	255.2	72.9	22.33	1.82	50.5	14.4
175	63	110	321.6	115.8	28.14	3.65	50.5	18.2
175	75	131	382.8	164.1	33.50	6.15	50.5	21.7
200	38	76	253.3	48.1	25.33	0.91	57.7	11.0
200	50	100	333.3	83.3	33.33	2.08	57.7	14.4
200	63	126	420.0	132.3	42.00	4.17	57.7	18.2
200	75	150	500.0	187.5	50.00	7.03	57.7	21.7
200	100	200	666.7	333.3	66.67	16.67	57.7	28.9
200	150	300	1000.0	750.0	100.00	56.25	57.7	43.3
200	200	400	1333.3	1333.3	133.33	133.33	57.7	57.7
225	38	85	320.6	54.2	36.07	1.03	65.0	11.0
225	50	112	421.9	93.8	47.46	2.34	65.0	14.4
225	63	141	531.6	148.8	59.80	4.69	65.0	18.2
225	75	168	632.8	210.9	71.19	7.91	65.0	21.7
250	50	125	520.8	104.2	65.10	2.60	72.2	14.4
250	75	187	781.3	234.4	97.66	8.79	72.2	21.7
250	100	250	1041.7	416.7	130.21	20.83	72.2	28.9
250	250	625	2604.2	2604.2	325.52	325.52	72.2	72.2
300	50	150	750.0	125.0	112.50	3.13	86.6	14.4
300	75	225	1125.0	281.3	168.75	10.55	86.6	21.7
300	100	300	1500.0	500.0	225.00	25.00	86.6	28.9
300	150	450	2250.0	1125.0	337.50	84.38	86.6	43.3
300	300	900	4500.0	4500.0	675.00	675.00	86.6	86.6

Note:
[a] Under dry exposure conditions. Please note the use of Eurocode convention for notation of the principle axes.

Source: BS 5268: Part 2:1991.

Tolerances on timber cross sections

BS EN 336 sets out the customary sizes of structural timber. Class 1 timbers are 'sawn' and Class 2 timbers are 'planed'. The permitted deviations for tolerance Class 1 are −1 mm to +3 mm for dimensions up to 100 mm and −2 mm to +4 mm for dimensions greater than 100 mm. For Class 2, the tolerance for dimensions up to 100 mm is ±1 mm and ±1.5 mm for dimensions over 100 mm. Structural design to BS EN 1995 allows for these tolerances and therefore analysis should be carried out for a 'target' section. It is the dimensions of the target section which should be included in specifications and on drawings.

Laminated timber products

Plywood

Plywood consists of veneers bonded together so that adjacent plies have the grain running in orthogonal directions. Plywoods in the United Kingdom generally come from the United States, Canada, Russia, Finland, or the Far East although only the Finnish have provided Eurocode-compatible data for structural applications. The type of plywood available is dependent on the import market. It is worthwhile calling around importers and stockists if a large or special supply is required. UK sizes are based on the imperial standard size of 8' × 4' (2.440 × 1.220 m). The main sources of imported plywood in the United Kingdom are:

Canada and America: The face veneer generally runs parallel to the longer side. Mainly imported as Douglas fir 18 mm ply used for concrete shuttering, although 9 and 12 mm are also available. Considered a specialist structural product by importers.

Finland: The face veneer can be parallel to the short or long side. Frequently spruce, birch or birch-faced ply. Birch plys are generally for fair-faced applications, while spruce 9, 12, 18 and 24 mm thick is for general building use, such as flooring and roofing.

Glue laminated timber

Timber layers, normally 43 mm thick, are glued together to build up deep beam sections. Long sections can be produced by staggering finger joints in the layers. Standard beam widths vary from 90 to 240 mm although widths up to 265 and 290 mm are available. Beam heights and lengths are generally limited to 2050 mm and 31 m, respectively. Column sections are available with widths of 90–200 mm and depths of 90–420 mm. Tapered and curved sections can also be manufactured. Loads are generally applied at 90° to the thickness of the layers.

Cross laminated timber

Kiln-dried softwood planks are finger jointed to form sheets, which are glued in perpendicular layers to form a thick panel product. Cross laminated timber is typically used to form wall or floor elements in timber panel building. In the United Kingdom, resulting boards are up to 13.5 m long and 2.95 m wide, and are typically 57–300 mm thick. Larger sizes are available by special order but these might require special transport considerations. Boards can be formed by three to nine layers but standard thicknesses are:

Three layers: 60, 63, 78, 90, 94, 98, 102, 108 mm.
Five layers: 95, 101, 117, 125, 128, 146, 158, 162, 170, 182 mm.
Seven layers: 202, 226, 256 mm.

Laminated veneered lumber

LVL (trade name Kerto®) is produced from rotary peeled 3 mm graded softwood veneers, which are glued to form continuous panels, similar to plywood. Veneers and can be laid in the same direction (Finnforest Kerto S) or with approximately 20% of veneer laid crossways (Finnforest Ketro Q), to give different strength and stability properties. Kerto S is ideal for beams, and for long spans with minimal deflections. Kerto Q should be selected where dimensional stability and/or compressive strength is required. Panels can be 1.8–2.5 m wide and up to a maximum of 25 m long, although standard sections are generally available up to 12 m in length. Design software is available on the Finnforest UK website. Standard sections are as follows:

Depth/width (mm)	Thickness of panel (mm)						
	27	33	39	45	57	75	90
200		•	•	•		•	•
220	•	•	•	•	•	•	•
240	•	•	•	•	•	•	•
300	•	•	•	•	•	•	•
360		•		•		•	•
400				•		•	
Kerto type	Q	Q	S	S	S	S	S

Special order thicknesses:
Kerto Q: 39, 45, 51, 57, 63, 69 mm.
Kerto S: 27, 33, 51, 63 mm can be manufactured as sanded on request.

Source: Metsawood. 2012.

Durability and fire resistance

Durability

Durability of timber depends on its resistance to fungal decay. Softwood is more prone to weathering and fungal attack than hardwood. Some timbers (such as oak, sweet chestnut, western red cedar and Douglas fir) are thought to be acidic and may need to be isolated from materials such as structural steelwork. The durability of timber products (such as plywood, LVL and glulam) normally depends on the stability and water resistance of the glue.

Weathering

On prolonged exposure to sunlight, wind and rain, external timbers gradually lose their natural colours and turn grey. Repeated wetting and drying cycles raise the surface grain, open up surface cracks and increase the risk of fungal attack, but weathering on its own generally causes few structural problems.

Fungal attack

For growth in timber, fungi need oxygen, a minimum moisture content of 20% and temperatures between 20°C and 30°C. Kiln drying at temperatures over 40°C will generally kill fungi, but fungal growth can normally be stopped by reducing the moisture content. Where structural damage has occurred, the affected timber should be cut away and replaced by treated timber. The remaining timber can be chemically treated to limit future problems. Two of the most common destructive fungi are:

'Dry rot' – *Serpula lacrymans*: Under damp conditions, white cotton wool strands form over the surface of the timber. Under drier conditions, a grey-white layer forms over the timber with occasional patches of yellow or lilac. Fruiting bodies are plate-like forms which disperse red spores. As a result of an attack, the timber becomes dry and friable (hence the name dry rot) and breaks up into cube-like pieces both along and across the grain.

'Wet rot' – *Coniophora puteana*: Known as cellar fungus, this fungus is the most common cause of timber decay in the United Kingdom. It requires high moisture contents of 40–50% which normally result from leaks or condensation. The decayed timber is dark and cracked along the grain. The thin strands of fungus are brown or black, but the green fruiting bodies are rarely seen in buildings. The decay can be hidden below the timber surface.

Insect attack

Insect attack on timber in the United Kingdom is limited to a small number of species and tends to be less serious than fungal attack. The reverse is generally true in hotter climates. Insects do not depend on damp conditions although some species prefer timber which has already suffered from fungal attack. Treatment normally involves removal of timber and treatment with pesticides. Some common insect pests in the United Kingdom are:

'Common furniture beetle' – *Anobium punctatum:* This beetle is the most widespread. It attacks hardwoods and softwoods, and can be responsible for structural damage in severe cases. The brown beetle is 3–5 mm long; leaves flight holes of approximately 2 mm in diameter between May and September, and is thought to be present in up to 20% of all buildings.

'Wood boring weevils' – *Pentarthrum huttoni and Euophryum confine:* Wood boring beetles attack timber previously softened by fungal decay. *P. huttoni* is the most common of the weevils and produces damage similar in appearance to the common furniture beetle. The beetles are 3–5 mm long and leave 1 mm diameter flight holes.

'Powder post beetle' – *Lyctus brunneus:* The powder post beetle attacks hardwoods, particularly oak and ash, until the sapwood is consumed. The extended soaking of vulnerable timbers in water can reduce the risk of attack but this is not normally commercially viable. The 4 mm reddish-brown beetle leaves flight holes of about 1.5 mm diameter.

'Death watch beetle' – *Xestobium rufovillosum:* The death watch beetle characteristically attacks partly decayed hardwoods, particularly oak, and is therefore responsible for considerable damage to old or historic buildings. The beetles typically make tapping noises during their mating season between March and June. Damp conditions encourage infestation. The brown beetle is approximately 8 mm long and leaves a flight hole of 3 mm diameter.

'Longhorn beetle' – *Hylotrupes bajulus:* The house long horn beetle is a serious pest, mainly present in parts of southern England. The beetle can infest and cause significant structural damage to the sapwood of seasoned softwood. Affected timbers bulge where tunnelling has occurred just below the surface caused by larvae that can be up to 35 mm long. The flight holes of the black beetle are oval and up to 10 mm across.

Source: BRE Digests 299, 307 and 345. Reproduced with permission by Building Research Establishment.

Fire resistance

Timber is an organic material and is therefore combustible. As timber is heated, water is driven off as vapour. By the time it reaches 230–250°C, the timber has started to break down into charcoal, producing carbon monoxide and methane (which cause flaming). The charcoal will continue to smoulder to carbon dioxide and ash. However, despite its combustibility, large sections of timber can perform better in fire than the equivalent sections of exposed steel or aluminium. Timber has a low thermal conductivity, which is further protected by the charred surface, preventing the interior of the section from burning.

BS EN 1995-1-2 details the predicted rates of charring for different woods which allows them to be 'fire engineered', if not protected. Most timbers have accepted charring rates of 0.7 mm/min except for hardwoods with characteristic density (ρ) of over 450 kg/m³, which char slower at 0.55 mm/min. The exception is beech which should be assumed to char at the same rate as softwood. These rates are based on charring from only one side, or for fire exposure of less than 20 min.

Design methods for fire design are given in BS EN 1995-1-2 using Equation 6.11b from BS EN 1990; $\gamma_m = 1.0$, $k_{mod} = 1.0$ and modified characteristic values for strength and stiffness using k_{fire} from clause 2.3, while considering reduced cross section and deflection effects.

Preliminary sizing of timber elements

Typical span/depth ratios for softwoods

Description	Typical depth (mm)
Domestic floor (50 mm wide joists at 400 mm c/c)	$L/24 + 25$ to 50
Office floors (50 mm wide joists at 400 mm c/c)	$L/15$
Rafters (50 mm wide joists at 400 mm c/c)	$L/24$
Beams/purlins	$L/10$ to 15
Independent posts	Minimum 100 mm^2
Triangular trusses	$L/5$ to 8
Rectangular trusses	$L/10$ to 15
Plywood stressed skin panels	$L/30$ to 40

Connections which rely on fixings (rather than dead bearing) to transfer the load in and out of the timber can often control member size and preliminary sizing. Highly stressed individual members should be kept at about 50% capacity until the connections can be designed in detail.

Plywood stress skin panels

Stress skin panels can be factory made using glue and screws or on site just using screws. The screws tend to be at close centres to accommodate the high longitudinal shear stresses. Plywood can be applied to the top, or top and bottom, of the internal softwood joists (webs). The webs are spaced according to the width of the panel and the point loads that the panel will need to carry. The spacing is normally about 600 mm for a UDL of 0.75 kN/m^2, about 400 mm for a UDL of 1.5 kN/m^2 or about 300 mm for a UDL of more than 1.5 kN/m^2. The direction of the face grain of the plywood skin will depend on the type of plywood chosen for the panel. The top ply skin will need to be about 9–12 mm thick for a UDL of 0.75 kN/m^2 or about 12–18 mm thick for a UDL of 1.5 kN/m^2. The bottom skin, if required, is usually 8–9 mm. The panel design is normally controlled by deflection and for economy the EI of the trial section should be about $4.4WL^2$.

Domestic floor joist capacity chart

See the graph below for an indication of the load carrying capacity of various joist sizes in grade C16 timber spaced at 400 mm centres.

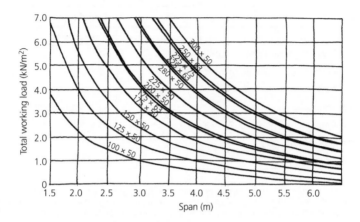

Span tables for solid timber tongued and grooved decking

UDL[a] (kN/m²)	Single span (m) for decking thickness				Double span (m) for decking thickness			
	38 mm	50 mm	63 mm	75 mm	38 mm	50 mm	63 mm	75 mm
1.0	2.2	3.0	3.8	4.7	3.0	3.9	5.2	6.3
1.5	1.9	2.6	3.3	4.0	2.6	3.4	4.5	5.4
2.0	1.7	2.3	3.0	3.6	2.4	3.1	4.0	4.9
2.5	1.6	2.2	2.8	3.4	2.2	2.9	3.7	4.6
3.0	1.5	2.1	2.6	3.2	2.1	2.7	3.5	4.3

Note:
[a] These loads limit the deflection to span/240 as the decking is not normally used with a ceiling.

Timber design to BS EN 1995

Design strength of timber

$$X_d = \eta \frac{X_k}{\gamma_m} \quad \text{Equation 6.3 from BS EN 1990}$$

where

X_d = Design value of material properties
X_k = Characteristic value of material properties
γ_m = Partial factor for material properties
η = Conversion factor (accounting for size, moisture, temperature effects and so on, made up in practice for timber by a number of different modification factors.)

$$\text{Design bending resistance,} \quad f_{m,y,d} = \frac{k_{crit}k_h k_{mod}k_{sys}}{\gamma_m} f_{m,k}$$

$$\text{Design shear resistance,} \ f_{v,d} = \frac{k_{mod}, f_{vk}}{\gamma_m}$$

$$\text{Design resistance check:} \frac{\sum \sigma}{\sum f} \leq 1.0$$

where

σ is the factored applied stress
f is the design resistance
For biaxial bending, see clause 6.16 for useful reduction factors

Recommended partial factors for material properties, γ_m

Solid timber	1.30
Glue laminated timber	1.25
LVL, plywood, OSB	1.20
Other wood-based materials	1.30
Connections (expect punched metal plates)	1.30
Punched metal plate fasteners	1.25
Accidental combinations	1.00

Source: BS EN 1990:2002; Equation 6.3; BS EN 1995-1-1:2004 + A1:2008. Table 2.3, Equations 2.17, 6.11 and 6.13. NA to BS EN 1995-1-1:2004 + A1:2008. Table NA.3.

Selected characteristic properties for timber

Strength class	Bending parallel to grain $f_{m,k}$ (N/mm²)	Tension parallel to grain $f_{t,0,k}$ (N/mm²)	Tension perpendicular to grain $f_{t,90,k}$ (N/mm²)	Compression parallel to grain $f_{c,0,k}$ (N/mm²)	Compression perpendicular to grain $f_{c,90,k}$ (N/mm²)	Shear parallel to grain $f_{v,k}$ (N/mm²)	Modulus of elasticity			Mean shear modulus G mean (N/mm²)	Density	
							Mean parallel to grain $E_{0,05}$ (N/mm²)	5th percentile parallel to grain $E_{0,05}$ (N/mm²)	Mean percentile to grain $E_{90,mean}$ (N/mm²)		Mean P mean (kg/m³)	Minimum P_k (kg/m³)
Softwood												
C16	16	10	0.50	17.0	2.2	1.8	8000	5400	270	500	370	310
C24	24	14	0.50	21.0	2.5	2.5	11000	7400	370	690	420	350
D40	40	24	0.60	26.0	8.3	4.0	13000	10900	860	810	550	660
D50	50	30	0.60	29.0	9.3	4.0	10900	11800	930	880	620	750
Glulam												
Gl28c	28	19.5	0.45	26.5	3.0	3.2	12600	10200	420	780	410	–
Gl28h	28	16.5	0.40	24.0	2.7	2.7	12600	10200	390	720	380	–
Finish ply Birch – Birch												
9 mm	45.6 (32.1)c	40.8	34.2	28.3	23.7	Planar 2.67 Panel 9.5	11395	–	6105	Planar 155f Panel 620	680	630
12 mm	42.9 (33.2)	40.0	35.0	27.7	24.3		10719	–	6781		680	630
18 mm	40.2 (34.1)	39.2	35.8	27.2	24.8		10048	–	7452		680	630
Conifer												
9 mm	26.0 (18.3)	15.2	12.8	19.6	16.4	Planar 2.67 Panel 9.5	8465	–	4535	Planar 52 Panel 530	520	460
12 mm	24.5 (19.0)	14.9	13.1	19.2	16.8		7963	–	5037		520	460
18 mm	23.0 (19.5)	14.6	13.4	18.8	17.2		7464	–	5536		520	460

OSB 2/3	16.4	9.4	7.0	15.4	12.7	Planar 1.0 Panel 6.8	4930	4190	1980	Planar 50 Panel 1080	550	—
LVL Kerto S	Planar 50.0 Panel 44.0	35.0	0.8	35.0	Planar 1.8 Panel 6.0	Planar 2.3 Panel 4.1	13800	11600	Planar 130 Panel 430	600	510	480
Kerto Q (27–69 mm)	Planar 36.0 Panel 32.0	26.0	6.0	26.0	Planar 1.8 Panel 9.0	Planar 1.3 Panel 4.5	10500	8800	Planar 130 Panel 2400	600	510	480
Cross-laminated timber Planar	24.0	—	0.12	—	2.7	2.7 (1.5)[i]	12000	—	370	690 (50)[i]	500	450
Panel	23.0	16.5	—	30.0	24.0	5.2	12000	—	—	250	500	450

Notes:

a The moisture contents for service classes are: class 1 ≤12%, class 2 ≤20% and class 3 >20%. Class 1 is generally warm/internal timber; class 2 is cold/sheltered/uninsulated timber and class 3 is external/unsheltered. Timber type, preservation and detailing should be carefully selected for class 3.

b Mean density is used to calculate weight and minimum density is used to calculate mechanical fastener strengths. Minimum is also referred to as 'characteristics' density in BS EN 1995.

c Plywood bending strengths in brackets refer to perpendicular to the (face) grain.

d Planar properties refer to the 'flat' dimension and concern rolling shear. Panel properties refer to the 'on edge' orientation.

e E values for tension and compression are available in manufacturer's literature but are omitted for clarity.

f Shear modulus, G, figures are quoted as perpendicular to the grain. Slightly higher values are available for parallel to the grain.

g Kerto LVL by Metsawood: Kerto S for beams/long spans or Kerto Q for stability/compressive strength.

h As at 2012 no characteristic values for North American plywood were available for use with Eurocodes. There is no timetable for their release. The North American Plywood Associations advised that the only design values available were for use with BS 5628.

i Cross-laminated timber values by KLH Ltd. Values in brackets refer to properties perpendicular to the grain of the outer lamella.

Source: BS EN 338:2009. Table 1; BS EN 12369. 2001. Part 1. Table 2; Metsawood (2012). Finnish Plywood Association. 2012.

Modification factors

Selected strength modification factors, k_{mod}

Load duration	Service class 1 + 2 Solid timber, glulam LUL and plywood parts 1, 2 and 3	Service class 1 + 2 OSB/3 or 4	Service class 3 Solid timber, glulam LVL and plywood part 3
Permanent (>10 years) (e.g. self weight)	0.6	0.3	0.5
Long term (6 months–10 years) (e.g. temporary structures, water tanks storage loads including lofts)	0.7	0.4	0.55
Medium term (1 week–6 months) (e.g. imposed floor loading)	0.8	0.55	0.65
Short term (<1 week) (e.g. snow, maintenance, after-effects of accident)	0.9	0.7	0.7
Instantaneous (e.g. wind, impact, explosion)	1.1	0.9	0.9

Notes:
a The moisture contents for the service classes are: class 1 ≤12%, class 2 ≤20% and class 3 >20%. Class 1 is generally warm/internal timber, while class 2 is cold/sheltered/uninsulated timber.
b For class 3 timbers, the type and preservation of the timber must be carefully specified.
c For unseasoned or saturated timbers (with minimum cross section >100 mm), use values for service class 3.

Effective length of bending members

Beam type	Loading condition	l_e/l
Simply supported	Constant moment	1.0
	Uniformly distributed load	0.9
	Concentrated load at midspan	0.8
Cantilever	Uniformly distributed load	0.5
	Concentrated load at free end	0.8

Notes:
a Effective lengths are valid for beams torsionally restrained at supports and loaded at the centre of gravity.
b For beams loaded at the compression edge, should have l_e increased by adding $2h$, while beams loaded at the tension edge can have l_e decreased by subtracting $0.5h$, where h is the depth of the beam.

Source: BS EN 1995-1-1:2004 + A1:2008. Tables 3.1 and 6.1.

Lateral torsional buckling modification factor, k_{crit}

If lateral displacement of the member's compression edge is prevented throughout its length, and torsional rotation is prevented at its supports (e.g. by decking, sheathing or blocking), it can be assumed that $k_{crit} = 1.0$; otherwise, k_{crit} is calculated from $E_{0,05}$ and the member relative stiffness, $\lambda_{rel,m}$ selected values are provided in the charts below.

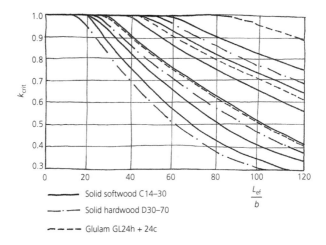

Solid softwood C14–30

Solid hardwood D30–70

Glulam GL24h + 24c

$\dfrac{L_{ef}}{b}$

Source: IStructE. 2007.

Depth factor for flexural members or width factor for tension members, k_h

k_h applies to small sections which are thought to have an enhanced capacity.

For solid timber, $h < 150$mm $k_h = \left(\dfrac{150}{h}\right)^{0.2} \leq 1.3$

For glulam timber, $h < 600$mm $k_h = \left(\dfrac{600}{h}\right)^{0.1} \leq 1.1$

For LVL $h < 300$ mm, k_h values should be adjusted for length using manufacturer's size effect factor, s, quoted with characteristic stresses.

$$k_h = \left(\frac{300}{h}\right)^{s} \leq 1.2$$

Load-sharing system factor, k_{sys}

This factor applies to all strength properties (not stiffness).

Where 'several' equally spaced, similar members, components or assemblies are connected in such a way that load can effectively be transferred between them. It is suggested that 'several' means 'four or more' adjacent members such as rafters, joists or studs or 'two or more', where items such as trimmers, are securely bolted together. In such conditions, $k_{sys} = 1.1$.

Shear modification factor, k_{CR}

k_{CR} is a shear modification factor to allow for the influence of during splits/cracks and glue line failure.

Material	Modification factor, k_{CR}
Solid timber	0.67
Glulam	0.67
LVL	1.00
Other wood-based panels	1.00

Source: BS EN 1995-1-1:2004 + A1: 2008. Clauses 3.2, 3.4, 6.6 and 6.1.7 and Table NA.4.

Biaxial bending modification factor, k_m

For rectangular sections in solid or glulam timber, or LVL, $k_m = 0.7$
For other cross sections in solid or glulam timber, or LVL, $k_m = 1.0$
For other wood-based products and cross sections, $k_m = 1.0$

where the following expression should be satisfied:

$$\frac{\sigma_{m,y}}{f_{m,y}} + k_m \frac{\sigma_{m,z}}{f_{m,z}} \leq 1.0 \quad \text{and} \quad k_m \frac{\sigma_{m,y}}{f_{m,y}} + \frac{\sigma_{m,z}}{f_{m,z}} \leq 1.0$$

where
  σ_m = design stresses and
  f_m = design bending strengths

Bearing modification factor, $k_{c,90}$

$k_{c,90}$ is generally 1.0 but compressive strength perpendicular to the grain can be enhanced for discrete supports as follows:

For solid softwood timber, $k_{c,90} = 1.5$
For glulam where bearing length ≤ 400 mn, $k_{c,90} = 1.75$

Notched bearings modification factor, k_v

It is a good practice to avoid notches in the underside of timber to avoid tensile stress concentrations. It is also conservative to assure $k_v = 1.0$; however, it is possible to calculate strength enhancement at notches according to clause 6.6.

Source: BS EN 1995-1-1:2004 + A1: 2008; clause 6.1.6.

Effective length of compression members

End conditions	$\dfrac{l_e}{l}$
Restrained at both ends in position and in direction	0.7
Restrained at both ends in position and one end in direction (typical case for a stud in a wall panel)	0.85
Restrained at both ends in position but not in direction	1.0
Restrained at one end in position and in direction and at the other end in direction but not in position	1.5
Restrained at one end in position and direction and free at the other end	2.0

Generally, the slenderness should be less than 180 for members carrying compression, or less than 250 where compression would only occur as a result of load reversal due to wind loading.

Compression buckling factor $k_{c,y}$ and $k_{c,z}$

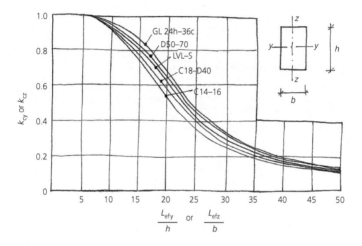

Source: BS EN 1995-1-1:2004 + A1: 2008. Clause 6.2.4; IStructE. 2007.

Deflections

When calculating the design values of actions or the effect of actions for serviceability limit state all partial load factors (γ_f) are set to 1.0, or 0.0 if the loading is favourable. However combination load factors are still applied.

While BS EN 1990 requires that 'a distinction shall be made between reversible and irreversible serviceability limit states', BS EN 1995 states that the 'characteristic combination' (normally used for irreversible state) should be used for all deflection calculations. The effects of instantaneous and final joint slip on members in assemblies (e.g. trusses, moment connections, portal frames etc.) can be considerable, affecting both member deflection and distribution of forces and moments, and should be included in relevant analyses.

In practice, the calculations generally need to be subdivided so that deformations are estimated for different actions and then combined – rather than the traditional approach of combining actions to calculate a combined deformation.

The calculation method in clause 2.2.3(5) can be simplified to:

Final deformation = Instantaneous deformation (elastic + shear) + Creep deformation

$$U_{fin} = U_{inst} + U_{creep}$$

where

$$U_{insr} = \sum U_{inst,G} + U_{inst,Q1} + \sum \Psi_{o,i} U_{inst,Qi}$$
$$U_{creep} = k_{def}\left[\sum U_{inst,G} + \sum \Psi_{2,i} U_{inst,Qi}\right]$$

and

U = Calculated
Ψ_0 = Combination load factor for variable actions from BS EN 1990
Ψ_2 = Combination load factor for quasi-permanent BS EN 1990

Deformation factor, k_{def}

Applied to deflection and connection calculations, it is not normally relevant to ultimate limit state calculations unless creep affects stress distribution. Selected values are as follows:

Material	Service class		
	1	**2**	**3**
Solid timber, glulam or LVL	0.60	0.80	2.00
Plywood	0.80	1.00	2.50
OSB/2	2.25	–	–
OSB/3 and OSB/4	1.50	2.25	–

Source: BS EN 1995-1-1:2004 + A1:2008. Clause 2.2.3(5) and Table 3.2.

Timber joints

BS EN 1995 deals with nailed, screwed, bolted, split ring toothed connectors; dowelled and glued joints. Joint positions and fixing edge distances can control member sizes, as a result of the reduced timber cross section at the joint positions. Joint slip (caused by fixings moving in pre-drilled holes) can cause rotations which will have a considerable effect on overall deflections.

Simplified rules for spacing of fasteners in timber-to-timber joints

The spacing rules in BS EN 1995 allow for almost infinite complexity in combination of spacing, hole drilling, load direction, direction of grain and so on. However, this does not make for a quick and easy design solution. Engineers can use the conservative (simplified) spacing rules in most cases. Close spacing or complex arrangement will require detailed consideration of clauses 8.3–8.7.

Nails – The values given for nailed joints are for nails made from steel wire driven at right angles to the grain without predrilling in softwood. In hardwood, holes normally need to be pre-drilled not bigger than 0.8d (where d is the fixing diameter). Minimum permitted pointside penetrations are 8d for smooth nails, 6d for other nails in side grain or 10d for other nails in end grain.

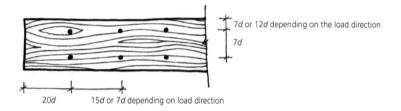

7d or 12d depending on the load direction

7d

20d 15d or 7d depending on load direction

Small screws ($\phi \leq 6$ mm) – The values given for screwed joints are for screws that conform to BS EN 14592 in pre-drilled holes. The holes should be drilled with a diameter equal to that of the screw shank (ϕ) for the part of the hole to contain the shank, reducing to a pilot hole (with a diameter of $\phi/2$) for the threaded portion of the screw. Where the standard headside thickness is less than the values in the table, the basic load must be reduced by the ratio: actual/standard thickness. The headside thickness must be greater than 2ϕ. The following tables give values for the UK screws rather than the European screws quoted in the latest British Standards and Eurocodes.

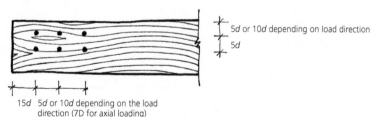

5d or 10d depending on load direction

5d

15d 5d or 10d depending on the load
       direction (7D for axial loading)

Bolts and large screws ($\phi > 6$ mm) – The values given for bolted joints are for black bolts which conform to BS EN 20898-1 with washers which conform to BS 4320. Bolt holes should not be drilled more than 2 mm larger than the nominal bolt diameter. Washers should have a diameter or width of three times the bolt diameter with a thickness of 0.25 times the bolt diameter and be fitted under the head and nut of each bolt. At least one complete thread should protrude from a tightened nut.

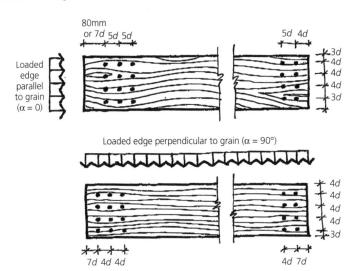

Design strength of a timber-to-timber connection parallel to the grain for service classes 1 and 2

$$F_{v_{ef} \, Rd \, o} = f_{v \, Rk_o} \times n_{sp} \times r_{pl} \times n_{ef} \times \frac{k_{mod}}{\gamma_m}$$

where

$F_{v \, Rk_o}$ is the minimum characteristic load capacity (calculated from clause 8.2.2(1) or tables below for single or double shear)

n_{sp} is the number of shear planes

r_{pl} is the number of rows of fasteners laterally loaded parallel to the grain, each row containing n equally sized and spaced fasteners

n_{ef} is the effective number of fixings per row where all or one component of loading is parallel to the grain. If fastener spacing is taken $\geq 14\phi$ then $n_{ef} = n$ based on simplified spacing rules

Design strength of a timber-to-timber connection perpendicular to the grain for service classes 1 and 2

Two modes of failure need to be checked.

Ductile yielding of the fastener

$$F_{90 \, Rd \, dy} = F_{v \, Rk90} \times n_{sp} \times r_{pr} \times n \times \frac{k_{mod}}{\gamma_m}$$

where

r_{pr} is the number of rows of fasteners laterally loaded perpendicular to the grain, each row containing n equally sized and spaced fasteners

$F_{v \, Rk90}$ is the minimum characteristic load capacity for nails ($\phi < 8 \, mm$) or small screws ($\phi \leq 6 \, mm$) $F_{v \, Rk_o}$, but this is modified for bolts ($\phi > 6 \, mm$) as follows:

$$F_{v \, Rk90} = \frac{F_{v \, Rk_o}}{k_{90}}$$

where

$k_{90} = (1.35 + 0.015\phi)$ for softwood
$= (1.30 + 0.015\phi)$ for LVL
$= (0.9 + 0.015\phi)$ for hardwood

Timber splitting parallel to the grain

$$F_{90Rdts} = 14\, b\omega \sqrt{\dfrac{h_c}{1 - \dfrac{h_c}{h}}} \times \dfrac{k_{mod}}{\gamma_m}$$

where
ω 1.0 is a modification factor for all fasteners except punched metal plates
b is the member thickness
h_c is the distance to the loaded edge from centre of most distant connector
h is the full depth of the timber member

Characteristic lateral load capacity $f_{v,rk}$ for selected fasteners in timber-to-timber connections

Fastener					Load parallel to the grain				Load perpendicular to the grain			
					Single shear		Double shear		Single shear		Double shear	
		Assumed	Minimum	Minimum	Load capacity		Load capacity		Load capacity		Load capacity	
Diameter mm	Reference	Effective diameter (mm)	Timber thickness (mm)	Fastener length (mm)	C16 (kN)	C24 (kN)	C16 (kN)	C24 (kN)	C16 (kN)	C24 (kN)	C16 (kN)	C24 (kN)
Nails												
3.0	swg 11	1.3	36	72	0.73	0.78	–	–	0.62	0.67	–	–
3.4	swg 10	1.5	41	82	0.89	0.97	–	–	0.77	0.83	–	–
4.5	swg 7	4.5	54	108	1.43	1.55	–	–	1.22	1.32	–	–
5.5	swg 5	5.5	66	132	2.01	2.17	–	–	1.71	1.85	–	–
Small wood screws												
3.48	no. 6	2.6	28	84	0.79	0.84	–	–	0.61	0.68	–	–
4.17	no. 8	3.1	33	100	0.94	1.03	–	–	0.71	0.80	–	–
4.88	no. 10	3.6	39	117	1.27	1.41	–	–	0.99	1.09	–	–
5.59	no. 12	4.1	45	134	1.59	1.68	–	–	1.35	1.43	–	–
Large wood screws												
6.3	no.14	4.6	50	101	2.57	2.74	–	–	1.93	2.17	–	–
7.01	no. 16	5.1	84	112	3.09	3.29	–	–	2.61	2.78	–	–
7.93	8 mm coach	7.9	47	94	3.46	3.89	–	–	2.45	2.76	–	–
7.93	8 mm coach	7.9	63	126	3.51	3.89	–	–	2.66	2.92	–	–
9.52	10 mm coach	9.5	47	94	4.56	5.02	–	–	3.47	3.83	–	–
9.52	10 mm coach	9.5	63	126	5.24	5.58	–	–	3.83	4.31	–	–
12.5	12 mm coach	12.5	47	94	6.57	7.19	–	–	4.97	5.54	–	–
12.5	12 mm coach	12.5	97	194	8.40	8.96	–	–	6.15	6.78	–	–

Bolts

M8	8	—	47	110	4.30	4.62	8.60	9.24	3.10	3.50	5.98	6.75
	8	—	63	142	4.30	4.62	8.60	9.24	3.66	3.94	7.33	7.88
	8	—	97	210	4.30	4.62	8.60	9.24	3.66	3.94	7.33	7.88
M10	10	—	47	114	6.41	6.89	10.75	12.14	3.71	4.19	7.17	8.09
	10	—	63	146	6.41	6.89	12.83	13.78	4.97	5.62	9.61	10.85
	10	—	97	214	6.41	6.89	12.83	13.78	5.43	5.83	10.85	11.66
M12	12	—	47	118	6.53	7.37	12.62	14.24	4.27	4.82	8.25	9.31
	12	—	63	150	8.50	9.54	16.91	19.08	5.72	6.46	11.05	12.48
	12	·	97	218	8.87	9.54	17.75	19.08	7.46	7.93	14.92	15.85
M16	16	—	47	126	8.31	9.39	16.06	18.13	5.23	5.90	10.10	11.40
	16	—	63	158	11.14	12.58	21.52	24.30	7.01	7.91	13.54	15.28
	16	—	97	226	14.76	15.88	29.52	31.75	10.79	12.18	20.84	23.53

Notes:

[a] Minimum headside thickness is $12d$ for nails and $4d$ for screws.

[b] Minimum coach screw pointside penetration is $6d$.

[c] Assumed fastener strength, $f_{uk} = 540 \text{ N/mm}^2$.

[d] Predrilling allowances made for large screws only.

[e] Values for service classes 1 and 2.

[f] Minimum spacings based on simplified rules.

[g] Round smooth shank nails/screws.

[h] Joint capacity $= F_{v,Rk} \times$ number of shear planes $\times$ (effective) number of fasteners/γ_m, where typically $\gamma_m = 1.3$ for connections (except metal punched plates).

Summary of differences with BS 5268: Structural timber

While the subtleties of the Eurocode will definitely allow designers to maximise the efficiency of an elemental timber frame, the design of a basic timber joist or rafter has been made considerably more complicated compared to BS 5268.

	BS EN 1995	BS 5268
Basis of design	Limit state principles	Permissible stress design
Methodology	More care is required when using BS EN 1995 as ultimate strengths need to be factored down to obtain safe design strengths	BS 5268 grade stresses are factored up to obtain design strengths and making a conservative design if any modification factors are inadvertently missed
Safety factors	Transparent and easy to modify	Opaque and not readily altered to suit different projects
Deflections	More detailed deflection calculations: Firstly, the deflections for permanent and variable actions need to be calculated separately and then added together. Secondly, instantaneous and permanent deflections are also calculated separately. Finally, a modification factor, k_{def}, is applied to account for the combined effects of creep and moisture content	Simplified deflection calculations provided quick but conservative designs
Vibration	Provides method for the calculation of the vibration characteristics of residential floors	Not covered
Notches	Complex method	Simplified notch calculations provided quick but conservative designs
Commercial	Impossible to design without the use of a computer and extra supporting documents. Efficient use of the available factors can yield more efficient material solutions	Simplified calculations provided quick but conservative designs ideal for small projects and refurbishments
Guidance	Practical guidance is patchy	BS does provide guidance on lateral restraint for beams, trussed rafter bracing, masonry wind shielding and methods of calculating spans for domestic members and certain glued connections which is missing from EC5

11
Masonry

Masonry, brought to the United Kingdom by the Romans, became a popular method of construction as the units could originally be lifted and placed with one hand. Masonry has orthotropic material properties relating to the bed or perpend joints of the masonry units. The compressive strength of the masonry depends on the strength of the masonry units and on the mortar type. Masonry is good in compression and has limited flexural strength. Where the flexural strength of masonry 'parallel to the bed joints' can be developed, the section is described as 'uncracked'. A cracked section (e.g. due to a damp-proof course or a movement joint) relies on the dead weight of the masonry to resist tensile stresses. The structure should be arranged to limit tension or buckling in slender members, or crushing of stocky structures.

Summary of material properties

Clay bricks: The wide range of clays in the United Kingdom results in a wide variety of available brick strengths, colours and appearance. Bricks can be hand or factory made. Densities range between 22.5 and 28 kN/m³. Clay bricks tend to expand due to water absorption. Engineering bricks have low water absorption, high strength and good durability properties (Class A strength >70 N/mm²; water absorption ≤4.5% by mass. Class B: strength >50 N/mm²; water absorption ≤7.0% by mass).

Calcium silicate bricks: Calcium silicates are low-cost bricks made from sand and slaked lime rarely used due to their tendency to shrink and crack. Densities range between 17 and 21 kN/m³.

Concrete blocks: Cement-bound blocks are available in densities ranging between 5 and 20 kN/m³. The lightest blocks are aerated; medium dense blocks contain slag, ash or pumice aggregate; dense blocks contain dense natural aggregates. Blocks can shrink by 0.01–0.09%, but blocks with shrinkage rates of no more than 0.03% are preferable to avoid the cracking of plaster and brittle finishes on the finished walls.

Stone masonry: Stone as rubble construction, bedded blocks, or facing to brick or blockwork is covered in BS EN 1996 in place of BS 5390. Thin stone used as cladding or facing is covered by BS 8298.

Cement mortar: Sand, dry hydrate of lime and cement are mixed with water to form mortar. The cement cures on contact with water. It provides a bond strong enough that the masonry can resist flexural tension, but structural movement will cause cracking.

Lime mortar: Sand and non-hydraulic lime putty form a mortar, to which some cement (or other pozzolanic material) can be added to speed up setting. The mortar needs air, and warm, dry weather to set. Lime mortar is more flexible than cement mortar and therefore can resist considerably more movement without visible cracking.

Typical unit strengths of masonry

Material (relevant BS)	Class	Water absorption	Typical unit compressive strength (N/mm^2)
Fired clay bricks (BS EN 771–1)	Engineering A	<4.5%	>70
	Engineering B	7.0%	>50
	Facings (bricks selected for appearance)	10–30%	10–50
	Commons (Class 1 to 15)	20–24%	
	Class 1		7
	Class 2		14
	Class 3		20
	Class 15		105
	Flettons	15–25%	15–25
	Stocks (bricks without frogs)	20–40%	3–20
Calcium silicate bricks (BS EN 771-2)	Classes 2 to 7		14–48.5
Concrete bricks (BS EN 771-3)			7–20
Concrete blocks (BS EN 771-3)	Dense solid		7, 10–30
	Dense hollow		3.5, 7, 10
	Lightweight		2.8, 3.5, 4, 7
Reconstituted stone (BS EN 771-5)	Dense solid		Typically as dense concrete blocks
Natural stone (BS EN 1996 and BS 8298)	Structural quality. Strength is dependent on the type of stone, the quality, the direction of the bed, the quarry location		15–100

Geometry and arrangement

Brick and block sizes

The standard UK brick is 215 × 102 × 65 mm which gives a co-ordinating size of 225 × 112 × 75 mm. The standard UK block face size is 440 × 215 mm in thicknesses from 75 to 215 mm, giving a co-ordinating size of 450 × 225 mm. This equates to two brick stretchers by three brick courses.

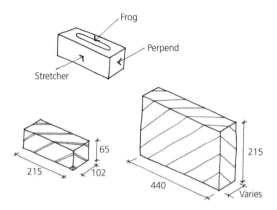

The Health and Safety Executive (HSE) require designers to specify blocks which weigh less than 20 kg to try to reduce repetitive strain injuries in bricklayers. Medium dense and dense blocks of 140 mm thick, or more, often exceed 20 kg. The HSE prefers designers to specify half blocks (such as Tarmac Topcrete) rather than rely on special manual handling (such as hoists) on site. In addition to this, the convenience and speed of block laying is reduced as block weight increases.

Group number

European masonry units vary greatly in terms of shape, size, voids, wall thickness etc. BS EN 1996 has had to be drafted to take account of this variation, alongside BS EN 771-1 to 6 which sets out material requirements for different masonry types; grouped into one of four groups or categories: 1, 2, 3 or 4. Group 1 covers superior structural units for use in unprotected masonry with no more than 25% voids. Group 2 includes calcium silicate bricks and hollow concrete blocks with 25–60% voids. All standard UK bricks manufactured to British standards (including frogged and perforated types) fall into Group 1; although some Group 2 units are becoming available. Most structural concrete blocks also fall into Group 1, but Group 2 can apply. Masonry in Groups 3 and 4 (e.g. continental extruded clay units etc.) has not traditionally been used in the United Kingdom. Engineers should check manufacturers' literature carefully.

Non-hydraulic lime mortar mixes for masonry

Mix constituents	Approximate proportions by volume	Notes on general application
Lime putty: coarse sand	1:2.5–3	Used where dry weather and no frost are expected for several months
Pozzolanic[a]: lime putty: coarse sand	1:2–3:12	Used where an initial mortar set is required within a couple of days

Note:
[a] Pozzolanic material can be cement, fired China dust or ground-granulated blast furnace slag (ggbfs).

The actual amount of lime putty used depends on the grading of the sand and the volume of voids. Compressive strength values for non-hydraulic lime mortar masonry can be approximated using the values for Class IV cement mortar. Owing to the flexibility of non-hydraulic lime mortar, thermal and moisture movements can generally be accommodated by the masonry without cracking of the masonry elements or the use of movement joints. This flexibility also means that resistance to lateral load relies on mass and dead load rather than flexural strength. The accepted minimum thickness of walls with non-hydraulic lime mortar is 215 mm and therefore the use of lime mortar in standard single-leaf cavity walls is not appropriate.

Cement mortar mixes for masonry

Mortar class	Compressive strength class	Type of mortar (proportions by volume)			Compressive strengths at 28 days (N/mm²)
		Cement: lime:sand	Masonry cement[d]:sand	Cement: sand	
Dry pack	–	1:0:3	–	–	
I	M12	1:0 to 1/4:3	–	–	12
II	M6	1:1/2:4 to 41/2	1:21/2 to 31/2	1:3 to 4	6
III	M4	1:1:5 to 6	1:4 to 5	1:5 to 6	4
IV	M2	1:2:8 to 9	1:51/2 to 61/2	1:7 to 8	2

Notes:
[a] Mix proportions are given by volume. Where sand volumes are given as variable amounts, use the larger volume for well-graded sand and the smaller volume for uniformly graded sand.
[b] As the mortar strength increases, the flexibility reduces and likelihood of cracking increases.
[c] Cement:lime:sand mortar provides the best bond and rain resistance, while cement:sand and plasticiser is more frost resistant.
[d] Masonry cement containing non-lime inorganic filler. Proportions vary if lime filler used.

Source: BS EN 1996-1-1:2005. Table NA.2.

Selected bond patterns

For strength, perpends should not be less than one-quarter of a brick from those in an adjacent course.

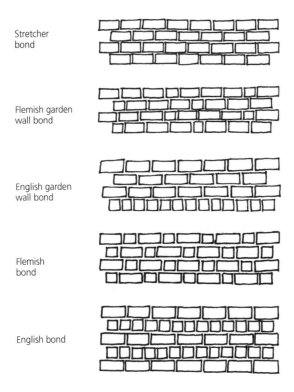

Stretcher bond

Flemish garden wall bond

English garden wall bond

Flemish bond

English bond

Durability and fire resistance

Durability

Durability of masonry relies on the selection of appropriate components detailed to prevent water and weather penetration. Wet bricks can suffer from spalling as a result of frost attack. Bricks of low porosity are required in positions where exposure to moisture and freezing is likely. In the absence of specific guidance in BS EN 1996-1-1 Section 4, BS 5268: Part 3 gives guidance on recommended combinations of masonry units and mortar for different exposure conditions as summarised below.

Durability issues for selection of bricks and mortar

Application		Minimum strength of masonry unit	Mortar class[a]	Brick frost resistance[b] and soluble salt content[c]
Internal walls generally/ external walls above DPC		Any block/15 N/ mm² brick	III / M4	FL, FN, ML, MN, OL or ON
External below DPC/ freestanding walls/ parapets		7 N/mm² dense block/20 N/mm² brick	III / M4	FL or FN (ML or MN if protected from saturation)
Brick damp proof courses in buildings (BS 743)		Engineering brick A	I / M12	FL or FN
Earth retaining walls		7 N/mm dense block/30 N/mm² brick	I or II / M12 or M6	FL or FN
Planter boxes		Engineering brick/20 N/mm² commons	I or II / M12 or M6	FL or FN
Sills and copings		Selected block/30 N/ mm² brick	I / M12	FL or FN
Manholes and inspection chambers	Surface water	Engineering brick/20 N/mm² commons	I or II / M12 or M6	FL or FN (ML or MN if more than 150 mm below ground level)
	Foul drainage	Engineering brick A	I or II / M12 or / M6	

Notes:
[a] Sulphate resisting mortar is advised where soluble sulphates are expected from the ground, saturated bricks or elsewhere.
[b] F indicates that the bricks are frost resistant, M indicates moderate frost resistance and 0 indicates no frost resistance.
[c] N indicates that the bricks have normal soluble salt content and L indicates low soluble salt content.
[d] Retaining walls and planter boxes should be waterproofed on their retaining faces to improve durability and prevent staining.

Fire resistance

As masonry units have generally been fired during manufacture, their performance in fire conditions is generally good. Perforated and cellular bricks have a lesser fire resistance than solid units of the same thickness. The fire resistance of blocks is dependent on the grading of the aggregate and cement content of the mix, but generally 100 mm solid blocks will provide a fire resistance of up to 2 h if load bearing and 4 h if non-load bearing. Longer periods of fire resistance may require a thicker wall than is required for strength. Specific product information should be obtained from masonry manufacturers. Design guidance for fire engineering is given in BS EN 1996-1-2.

Movement joints in masonry with cement-based mortar

Movement joints to limit the lengths of walls built in cement mortar are required to minimise cracking due to deflection, differential settlement, temperature change and shrinkage or expansion. In addition to long wall panels, movement joints are also required at points of weakness, where stress concentrations might be expected to cause cracks (such as at steps in height or thickness or at the positions of large chases). Typical movement joint spacings are as follows:

Material	Approximate horizontal joint spacing[b] and reason for provision	Typical joint thickness (mm)	Maximum suggested panel length: Height ratio[a]
Clay bricks	12 m for expansion	16	3:1
	15–18 m with bed joint reinforcement at 450 mm c/c	22	
	18–20 m with bed joint reinforcement at 225 mm c/c	25	
Calcium silicate bricks	7.5–9 m for shrinkage	10	3:1
Concrete bricks	6 m for shrinkage	10	2:1
Concrete blocks	6–7 m for shrinkage	10	2:1
	15–18 m with bed joint reinforcement at 450 mm c/c	22	
	18–20 m with bed joint reinforcement at 225 mm c/c	25	
Natural stone cladding	6 m for thermal movements	10	3:1

Notes:
[a] Consider bed joint reinforcement for ratios beyond the suggested maximum.
[b] The horizontal joint spacing should be halved for joints which are spaced off corners.

Vertical joints are required in cavity walls every 9 m or three storeys for buildings over 12 m or four storeys. This vertical spacing can be increased if special precautions are taken to limit the differential movements caused by the shrinkage of the internal block and the expansion of the external brick. The joint is typically created by supporting the external skin on a proprietary stainless steel shelf angle fixed back to the internal structure. Normally, 1 mm of joint width is allowed for each metre of masonry (with a minimum of 10 mm) between the top of the masonry and underside of the shelf angle support.

Preliminary sizing of masonry elements

Typical span/thickness ratios

Description	Typical thickness		
	Freestanding/ cantilever	Element supported on two sides	Panel supported on four sides
Lateral loading			
Solid wall with no piers – uncracked section	H/6–8.5[a]	H/20 or L/20	H/22 or L/25
Solid wall with no piers – cracked section	H/4.5–6.4[a]	H/10 or L/20	H/12 or L/25
External cavity wall[b] panel	–	H/20 or L/30	H/22 or L/35
External cavity wall[b] panel with bed joint reinforcement	–	H/20 or L/35	H/22 or L/40
External diaphragm wall panel	H/10	H/14	–
Reinforced masonry retaining wall (bars in pockets in the walls)	H/10–15	–	–
Solid masonry retaining wall (thickness at base)	H/2.5–4	–	–
Vertical loading			
Solid wall	H/8	H/18–22	–
Cavity wall	H/11	H/5.5	–
Masonry arch/vault	–	L/20–30	L/30–60
Reinforced brick beam depth	–	L/10–16	–

Notes:
[a] Depends on the wind exposure of the wall.
[b] The spans or distances between lateral restraints are L in the horizontal direction and H in the vertical direction.
[c] In cavity walls, the thickness is the sum of both leaves excluding the cavity width.

Vertical load capacity wall charts

Vertical load capacity of selected walls less than 150 mm thick

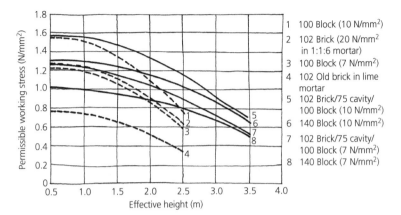

1 100 Block (10 N/mm²)
2 102 Brick (20 N/mm² in 1:1:6 mortar)
3 100 Block (7 N/mm²)
4 102 Old brick in lime mortar
5 102 Brick/75 cavity/ 100 Block (10 N/mm²)
6 140 Block (10 N/mm²)
7 102 Brick/75 cavity/ 100 Block (7 N/mm²)
8 140 Block (7 N/mm²)

Vertical load capacity of selected solid walls greater than 150 mm thick

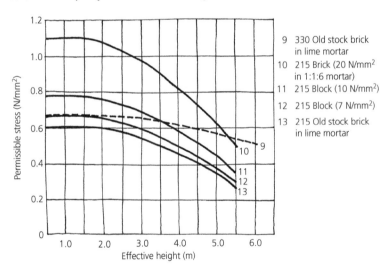

9   330 Old stock brick in lime mortar
10 215 Brick (20 N/mm² in 1:1:6 mortar)
11 215 Block (10 N/mm²)
12 215 Block (7 N/mm²)
13 215 Old stock brick in lime mortar

Preliminary sizing of external cavity wall panels

The following approach can be considered for cavity wall panels in non-load-bearing construction up to about 3.5 m tall in buildings of up to four storeys high, in areas which have many windbreaks. Without major openings, cavity wall panels can easily span up to 3.5 m if spanning horizontally, while panels supported on four sides can span up to about 4.5 m. Load-bearing wall panels can be larger as the vertical loads pre-compress the masonry and give it much more capacity to span vertically. Gable walls can be treated as rectangular panels and their height taken as the average height of the wall.

Detailed calculations for masonry around openings can sometimes be avoided if:

1. The openings are completely framed by lateral restraints.
2. The total area of openings is less than the lesser of 10% of the maximum panel area (see the section 'Lateral Load' on the design of external wall panels later in this chapter) or 25% of the actual wall area.
3. The opening is more than half its maximum dimension from any edge of the wall panel (other than its base) and from any adjacent opening.

Internal non-load-bearing masonry partition chart

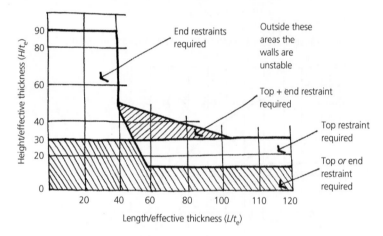

Source: BS 5628: Part 3: 2005.

Typical ultimate strength values for stone masonry

	Crushing (N/mm²)	Tension (N/mm²)	Shear (N/mm²)	Bending (N/mm²)
Basalt	8.5	8.6	4.3	–
Chalk	1.1	–	–	–
Granite	96.6	3.2	5.4	10.7
Limestone	53.7	2.7	4.3	6.4
Limestone soft	10.7	1.0	3.8	5.4
Marble	64.4	3.2	5.4	–
Sandstone	53.7	1.1	3.2	5.4
Sandstone soft	21.5	0.5	1.1	2.1
Slate	85.8	1.1	3.2	5.4

The strength values listed above assume that the stone is of good average quality and that the factor of safety commonly used will be 10 for working stresses. While this seems sensible for tension, shear and bending, it does seem conservative for crushing strength. Better values can be achieved on the basis of strength testing. These values can be used in preliminary design, but where unknown stones or unusual uses are proposed, strength testing is advised. The strength of stone varies between sources and samples, and also depends on the mortar and the manner of construction. Strength properties also vary with time depending on which part of the quarry is being worked. The British Stone website has listings of stone tests carried out by the Building Research Establishment.

As compressive load can be accompanied by a shear stress of up to half the compressive stress, shear stresses normally control the design of slender items such as walls and piers. Safe wall and pier loads are generally obtained by assuming a safe working compressive stress equal to twice the characteristic shear stress.

Source: Howe, J.A. 1910.

Masonry design to BS EN 1996

Design strength of masonry

$$f_d = \frac{f_k}{\gamma_m}$$

Partial factors for selected unreinforced masonry

	Class of execution control	
	Class 1	Class 2
Direct or flexural tension		
Manufacturing category I	2.3	2.7
Manufacturing category II	2.6	**3.0**
Flexural tension		
Manufacturing category I or II	2.3	2.7
Shear		
Manufacturing category I or II	2.5	2.5

Notes:
[a] Execution control should follow the requirements for workmanship, suspension and inspection in BS EN 1996-2. Class 1 is a high level of control using factory-made mortar and regular testing in addition. Class 2 is probably appropriate for most general site-based work in the United Kingdom, i.e. 'normal' controls.
[b] Manufacturing control categories should be specified by manufacturers.
[c] When considering the effects of misuse or accidents, these values should be halved.
[d] γ_m would typically be taken as 3.0 for normal UK situations.

Source: NA to BS EN. 1996. 2005. Part 1-1. Table NA.1.

Axial load

Selected characteristic compressive masonry strengths for Group 1 standard format brick masonry (N/mm²)

		Mean compressive strength of unit									
		5	10	15	20	30	40	50	75	100	125
Mortar class					Stock/Fletton				Class B		Class A	
		Normalised mean compressive strength of unit, f_b										
		5.1	10.2	15.3	20.4	30.6	40.8	51	76.5	102	127.5
i	M12	3.30	5.35	7.11	8.70	11.55	14.13	16.52	21.94	26.84	31.37
ii	M6	2.68	4.35	5.78	7.07	9.38	11.48	13.42	17.82	21.80	25.48
iii	M4	2.37	3.85	5.12	6.26	8.31	10.16	11.88	15.78	19.30	22.57
iv	M2	1.93	3.13	4.15	5.08	6.75	8.26	9.65	12.82	15.68	18.33

Selected characteristic compressive strengths for Group 1 concrete block masonry, f_k (N/mm²)

Mortar class and type of unit		Mean compressive strength of unit									
		2.9	3.6	4.0	5.2	7.3	10.4	15.0	17.5	22.5	30.0
100 mm solid or concrete-filled hollow blocks		Normalised mean compressive strength of unit, f_b									
		4.8	6.0	6.6	8.6	12.1	17.2	24.8	29.0	37.3	49.7
		Characteristic strength, f_k									
ii	M6	2.57	2.99	3.22	3.86	4.90	6.28	8.11	9.03	10.77	13.17
iii	M4	2.27	2.64	2.85	3.42	4.34	5.56	7.18	8.00	9.54	11.67
140 mm solid or concrete-filled hollow blocks		Normalised mean compressive strength of unit, f_b									
		4.5	5.6	6.2	8.1	11.4	16.2	23.4	27.3	35.1	46.8
		Characteristic strength, f_k									
ii	M6	2.46	2.86	3.08	3.71	4.70	6.02	7.78	8.66	10.33	12.64
iii	M4	2.18	2.54	2.73	3.28	4.16	5.33	6.89	7.67	9.15	11.19
215 mm solid concrete block wall (140 mm block laid on side)		Normalised mean compressive strength of unit, f_b									
		3.3	4.1	4.5	5.9	8.2	11.7	16.9	19.7	25.4	33.8
		Characteristic strength, f_k									
ii	M6	1.96	2.28	2.46	2.95	3.74	4.80	6.20	6.90	8.23	10.07
iii	M4	1.74	2.02	2.18	2.61	3.32	4.25	5.49	6.11	7.29	8.92

Notes to both tables:
1. Values apply to Group 1 masonry units in general-purpose mortar.
2. Characteristic strengths for Equation 3.1 (clause 3.6.1.2 (1) (i))

$$f_k = k \, f_b^\alpha f_m^\beta$$

where
   f_k is the characteristic compressive strength of the masonry
   k is a constant (taken as 0.5 for simplicity see Table NA.4 for more details)
   f_b is the normalised mean compressive strength of the masonry unit
   f_m is the compressive strength of the mortar
   $\alpha = 0.7$ and $\beta = 0.3$ for general-purpose mortar (see clause NA 2.4)
3. f_b = conditioning factor (γ_c) × shape factor (γ_s) × declared mean compressive stress.
   a.   BS EN 772-1 suggests $\gamma_c = 1.0$ for air-dried units and $\gamma_c = 1.2$ for wet strengths. As masonry units are generally manufacturing and classified using wet strengths, this table assumes $\gamma_c = 1.2$.
   b.   BS EN 772-1, Table A1 gives shape factors for standard UK masonry units as follows:
      i.   Standard UK clay brick, $\gamma_s = 0.85$
      ii.   Standard UK concrete block, $\gamma_s = 1.38$ for 100 mm, $\gamma_s = 1.30$ for 140 mm
      iii. Standard UK 140 concrete block laid on side, $\gamma_s = 0.94$
4. For columns or piers with a cross-sectional area, $A < 0.1$ m², f_k should be multiplied by $\gamma_{col} = (0.7 + 3A)$, where A is the load-bearing cross-sectional area of the wall in square metres.
5. Values for stone masonry can be approximated using the values for concrete blocks of similar strength. The strength of random rubble masonry in cement mortar can be approximated as 75% of the concrete block values. Much lower values are achieved with lime mortar. Detailed calculations should be carried out using the full BS EN 1996 design method.
6. For collar jointed walls the given values of f_k should be multiplied by a factor of 0.8.

Source: Based on data as listed above from BS EN 1996-1-1:2005.

Effective thickness

Stiffness coefficient p_t for walls stiffened by piers – Minimum wall thicknesses should be 90 mm for a single leaf and 75 mm for one leaf of a cavity wall.

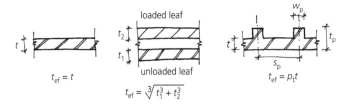

$$t_{ef} = t$$

$$t_{ef} = \sqrt[3]{t_1^3 + t_2^3}$$

$$t_{ef} = p_t t$$

Ratio of pier spacing (centre to centre) to pier width S_p/W_p	Ratio of pier depth to wall thickness t_p/t		
	1	**2**	**3**
6	1.0	1.4	2.0
10	1.0	1.2	1.4
20	1.0	1.0	1.0

Note: Linear interpolation (but not extrapolation) is permitted.

Source: BS EN 1996-1-1:2005. Clause 5.5.1.3 and T5.1. NA to BS EN. 1996. 2005. Part 1-1. Clause NA 2.13 and NA2.15.

Effective height or length

Lateral supports should be able to carry any applied horizontal loads or reactions plus 2.5% of the total vertical design load in the element to be laterally restrained. The effective height of a masonry wall depends on the horizontal lateral restraint provided by the floors or roofs supported on the wall. The effective length of a masonry wall depends on the vertical lateral restraint provided by cross walls, piers or returns. BS EN 1996 clause 5.5.1.2 defines types of arrangement that provide simple or enhanced lateral restraint. The slenderness ratio of a masonry element should generally not exceed 27. For walls of less than 90 mm thick in buildings of two storeys or more, the slenderness should not exceed 20.

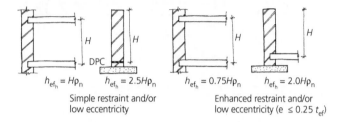

$h_{ef_h} = H\rho_n$ $h_{ef_h} = 2.5H\rho_n$ $h_{ef_h} = 0.75H\rho_n$ $h_{ef_h} = 2.0H\rho_n$

Simple restraint and/or   Enhanced restraint and/or
low eccentricity   low eccentricity (e ≤ 0.25 t_{ef})

Enhanced restraint can be assumed if stiffening cross walls are present. Cross walls being taken into account should have a length of at least 0.2 H and thickness of at least 0.3 t_{ef} using the properties of the wall to be stiffened. If 0.2 H length of cross wall is interrupted by openings, then a minimum of 0.2 H wall should extend beyond each opening. Cross walls can only be considered to restrain a wall of length, L, when L ≤ 15 t_{ef} if restrained vertically on one edge or L ≤ 30 t_{ef} if restrained vertically on two edges. For larger panels, the cross walls are considered insignificant and should be ignored.

For walls stiffened on one vertical edge in addition to top and bottom restraints:

$$\rho_n = \frac{1}{1 + \left[\dfrac{h_{ef_h}}{3L}\right]^2} \quad \text{where } H \le 3.5L$$

$$\rho_n = \frac{1.5L}{H} \ge 0.3 \quad \text{where } H > 3.5L$$

For walls stiffened on two vertical edges in addition to top and bottom restraint:

$$\rho_n = \frac{1}{1 + \left[\dfrac{h_{ef_h}}{L}\right]^2} \quad \text{where } H \le 1.15L$$

$$\rho_n = \frac{0.5L}{H} \quad \text{where } H > 1.15L$$

Source: BS EN 1996-1-1: 2005. Clause 5.5.1.2.

Horizontal resistance to lateral movement

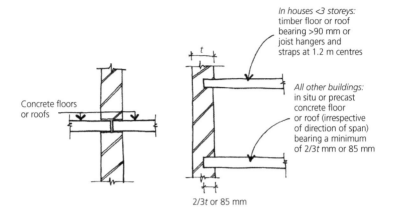

Concrete floors or roofs

In houses <3 storeys: timber floor or roof bearing >90 mm or joist hangers and straps at 1.2 m centres

All other buildings: in situ or precast concrete floor or roof (irrespective of direction of span) bearing a minimum of 2/3t mm or 85 mm

2/3t or 85 mm

Vertical resistance to lateral movement

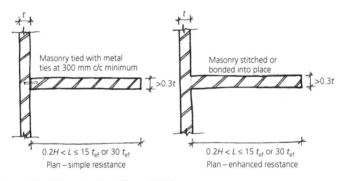

Masonry tied with metal ties at 300 mm c/c minimum

>0.3t

0.2H < L ≤ 15 t_{ef} or 30 t_{ef}
Plan – simple resistance

Masonry stitched or bonded into place

>0.3t

0.2H < L ≤ 15 t_{ef} or 30 t_{ef}
Plan – enhanced resistance

Source: BS EN 1996-1-1:2005. Clause 5.5.1.2.

Eccentric loading

BS EN 1996 uses a more detailed method of assessing eccentricity of load in masonry walls and encourages the use of subframe analysis to calculate loads at the top, bottom and middle of the wall. A slenderness reduction factor is calculated for each of these cases, with only the smallest of the three being applied.

Capacity reduction factor for eccentricity at the top or bottom of the wall, ϕ_i

$$\phi_i = 1 - \frac{2e_i}{t}$$

Eccentricity, e_i, can be approximated for quick calculation from:

$$e_i = \frac{M}{N} \pm \frac{h_{ef}}{450} \leq 0.05\,t$$

where M is the moment and N the axial load (for either the top or bottom of the wall). For a symmetrical arrangement with axial load and minimum eccentricity ($e = 0.05\,t$), $\phi_i = 0.9$. For all other arrangements, see BS EN 1996, clause 6.1.2.2.

Capacity reduction for eccentrically at mid-height of the wall, ϕ_m

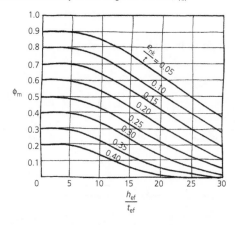

Vertical design resistance of solid walls and columns

$$\text{Wall:} \quad N_{RD} = \phi_{(m\,or\,i)} t \frac{f_k}{\gamma_m}$$

$$\text{Column:} \quad N_{RD} = \phi_{(m\,or\,i)} 6t \frac{f_k}{\gamma_m} \gamma_{col}$$

where t is the actual thickness of the wall, b is the column width, ϕ_m or ϕ_i is the smallest of the possible three slenderness reduction factors, γ_{col} is the column reduction factor, γ_m is the partial factor for material and workmanship and f_k is the characteristic compressive strength of the masonry.

Source: BS EN 1996-1-1: 2005. Clause 6.1.2.1 and Figure G.1., Annex G.

Shear strength

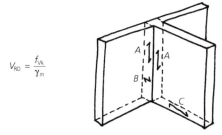

$$V_{RD} = \frac{f_{Vk}}{\gamma_m}$$

A Complementary shear acting in the vertical direction of the vertical plane: $f_{Vk} = f_{Vk_o}$
B Complementary shear acting in the horizontal direction of the vertical plane
C Shear acting in the horizontal direction of the horizontal plane:

$$f_{Vk} = f_{Vk_o} + 0.4\,\sigma_{Ed} \le 0.065\,f_b \text{ for fully filled perpend joints}$$

where
  f_{Vk_o} is the characteristic initial shear strength of the masonry (from test or manufacturer's data). Typically, 0.2 N/mm² for bricks or 0.15 for blocks in class II or III (M6 or M4) general-purpose mortar with filled perpend joints
  f_{vk} is the characteristic shear strength
  f_b is the normalised mean compressive strength of the masonry units

BS EN 1996 part 3 gives a simplified calculation method for the design resistance of shear walls (with fully filled perpends) accounting for applied moment and eccentricity (where $e = (M/N) \not> (L/b)$ and applied axial load is half the axial load capacity of the wall):

$$V_{Rd} = 3e_{sw}t\frac{f_{Vk_o}}{\gamma_m} + 0.4\frac{N_{Ed}}{\gamma_m} \le 0.195\,e_{sw}\,t, \quad \text{where } e_{sw} = \frac{1}{2} - e$$

Source: BS EN 1996-1-1: 2005; 3.6.2; NA to BS EN 1996-1-1: 2005; Table NA.5: BS EN 1996-3: 2006; 4.5.

Lateral load

Wind zone map – The sizes of external wall panels are limited depending on the elevation of the panel above the ground and the typical wind speeds expected for the site.

Maximum permitted panel areas for external cavity wall panels[a] (m²) – While BS EN 1996-1-1 gives no specific limits on the sites of external wall panels, the guidance from BS 5628 part 3 is still useful design guidance.

Wind zone	Height above ground (m)	Panel type								
		A	B	C	D	E	F	G	H	I
1	5.4	11.0	17.5	26.5	20.5	32.0	32.0	8.5	14.0	19.5
	10.8	9.0	13.0	17.5	15.5	24.0	32.0	7.0	10.0	15.5
2	5.4	9.5	14.0	21.0	17.5	27.0	32.0	7.5	10.5	17.0
	10.8	8.0	11.5	13.5	13.0	19.0	28.0	6.0	9.0	13.0
3	5.4	8.5	12.5	15.5	14.5	22.0	30.5	6.5	9.5	14.5
	10.8	7.0	10.0	11.5	11.0	14.5	24.5	5.0	7.5	11.5
4	5.4	8.0	11.0	13.0	12.5	18.0	27.0	6.0	8.5	12.5
	10.8	6.5	9.0	10.5	9.5	12.5	21.5	4.0	6.5	10.0

Note:
[a] The values in the table are given for cavity walls with leaves of 100 mm block inner skin. If either leaf is increased to 140 mm, the maximum permitted areas in the table can be increased by 20%.

No dimension of unreinforced masonry panels should generally exceed 50 × effective thickness. Reinforced masonry panel areas can be about 20% bigger than unreinforced panels and generally no dimension should exceed 60 × effective thickness.

Source: BS 5628: Part 3: 2005.

Characteristic flexural strength of masonry – The characteristic flexural strength of masonry works on the assumption that the masonry section is uncracked and therefore can resist some tensile stresses. If the wall is carrying some compressive load, then the wall will have an enhanced resistance to tensile stresses. Where tension develops which exceeds the tensile resistance of the masonry (e.g. at a DPC or crack location), the forces must be resisted by the dead loads alone and therefore the wall will have less capacity than an uncracked section.

Type of masonry unit	Plane of failure				Basic orthogonal ratio, μ^a	
	Parallel to the bed joints, f_{kx1}		Perpendicular to the bed joints, f_{kx2}			
	Class I M12 (N/mm²)	I and III M6 and M4 (N/mm²)	Class I M12 (N/mm²)	II and III M6 and M4 (N/mm²)	Class I M12	II and III M6 and M4
Clay bricks having a water absorption of:						
Less than 7%	0.7	0.5	2.0	1.5	0.35	0.33
Between 7% and 12%	0.5	0.4	1.5	1.1	0.33	0.36
Over 12%	0.4	0.3	1.1	0.9	0.36	0.33
Calcium silicate or concrete bricks		0.3		0.9		0.33
7.3 N/mm² concrete blocks:						
≤100 mm wide wall		0.25		0.60		0.42
140 mm wide wall		0.22		0.53		0.42
215 mm wide wall		0.17		0.41		0.41
>250 mm wide wall		0.15		0.35		0.43
Any thickness concrete block walls:						
10.4 N/mm²		0.25		0.75		0.33
≥14.0 N/mm²		0.25		0.90^d		0.28

Notes:

[a] The orthogonal ratio $\mu = f_{kx1}/f_{kx2}$ can be improved if f_{kx1} is enhanced by the characteristic dead load: $f_{kx1enhanced} = f_{kx1}/\gamma_m + 0.9\ G_k$.

[b] The wall thickness is taken as the thickness of a solid wall or leaf of a cavity wall.

[c] Linear interpolation is permitted for walls >100 mm and <250 mm thick.

[d] When used with flexural strength parallel to the bed joints, assume $\rho = 0.3$ for this particular value of K_{kx2}.

Source: NA to BS EN. 1996. 2005. Part 1-1. Table NA.6.

Ultimate moments applied to wall panels – The flexural strength of masonry parallel to the bed joints is about a third of the flexural strength perpendicular to the bed joints. The overall flexural capacity of a panel depends on the dimensions, orthogonal strength ratio and support conditions of that panel. Like BS 5628, BS EN 1996 uses α (which is a bending moment coefficient based on experimental data and yield line analysis) to estimate how a panel will combine these different orthogonal properties to distribute the applied lateral loads and express this as a moment in one direction:

$M_{Ed1} = \mu\alpha\gamma_f W_k L^2$ per unit length (parallel to bed joints) or
$M_{Ed2} = \alpha\gamma_f W_k L^2$ per unit height (perpendicular to bed joints)

W_k is the applied distributed lateral load panel, L is the horizontal panel length and $M_{Ed2} = M_{Ed1}/\mu$. Therefore the applied moment need only be checked against the flexural strength in one direction-normally parallel to the bed joints as this is the weaker case. γ_f is the combined partial load factor for actions.

Ultimate flexural strength of an uncracked wall spanning horizontally

$$M_{Rd2} = \left(\frac{f_{kx2}}{\gamma_M}\right)Z \quad \text{(perpendicular to bed joints)}$$

Ultimate flexural strength of an uncracked wall spanning vertically

$$M_{Rd1} = \left(\frac{f_{kx1}}{\gamma_M} + 0.9G_k\right)Z \quad \text{(parallel to bed joints)}$$

The flexural strength parallel to the bed joints varies with the applied dead load and the height of the wall. Therefore, the top half of non-load-bearing walls are normally the critical case. There are published tables for non-load-bearing and freestanding walls, which greatly simplify calculations.

Ultimate flexural strength of a cracked wall spanning vertically

$$M_{Rd1} = \frac{\gamma_w t^2 h}{2\gamma_F}$$

The dead weight of the wall is used to resist lateral loads. Tension can be avoided if the resultant force is kept within the middle third of the wall. γ_F is an appropriate factor of safety against overturning and γ_w is the weight of the wall.

Selected bending moment coefficients in laterally loaded wall panels

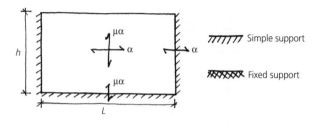

Panel support conditions	Orthogonal ratio (μ)	Values of panel factor (α)						
		H/L						
		0.35	0.50	0.75	1.00	1.20	1.50	1.75
A	1.00	0.031	0.045	0.059	0.071	0.079	0.085	0.090
	0.90	0.032	0.047	0.061	0.073	0.081	0.087	0.092
	0.80	0.034	0.049	0.064	0.075	0.083	0.089	0.093
	0.70	0.035	0.051	0.066	0.077	0.085	0.091	0.095
	0.60	0.038	0.053	0.069	0.080	0.088	0.093	0.097
	0.50	0.040	0.056	0.073	0.083	0.090	0.095	0.099
	0.40	0.043	0.061	0.077	0.087	0.093	0.098	0.101
	0.35	0.045	0.064	0.080	0.089	0.095	0.100	0.103
	0.30	0.048	0.067	0.082	0.091	0.097	0.101	0.104
B	1.00	0.024	0.035	0.046	0.053	0.059	0.062	0.065
	0.90	0.025	0.036	0.047	0.055	0.060	0.063	0.066
	0.80	0.027	0.037	0.049	0.056	0.061	0.065	0.067
	0.70	0.028	0.039	0.051	0.058	0.062	0.066	0.068
	0.60	0.030	0.042	0.053	0.059	0.064	0.067	0.069
	0.50	0.031	0.044	0.055	0.061	0.066	0.069	0.071
	0.40	0.034	0.047	0.057	0.063	0.067	0.070	0.072
	0.35	0.035	0.049	0.059	0.065	0.068	0.071	0.073
	0.30	0.037	0.051	0.061	0.066	0.070	0.072	0.074
C	1.00	0.020	0.028	0.037	0.042	0.045	0.048	0.050
	0.90	0.021	0.029	0.038	0.043	0.046	0.048	0.050
	0.80	0.022	0.031	0.039	0.043	0.047	0.049	0.051
	0.70	0.023	0.032	0.040	0.044	0.048	0.050	0.051
	0.60	0.024	0.034	0.041	0.046	0.049	0.051	0.052
	0.50	0.025	0.035	0.043	0.047	0.050	0.052	0.053
	0.40	0.027	0.038	0.044	0.048	0.051	0.053	0.054
	0.35	0.029	0.039	0.045	0.049	0.052	0.053	0.054
	0.30	0.030	0.040	0.046	0.050	0.052	0.054	0.055

(continued)

Panel support conditions	Orthogonal ratio (μ)	Values of panel factor (α)						
		H/L						
		0.35	0.50	0.75	1.00	1.20	1.50	1.75
D	1.00	0.013	0.021	0.029	0.035	0.040	0.043	0.045
	0.90	0.014	0.022	0.031	0.036	0.040	0.043	0.046
	0.80	0.015	0.023	0.032	0.038	0.041	0.044	0.047
	0.70	0.016	0.025	0.033	0.039	0.043	0.045	0.047
	0.60	0.017	0.026	0.035	0.040	0.044	0.046	0.048
	0.50	0.018	0.028	0.037	0.042	0.045	0.048	0.050
	0.40	0.020	0.031	0.039	0.043	0.047	0.049	0.051
	0.35	0.022	0.032	0.040	0.044	0.048	0.050	0.051
	0.30	0.023	0.034	0.041	0.046	0.049	0.051	0.052
E	1.00	0.008	0.018	0.030	0.042	0.051	0.059	0.066
	0.90	0.009	0.019	0.032	0.044	0.054	0.062	0.068
	0.80	0.010	0.021	0.035	0.046	0.056	0.064	0.071
	0.70	0.011	0.023	0.037	0.049	0.059	0.067	0.073
	0.60	0.012	0.025	0.040	0.053	0.062	0.070	0.076
	0.50	0.014	0.028	0.044	0.057	0.066	0.074	0.080
	0.40	0.017	0.032	0.049	0.062	0.071	0.078	0.084
	0.35	0.018	0.035	0.052	0.064	0.074	0.081	0.086
	0.30	0.020	0.038	0.055	0.068	0.077	0.083	0.089
F	1.00	0.008	0.016	0.026	0.034	0.041	0.046	0.051
	0.90	0.008	0.017	0.027	0.036	0.042	0.048	0.052
	0.80	0.09	0.018	0.029	0.037	0.044	0.049	0.054
	0.70	0.010	0.020	0.031	0.039	0.046	0.051	0.055
	0.60	0.011	0.022	0.033	0.042	0.048	0.053	0.057
	0.50	0.013	0.024	0.036	0.044	0.051	0.056	0.059
	0.40	0.015	0.027	0.039	0.048	0.054	0.058	0.062
	0.35	0.016	0.029	0.041	0.050	0.055	0.060	0.063
	0.30	0.018	0.031	0.044	0.052	0.057	0.062	0.065
G	1.00	0.007	0.014	0.022	0.028	0.033	0.037	0.040
	0.90	0.008	0.015	0.023	0.029	0.034	0.038	0.041
	0.80	0.008	0.016	0.024	0.031	0.035	0.039	0.042
	0.70	0.009	0.017	0.026	0.032	0.037	0.040	0.043
	0.60	0.010	0.019	0.028	0.034	0.038	0.042	0.044
	0.50	0.011	0.021	0.030	0.036	0.040	0.043	0.046
	0.40	0.013	0.023	0.032	0.038	0.042	0.045	0.047
	0.35	0.014	0.025	0.033	0.039	0.043	0.046	0.048
	0.30	0.016	0.026	0.035	0.041	0.044	0.047	0.049

continued

(continued)

Panel support conditions	Orthogonal ratio (μ)	Values of panel factor (α)						
		H/L						
		0.35	0.50	0.75	1.00	1.20	1.50	1.75
H	1.00	0.005	0.011	0.018	0.024	0.029	0.033	0.036
	0.90	0.006	0.012	0.019	0.025	0.030	0.034	0.037
	0.80	0.006	0.013	0.020	0.027	0.032	0.035	0.038
	0.70	0.007	0.014	0.022	0.028	0.033	0.037	0.040
	0.60	0.008	0.015	0.024	0.030	0.035	0.038	0.041
	0.50	0.009	0.017	0.025	0.032	0.036	0.040	0.043
	0.40	0.010	0.019	0.028	0.034	0.039	0.042	0.045
	0.35	0.011	0.021	0.029	0.036	0.040	0.043	0.046
	0.30	0.013	0.022	0.031	0.037	0.041	0.044	0.047
I	1.00	0.004	0.009	0.015	0.021	0.026	0.030	0.033
	0.90	0.004	0.010	0.016	0.022	0.027	0.031	0.034
	0.80	0.005	0.010	0.017	0.023	0.028	0.032	0.035
	0.70	0.005	0.011	0.019	0.025	0.030	0.033	0.037
	0.60	0.006	0.013	0.020	0.026	0.031	0.035	0.038
	0.50	0.007	0.014	0.022	0.028	0.033	0.037	0.040
	0.40	0.018	0.016	0.024	0.031	0.035	0.039	0.042
	0.35	0.09	0.017	0.026	0.032	0.037	0.040	0.043
	0.30	0.010	0.019	0.028	0.034	0.038	0.042	0.044
J	1.00	0.009	0.023	0.046	0.071	0.096	0.122	0.151
	0.90	0.010	0.026	0.050	0.076	0.103	0.131	0.162
	0.80	0.012	0.028	0.054	0.083	0.111	0.142	0.175
	0.70	0.013	0.032	0.060	0.091	0.121	0.156	0.191
	0.60	0.015	0.036	0.067	0.100	0.135	0.173	0.211
	0.50	0.018	0.042	0.077	0.133	0.153	0.195	0.237
	0.40	0.021	0.050	0.090	0.131	0.177	0.225	0.272
	0.35	0.024	0.055	0.098	0.144	0.194	0.244	0.296
	0.30	0.027	0.062	0.108	0.160	0.214	0.269	0.325
K	1.00	0.009	0.021	0.038	0.056	0.074	0.091	0.108
	0.90	0.010	0.023	0.041	0.060	0.079	0.097	0.113
	0.80	0.011	0.025	0.045	0.065	0.084	0.103	0.120
	0.70	0.012	0.028	0.049	0.070	0.091	0.110	0.128
	0.60	0.014	0.031	0.054	0.077	0.099	0.119	0.138
	0.50	0.016	0.035	0.061	0.085	0.109	0.130	0.149
	0.40	0.019	0.041	0.069	0.097	0.121	0.144	0.164
	0.35	0.021	0.045	0.075	0.104	0.129	0.152	0.173
	0.30	0.024	0.050	0.082	0.112	0.139	0.162	0.183

continued

(continued)

Panel support conditions	Orthogonal ratio (μ)	Values of panel factor (α)						
		H/L						
		0.35	0.50	0.75	1.00	1.20	1.50	1.75
L	1.00	0.006	0.015	0.029	0.044	0.059	0.073	0.088
	0.90	0.007	0.017	0.032	0.047	0.063	0.078	0.093
	0.80	0.008	0.018	0.034	0.051	0.067	0.084	0.099
	0.70	0.009	0.021	0.038	0.056	0.073	0.090	0.106
	0.60	0.010	0.023	0.042	0.061	0.080	0.098	0.115
	0.50	0.012	0.027	0.048	0.068	0.089	0.108	0.126
	0.40	0.014	0.032	0.055	0.078	0.100	0.121	0.139
	0.35	0.016	0.035	0.060	0.084	0.108	0.129	0.148
	0.30	0.018	0.039	0.066	0.092	0.116	0.138	0.158

Notes:

[a] Linear interpolation of μ and h/L is permitted.

[b] When the dimensions of a wall are outside the range of h/L given in this table, it will usually be sufficient to calculate the moments on the basis of a simple span. For example, a panel of type A having h/L less than 0.3 will tend to act as a freestanding wall, while the same panel having h/L greater than 1.75 will tend to span horizontally.

Source: BS EN 1996-1-1:2005. Appendix E.

Concentrated loads

Increased stresses are permitted under and close to the bearings of concentrated loads. The load is assumed to be spread uniformly beneath the bearing. The effect of this bearing pressure in combination with the stresses in the wall due to other loads should be less than the design bearing strength.

BS EN 1996-3 gives a simplified method for the assessment of concentrated loads which allow normal design stresses to be exceeded by up to 50% where $N_{Edc} < N_{Rdc}$.

Group 1 masonry units: $N_{Rdc} = \dfrac{f_k}{\gamma_M}\left(1.2 + \dfrac{0.4a_1}{h_c}\right)A_b \not> 1.5\dfrac{f_k}{\gamma_m}A_b$

Group 2 masonry units: $N_{Rdc} = \dfrac{f_k}{\gamma_M}A_b$

where
    a_1 is the shortest distance from the concentrated load to a wall edge
    h_c is the height of the concentrated load about the floor
    A_b is the loaded area

Summary of differences with BS 5628: Structural masonry

	BS EN 1996	BS 5628
Basis of design	Methods designed to deal with the huge array of masonry units and mortars used across the European Union. One of the main challenges of using BS EN 1996 is to identify where the standard UK bricks and blocks sit within the Eurocode classifications. After that, the British Standard and Eurocode design methods are reasonably similar	Methods only deal with traditional British masonry elements and materials
Masonry products	BS EN 1996 classifies masonry products into 'Groups' depending on the volume of voids. At present, all UK bricks fall into Group 1, although it is thought that some Group 2 products might appear in due course. Most UK blocks will fall into either Group 1 or 2. Product literature should be checked for this before design begins	–
Compressive strength	Method uses normalised compressive strength to allow for the variety of testing regimes used. f_b = conditioning factor × shape factor × declared mean compressive strength	Method uses mean compressive strength for each unit
Construction control and safety factors	Allows discretion over the construction environment (Class 1 (factory-type conditions) and Class 2 ('normal' conditions)) and manufacturing control (either Category I or Category II which should be quoted in product literature), with factors further varied depending on whether the design is being checked for compression or tension. The Eurocode values are lower than those in BS 5628	Allows discretion over the construction environment and manufacturing control
Slenderness	Complex method requiring a sub-frame analysis to assess induced moments. Although designers need to calculate values for the top, middle and bottom of a given element, only the lowest modification factor should be used when calculating the design vertical load resistance. More sophisticated method for assessing vertical and/or horizontal restraints	Simplified method based on slenderness ratios and buckling curves
Column modification factor	$(0.7 + 3A)$ for a cross-sectional area of less than 0.1 m^2	$(0.7 + 1.5A)$ for a cross-sectional area of less than 0.2 m^2
Collar jointed walls	Reduction factor of 0.8	No guidance
≤ 100 mm walls	No guidance	Enhanced capacity permitted
Panel design	Wider range of bending moment coefficient, α, factors available than original BS 5628 tables	–

Masonry design to CP111

CP111 is the 'old' brick code which uses permissible stresses and has been withdrawn. Although it is now not appropriate for new construction, it can be helpful when refurbishing old buildings as the ultimate design methods used in BS EN 1996 are not appropriate for use with historic materials.

Basic compressive masonry strengths for standard format bricks (N/mm²)

Description of mortar proportions by volume	Hard-ening time	Basic compressive stress of unit[a] (N/mm²)								
Cement:lime:sand (BS 5628 mortar class)		2.8	7.0	10.5 Stock	20.5 Fletton	27.5	34.5	52.0	69.0 Class B	≥ 96.5 Class A
Dry pack – 1:0:3 (I)	7	0.28	0.70	1.05	1.65	2.05	2.50	3.50	4.55	5.85
Cement lime – 1:1:6 (III)	14	0.28	0.70	0.95	1.30	1.60	1.85	2.50	3.10	3.80
Cement lime – 1:2:9 (IV)	14	0.28	0.55	0.85	1.15	1.45	1.65	2.05	2.50	3.10
Non-hydraulic lime putty with pozzolanic/cement additive – 0:1:3	14	0.21	0.49	0.70	0.95	1.15	1.40	1.70	2.05	2.40
Hydraulic lime – 0:1:2	14	0.21	0.49	0.70	0.95	1.15	1.40	1.70	2.05	2.40
Non-hydraulic lime mortar – 0:1:3	28[b]	0.21	0.42	0.55	0.70	0.75	0.85	1.05	1.15	1.40

Notes:
[a] For columns or piers of cross-sectional area $A < 0.2$ m², the basic compressive strength should be multiplied by $\gamma = (0.7 + 1.5\,A)$.
[b] Longer time may be required if the weather is not warm and dry.

Source: CP111. 1970.

Capacity reduction factors for slenderness and eccentricity

Slenderness ratio	Slenderness reduction factor, β				
	Axially loaded	Eccentricity of loading			
		$t/6$	$t/4$	$t/3$	$t/3$ to $t/2$
6	1.000	1.000	1.000	1.000	1.000
8	0.950	0.930	0.920	0.910	0.885
10	0.890	0.850	0.830	0.810	0.771
12	0.840	0.780	0.750	0.720	0.657
14	0.780	0.700	0.660	0.620	0.542
16	0.730	0.630	0.580	0.530	0.428
18	0.670	0.550	0.490	0.430	0.314
20	0.620	0.480	0.410	0.340	0.200
22	0.560	0.400	0.320	0.240	–
24	0.510	0.330	0.240	–	–
26	0.450	0.250	–	–	–
27	0.430	0.220	–	–	–

Concentrated loads

Although CP111 indicates that concentrated compressive stresses up to 1.5 times the permissible compressive stresses are acceptable, it is now thought that this guidance is not conservative as it does not take account of the bearing width or position.

Therefore, it is generally accepted that bearing stresses should be kept within the basic permissible stresses. For historic buildings, this typically means maximum bearing stresses of 0.42 N/mm^2 for stock bricks in traditional lime mortar or 0.7 N/mm^2 where the structure has been 'engineered', perhaps with flettons in arches or vaults.

Source: CP111. 1970.

Lintel design to BS 5977

BS 5977 sets out the method for load assessment of lintels in masonry structures for openings up to 4.5 m in single-storey construction or up to 3.6 m in normal domestic two- to three-storey buildings. The method assumes that the masonry over an opening in a simple wall will arch over the opening. The code guidance must be applied with common sense as building elevations are rarely simple and load will be channelled down piers between openings. Typically, there should be not less than 0.6 m or 0.2L of masonry to each side of the opening (where L is the clear span), not less than 0.6L of masonry above the lintel at midspan and not less than 0.6 m of masonry over the lintel supports. When working on existing buildings, the effect of new openings on existing lintels should be considered.

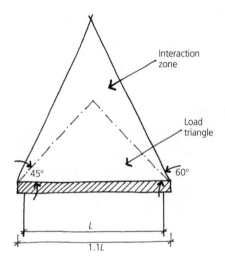

Loading assumptions:

1. The weight of the masonry in the loaded triangle is carried on the lintel – not the masonry in the zone of interaction.
2. Any point load or distributed load applied within the load triangle is dispersed at 45° and carried by the lintel.
3. Half of any point, or distributed, load applied to the masonry within the zone of interaction is carried by the lintel.

Where there are no openings above the lintel, and the loading assumptions apply, no loads outside the interaction zone need to be considered. Openings that are outside the zone of interaction, or which cut across the zone of interaction completely, need not be considered and do not add load to the lintel. However, openings which cut into (rather than across) the zone of interaction can have a significant effect on lintel loading as all the self-weight of the wall and applied loads above the line X–Y are taken into account. As for the other loads applied in the zone of interaction, they are halved and spread out at 45° from the line X–Y to give a line load on the lintel below.

All the loading conditions are illustrated in the following example:

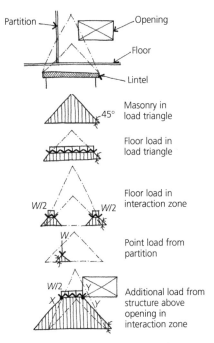

Source: BS 5977: Part 1:1981. *Lintels. Methods for Assessment of Load.* BSI. Figures 1 and 4.

Masonry accessories

Joist hangers

Joist hangers provide quick, economic and reliable timber-to-timber, timber-to-masonry and timber-to-steel junctions. Joist hangers should comply with BS 6178 and be galvanised for general use or stainless steel for special applications. Normally, 600 mm of masonry over the hangers is required to provide restraint and ensure full load-carrying capacity. Coursing adjustments should be made in the course below the course carrying the joist hanger to avoid supporting the hangers on cut blocks. The end of the joist should be packed tight to the back of the hanger, have enough bearing on the hanger and be fixed through every provided hole with 3.75 × 30 mm square twist sheradised nails. The back of the hanger must be tight to the wall and should not be underslung from beam supports. If the joist needs to be cut down to fit into the joist hanger, it may exceed the load capacity of the hanger. If the joist hangers are not installed to the manufacturer's instructions, they can be overloaded and cause collapse.

Straps for robustness

Masonry walls must be strapped to floors and roofs for robustness in order to allow for any out-of-plane forces, accidental loads and wind suction around the roof line. The traditional strap is 30 × 5 mm with a characteristic tensile strength of 10 kN. Straps are typically galvanised mild steel or austenitic stainless steel, fixed to three joists with four fixings and built into the masonry wall at a maximum spacing of 2 m. A typical strap can provide a restraining force of 5 kN/m depending on the security of the fixings. Compressive loads are assumed to be transferred by direct contact between the wall and floor/roof structures. Building Regulations and BS 8103: Part 1 set out recommendations for the fixing and spacing of straps.

Padstones

These are used to spread the load at the bearings of steel beams on masonry walls. The plan area of the padstone is determined by the permissible concentrated bearing stress on the masonry. The depth of the padstone is based on a 45° load spread from the edges of the steel beam on the padstone until the padstone area is sufficient that the bearing stresses are within permissible values.

Proprietary pre-stressed concrete beam lintels

The following values are working loads for beam lintels which do not act compositely with the masonry above the opening manufactured by Supreme Concrete. Supreme Concrete stock pre-stressed concrete lintels from 0.6 to 3.6 m long in 0.15 m increments but can produce special lintels up to about 4.2 m long. The safe loads given below are based on 150 mm end bearing. A separate range of fair faced lintels are also available.

Maximum uniformly distributed service load (kN)									
Lintel width (mm)	Lintel depth (mm)	Reference (mm)	Clear span between supports (m)						
			0.6	0.9	1.5	2.1	2.7	3.3	3.9
100	65	P100	8.00	5.71	3.64	2.67	2.11	1.74	–
	100	S10	17.92	12.80	8.15	5.97	4.72	3.90	–
	140	R15A	35.95	25.68	16.34	11.98	9.46	7.81	6.66
	215	R22A	59.42	53.71	34.18	25.07	19.79	16.35	13.93
140	65	P150	10.99	7.85	4.99	3.66	2.89	2.39	–
	100	R1 5	27.41	19.58	12.46	9.14	7.21	5.96	5.08
	215	R21A	73.90	73.90	52.65	38.61	30.48	25.18	21.45
215	65	P220	26.24	18.74	11.93	8.75	6.91	5.70	–
	100	R22	38.61	27.58	17.55	12.87	10.16	8.39	7.15
	140	R21	73.90	53.26	33.89	24.85	19.62	16.21	13.81

Source: Supreme Concrete Lintels. 2008. Note that this information is subject to change at any time. Consult the latest Supreme Concrete literature for up-to-date information.

Profiled steel lintels

Profiled galvanised steel lintels are particularly useful for cavity wall construction, as they can be formed to support both leaves and incorporate insulation. Profiled steel lintels are supplied to suit cavity widths from 50 to 165 mm, single leaf walls, standard and heavy-duty loading conditions, wide inner or outer leaves and timber construction. Special lintels are also available for corners and arches. The following lintels are selected from the range produced by I.G. Ltd.:

Lintel reference	Ext. leaf (mm)	Cavity (mm)	Int. leaf (mm)	Height (mm)	Available lengths (mm) (in 150 mm increments)	Gauge (mm)	Total uniformly distributed load for load ratio 1 (kN)	Total uniformly distributed load for load ratio 2 (kN)
L1/S 100	102	90–105	125	88	0.60–1.20	1.6	12	10
				88	1.35–1.50	2.0	16	13
				107	1.65–1.80	2.0	19	16
				125	1.95–2.10	2.0	21	17
				150	2.25–2.40	2.0	23	18
				162	2.55–2.70	2.6	27	22
				171	2.85–3.60	2.6	27	20
				200	3.15–4.05	3.2	27	20
				200	3.75–4.05	3.2	26	19
				200	4.20–4.80	3.4	27	22
L1/HD 100	102	90–105	125	110	0.60–1.20	2.9	30	22
				135	1.35–1.50	2.9	30	22
				163	1.65–2.10	2.9	40	35
				203	2.25–2.55	2.9	40	35
				203	2.70–3.00	3.2	40	35
				203	3.15–3.60	3.2	35	32
				203	3.75–4.20	3.2	33	28

L1/S 100

95 88 100

L1/HD 100

95 88 100

L1/XHD 100	102	90–105	125	163ª	0.60–1.50	3.2	50	45
				163ª	1.65–1.80	3.2	50	45
				203ª	1.95–2.10	3.2	55	45
				203ª	2.25–2.70	3.2	50	40
L1/S 100 WIL	102	90–105	150	82	0.60–1.20	1.6	13	11
				107	1.35–1.80	2.0	17	14
				142	1.95–2.40	2.0	23	18
				177	2.55–3.00	2.6	24	18
				191	3.15–3.60	3.2	30	26
				187	3.75–4.05	3.2	27	25
				187	4.20	3.4	26	21
L1/HD 100 WIL	102	90–105	150	113	0.60–1.35	2.9	20	17
				144	1.50–1.80	2.9	35	27
				165	1.95–2.10	2.9	30	25
				188	2.25–2.70	3.2	36	32

L1/XHD 100 — 95 88 100

L1/S 100 WIL — 95 88 125

L1/HD 100 WIL — 95 88 125

continued

(continued)

Lintel reference	Ext. leaf (mm)	Cavity (mm)	Int. leaf (mm)	Height (mm)	Available lengths (mm) (in 150 mm increments)	Gauge (mm)	Total uniformly distributed load for load ratio 1 kN	Total uniformly distributed load for load ratio 2 kNL1/S 100
L5/100[b]	102	90–105	120 max	229	0.60–2.40	2.5	48	
					2.55–3.60	3.0	50	
					3.75–4.80	3.0	38	
L6/100[c]	102	90–105	120 max	207	0.60–4.80	2.5	80	
					5.20	3.0	70	
					5.40		62	
					5.80		55	
					6.20		43	
					6.60		40	
L9i	200–215			55	0.60–1.50	2.5	6	
				55	1.65–1.80	3.0	6	
				100	1.95–2.70	3.0	10	

L10	102	60	0.60–1.20	3.0	4
		110	1.35–1.80	3.0	8
		210	1.95–2.70	3.0	10
L11	102	150	0.60–1.80	2.5	16
		225	1.95–2.40	2.5	20
		225	2.55–3.00	3.0	22

Notes:

a Indicates that a continuous bottom plate is added to lintel.
b L5 and L11 lintels are designed assuming composite action with the masonry over the lintel.
c L6 lintels are made with 203 × 133UB30 supporting the inner leaf. Minimum end bearing of 200 mm required.
d The L1/S lintel is also available as L1/S 110 for cavities 110–125 mm, L1/S 130 for 130–145 mm and L1/S 150 for 150–165 mm.
e IG can provide details for wide inner leaf lintels for cavities greater than 100 mm on request.
f Loads in tables are unfactored. A lintel should not exceed a maximum deflection of L / 325 when subject to the safe working load.
g Load ratio 1 – applies to walls with an inner to outer load ratio of between 1:1 and 3:1. This ratio is normally applicable to lintels that support masonry, or masonry and timber floors.
h Load ratio 2 – applies to walls with an inner to outer load ratio of between 4:1 and 19:1. This ratio is normally applicable to lintels that support concrete floors or are at eaves details.
i L9 lintels must be built in as shown to manufacturer's details. Prop during construction.

Source: IG Lintels Ltd. 2007.

12
Geotechnics

Geotechnics is the engineering theory of soils, foundations and retaining walls. The chapter is intended as a guide that can be used alongside information obtained from local building control officers; for feasibility purposes and for the assessment of site investigation results, scheme design should be carried out on the basis of a full site investigation designed specifically for the site and structure under consideration.

The relevant codes of practice are

- BS EN 1997-1 for Geotechnical Design
- BS EN 1997-2 for Ground Investigation and Testing

The following issues should be considered for all geotechnical problems:

- Like the rest of the Eurocode suite, Eurocodes use factored loads and characteristic strengths.
- Historic UK values are based on unfactored loads.
- Engineers not familiar with site investigation tests and their implications, soil theory and bearing capacity equations, should not use the information in this chapter without also using the sources listed in 'Further Reading' for information on theory and definitions.
- The foundation information included in this chapter allows for simplified or idealised soil conditions for preliminary sizing or scheme design purposes. In practice, soil layers and variability should be allowed for in the foundation design.
- The preliminary values in this chapter are based on unfactored loads and must have an adequate factor of safety (normally $\gamma_f = 2$–3) applied to the ultimate capacity to provide the allowable bearing pressure for design purposes.
- Settlement normally controls the design and allowable bearing pressures typically limit settlement to 25 mm. Differential settlements should be considered. Cyclic or dynamic loading can cause higher settlements to occur and therefore require higher factors of safety.
- Foundations in fine-grained soils (such as clay, silt and chalk) need to be taken down to a depth below which they will not be affected by seasonal changes in the moisture content of the soil, frost action and the action of tree roots. Frost action is normally assumed to be negligible from 450 mm below ground level. Guidelines on trees and shallow foundations are covered later in the chapter.
- Groundwater control is key to the success of ground and foundation works and its effects must be considered, both during and after construction. Dealing with water within a site may reduce the water table of surrounding areas and affect adjoining structures.
- It is nearly always cheaper to design wide shallow foundations to a uniform and predetermined depth than to excavate narrow foundations to a depth which might be variable on site.

Selection of foundations and retaining walls

The likely foundation arrangement for a structure needs to be considered so that an appropriate site investigation can be specified, but the final foundation arrangement will normally only be decided after the site investigation results have been returned.

Foundations for idealised structure and soil conditions

Foundations must always follow the building type, that is, a large-scale building needs large-scale/deep foundation. Pad and strip foundations cannot practically be taken beyond 3 m depth and these are grouped with rafts in the classification 'shallow foundation', while piles are called deep foundations. They can have diameters from 75 to 2000 mm and be 5 to 100 m in length. The smaller diameters and lengths tend to be bored cast in situ piles, while larger diameters and lengths are often driven steel piles.

Idealised extremes of structure type	Idealised soil conditions				
	Firm, uniform soil in an infinitely thick stratum	Firm stratum of soil overlying an infinitely thick stratum of soft soil	Soft uniform soil in an infinitely thick stratum	High water table and/or made ground	Soft stratum of soil overlying an infinitely thick stratum of firm soil or rock
Light flexible structure	Pad or strip footings	Pad or strip footings	Friction piles or surface raft	Piles or surface raft	Bearing piles or piers
Heavy rigid structure	Pad or strip footings	Buoyant raft or friction piles	Buoyant raft or friction piles	Buoyant raft or piles	Bearing piles or piers

Retaining walls for idealised site and soil conditions

Idealised site conditions	Idealised soil types		
	Dry sand and gravel	Saturated sand and gravel	Clay and silt
Working space[a] available	Gravity or cantilever retaining wall

Reinforced soil, gabion or crib wall | Dewatering during construction of gravity or cantilever retaining wall | Gravity or cantilever retaining wall |
| Limited working space | King post or sheet pile as temporary support

Contiguous piled wall

Diaphragm wall

Soil nailing | Sheet pile and dewatering

Secant bored piled wall

Diaphragm wall | King post or sheet pile as temporary support

Contiguous piled wall

Soil nailing

Diaphragm wall |
| Limited working space and special controls on ground movements | Contiguous piled wall

Diaphragm wall | Secant bored piled wall

Diaphragm wall | Contiguous piled wall

Diaphragm wall |

Note:
[a] Working space available to allow the ground to be battered back during wall construction.

Site investigation

In order to decide on the appropriate form of site investigation, the engineer must have established the position of the structure on the site, the size and form of the structure and the likely foundation loads.

It is good practice for the investigation to be taken to a depth of 1.5 times the width of the loaded area for shallow foundations. A loaded area can be defined as the width of an individual footing area, the width of a raft foundation, or the width of the building (if the foundation spacing is less than three times the foundation breadth). An investigation must be conducted to prove bedrock, should be taken down 3 m beyond the top of the bedrock to ensure that rock layer is sufficiently thick.

BS EN 1997-2 requires the preparation of a 'geotechnical investigation report' and that this should be included in a 'geotechnical design report'.

BS EN 1997-1 requires that a copy of the 'geotechnical design report' be provided to the client.

Summary of typical site investigation requirements for idealised soil types

| Soil type | Type of geotechnical work | | |
	Excavations	Shallow footings and rafts	Deep foundations and piles
Sand	Permeability for dewatering and stability of excavation bottom	Shear strength for bearing capacity calculations	Test pile for assessment of allowable bearing capacity and settlements
	Shear strength for loads on retaining structures and stability of excavation bottom	Site loading tests for assessment of settlements	Deep boreholes to probe zone of influence of piles
Clay	Shear strength for loads on retaining structures and stability of excavation bottom	Shear strength for bearing capacity calculations	Long-term test pile for assessment of allowable bearing capacity and settlements
	Sensitivity testing to assess strength and stability and the possibility of reusing material as backfill	Consolidation tests for assessment of settlements	Shear strength and sensitivity testing to assess bearing capacity and settlements
		Moisture content and plasticity tests to predict heave potential and effects of trees	Deep boreholes to probe zone of influence of piles

Soil classification

Soil classification is based on the sizes of particles in the soil as divided by the European Standard sieves. Clay is classified as passing the 0.002 mm sieve.

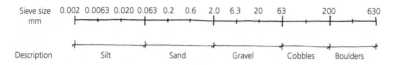

Soil description by particle size

As soils are not normally uniform, standard descriptions for mixed soils have been defined by BS EN 14688-2. The basic components are boulders, cobbles, gravel, sand, silt and clay and these are written in capital letters where they are the main component of the soil. Typically, soil descriptions are as follows:

	% Weight of fraction		Name of soil	
Fraction	≤63 mm	≤0.063 mm	Modifying	Main term
Gravel	>40	–	–	Gravel
	20–40	–	Gravelly	–
Sand	>40	–	–	Sand
	20–40	–	Sandy	–
Silt + clay	>40	>40	–	Clay
		20–40	Silty	Clay
		10–20	Clayey	Silt
		<10	–	Silt
	15–40	≥20	Clayey	–
		<20	Silty	–
	5–15	≥20	Slightly clayey	–
		<20	Slightly silty	–

Soil description by consistency

Homogeneous	A deposit consisting of one soil type
Heterogeneous	A deposit containing a mixture of soil types
Interstratified	A deposit containing alternating layers, bands or lenses of different soil types
Weathered	Coarse soils may contain weakened particles and/or particles sorted according to their size. Fine soils may crumble or crack into a 'column'-type structure
Fissured clay	Breaks into multifaceted fragments along fissures
Intact clay	Uniform texture with no fissures
Fibrous peat	Recognisable plant remains present, which retains some strength
Amorphous peat	Uniform texture, with no recognisable plant remains

Source: BS EN 14688. 2004. Table B1.

Typical soil properties

The presence of water is critical to the behaviour of soil and the choice of shear strength parameters (internal angle of shearing resistance ϕ and cohesion c) are required for geotechnical design.

If water is present in soil, applied loads are carried in the short term by pore water pressures. For granular soils above the water table, pore water pressures dissipate almost immediately as the water drains away and the loads are effectively carried by the soil structure. However, for fine-grained soils, which are not as free draining, pore water pressures take much longer to dissipate. Water and pore water pressures affect the strength and settlement characteristics of soil.

The engineer must distinguish between undrained conditions (short-term loading, where pore water pressures are present and design is carried out for total stresses on the basis of ϕ_u and C_u) and drained conditions (long-term loading, where pore water pressures have dissipated and design is carried out for effective stresses on the basis of ϕ' and c').

Drained conditions, $\phi' > 0$

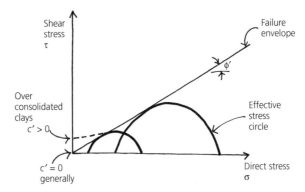

Approximate correlation of properties for drained granular soils

Description	SPT[a] N blows	Effective internal angle of shearing resistance ϕ'	Bulk unit weight γ_{bulk} (kN/m³)	Dry unit weight γ_{dry} (kN/m³)
Very loose	0–4	26–28	<16	<14
Loose	4–10	28–30	16–18	14–16
Medium dense	10–30	30–36	18–19	16–17
Dense	30–50	36–42	19–21	17–19
Very dense	>50	42–46	21	19

Note:
[a] An approximate conversion from the standard penetration test (SPT) to the Dutch cone penetration test: $C_r \approx 400\,N$ kN/m².

For saturated, dense, fine or silty sands, measured N values should be reduced by $N = 15 + 0.5(N-15)$.

Approximate correlation of properties for drained cohesive soils – The cohesive strength of fine-grained soils normally increases with depth. Drained shear strength parameters are generally obtained from very slow triaxial tests in the laboratory. The effective internal angle of shearing resistance, ϕ', is influenced by the range and distribution of fine particles, with lower values being associated with higher plasticity. For a normally consolidated clay, the effective (or apparent) cohesion, c', is zero but for an overconsolidated clay it can be up to 30 kN/m².

Soil description	Typical shrinkability	Plasticity index (PI%)	Bulk unit weight γ_{bulk} (kN/m³)	Effective internal angle of shearing resistance ϕ'
Clay	High	>35	16–22	18–24
Silty clay	Medium	25–35	16–20	22–26
Sandy clay	Low	10–25	16–20	26–34

Undrained conditions, $\phi_u = 0$

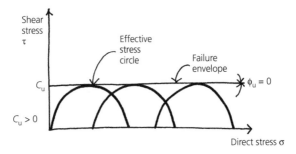

Approximate correlation of properties for undrained cohesive soils

Description	Undrained shear strength C_u (kN/m²)	Bulk unit weight γ_{bulk} (kN/m³)
Very stiff and hard clays	> 150	19–22
Stiff clays	100–150	
Firm to stiff clays	75–100	17–20
Firm clays	50–75	
Soft to firm clays	40–50	
Soft clays and silts	20–40	16–19
Very soft clays and silts	20	

It can be assumed that $C_u \approx 4.5N$ if the clay plasticity index is greater than 30, where N is the number of standard penetration test blows.

Typical values of Californian Bearing Ratio (CBR)

Type of soil	Plasticity index	Predicted CBR (%)
Heavy clay	70	1.5–2.5
	60	1.5–2.5
	50	1.5–2.5
	40	2.0–3.0
Silty clay	30	2.5–6.0
Sandy clay	20	2.5–8.0
	10	1.5–8.0
Silt	–	1.0–2.0
Sand (poorly graded)	–	20
Sand (well graded)	–	40
Sandy gravel (well graded)	–	60

Source: Highways Agency.

Typical angle of repose for selected soils – The angle of repose is very similar to, and often confused with, the internal angle of shearing resistance. The internal angle of shearing resistance is calculated from laboratory tests and indicates the theoretical internal shear strength of the soil for use in calculations while the angle of repose relates to the expected field behaviour of the soil. The angle of repose indicates the slope which the sides of an excavation in the soil might be expected to stand at. The values given below are for short-term, unweathered conditions:

Soil type	Description	Typical angle of repose	Description	Typical angle of repose
Top soil	Loose and dry	35–40	Loose and saturated	45
Loam	Loose and dry	40–45	Loose and saturated	20–25
Peat	Loose and dry	15	Loose and saturated	45
Clay/silt	Firm to moderately firm	17–19	Puddle clay	15–19
	Sandy clay	15	Silt	19
	Loose and wet	20–25	Solid naturally moist	40–50
Sand	Compact	35–40	Loose and dry	30–35
	Sandy gravel	35–45	Saturated	25
Gravel	Uniform	35–45	Loose shingle	40
	Sandy compact	40–45	Stiff boulder/hard shale	19–22
	Med coarse and dry	30–45	Med coarse	25–30
Broken rock	Dry	35	Wet	45

Preliminary sizing

Typical allowable bearing pressures under static loads

Description	Safe bearing capacity[a] (kN/m²)	Field description/notes
Strong igneous rocks and gneisses	10000	Footings on unweathered rock
Strong limestones and hard sandstones	4000	
Schists and slates	3000	
Strong shales and mudstones	2000	
Hard block chalk	80–600	Beware of sink holes and hollowing as a result of water flow
Compact gravel and sandy gravel[b]	>600	Requires pneumatic tools for excavation
Medium dense gravel and sandy gravel[b]	200–600	Hand pick – Resistance to shoveling
Loose gravel and sandy gravel[b]	<200	Small resistance to shoveling
Compact sand[b]	>300	Hand pick – Resistance to shoveling
Medium dense sand[b]	100–300	Hand pick – Resistance to shoveling
Loose sand[b]	<100	Small resistance to shoveling
Very stiff and hard clays	300–600	Requires pneumatic spade for excavation but can be indented by the thumbnail
Stiff clays	150–300	Hand pick – Cannot be moulded in hand but can be indented by the thumb
Firm clays	75–150	Can be moulded with firm finger pressure
Soft clays and silts	<75	Easily moulded with firm finger pressure
Very soft clays and silts	Nil	Extrudes between fingers when squeezed
Firm organic material/medieval fill	20–40	Can be indented by thumbnail. Only suitable for small-scale buildings where settlements may not be critical
Unidentifiable made ground	25–50	Bearing values depend on the likelihood of voids and the compressibility of the made ground
Springy organic material/peats	Nil	Very compressible and open structure
Plastic organic material/peats	Nil	Can be moulded in the hand and smears the fingers

Notes:
[a] This table should be read in accordance with the limitations of BS 8004.
[b] Values for granular soil assume that the footing width, B, is not less than 1 m and that the water table is more than B below the base of the foundation.

Source: BS 8004: 1986.

Quick estimate design methods for shallow foundations

General equation for allowable bearing capacity after Brinch Hansen

$$\text{Strip footings: } q_{allowable} = \frac{cN_c + q'_o N_q + 0.5\gamma BN_\gamma}{\gamma_f}$$

$$\text{Pad footings: } q_{allowable} = \frac{1.3cN_c + q'_o N_q + 0.4\gamma BN_\gamma}{\gamma_f}$$

Factor of safety against bearing capacity failure, $\gamma_f = 2.0$ to 3.0, q'_o is the effective overburden pressure, γ is the unit weight of the soil, B is the width of the foundation, c is the cohesion (for the drained or undrained case under consideration) and N_c, N_q and N_γ are shallow bearing capacity factors. Approximate values for the bearing capacity factors N_c, N_q and $N\gamma$ are set out below in relation to ϕ:

Internal angle of shear ϕ	Bearing capacity factors[a]		
	N_c	N_q	N_γ
0	5.0	1.0	0.0
5	6.5	1.5	0.0
10	8.5	2.5	0.0
15	11.0	4.0	1.4
20	15.5	6.5	3.5
25	21.0	10.5	8.0
30	30.0	18.5	17.0
35	45.0	34.0	40.0
40	75.0	65.0	98.0

Note:
[a] Values from charts by Brinch Hansen (1961).

Simplified equations for allowable bearing capacity after Brinch Hansen

For very preliminary design, Terzaghi's equation can be simplified for uniform soil in thick layers.

Spread footing on clay – $q_{allowable} = 2C_u$ spread footing on undrained cohesive soil ($\gamma_f = 2.5$).

Spread footing on gravel

$q_{allowable} = 10N$	Pad footing on dry soil ($\gamma_f = 3$)
$q_{allowable} = 7N$	Strip footing on dry soil ($\gamma_f = 3$)
$q_{wet\ allowable} = q_{allowable}/2$	Spread foundation at or below the water table

Where N is the standard penetration test (SPT) value.

Quick estimate design methods for deep foundations

Concrete and steel pile capacities – Concrete piles can be cast in situ or precast, prestressed or reinforced. Steel piles are used where long or lightweight piles are required. Sections can be butt welded together and the excess can be cut away. Steel piles have good resistance to lateral forces, bending and impact, but they can be expensive and need corrosion protection.

Typical maximum allowable pile capacities can be 300–1800 kN for bored piles (diameter 300–600 mm), 500–2000 kN for driven piles (275–400 mm square precast or 275–2000 mm diameter steel), 300–1500 kN for continuous flight auger (CFA) piles (diameter 300–600 mm) and 50–500 kN for mini piles (diameter 75–280 mm and length up to 20 m). The minimum pile spacing achievable is normally about three diameters between the pile centres.

Working pile loads for CFA piles in granular soil ($N = 15$), $\phi = 30°$

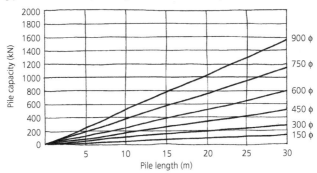

Working pile loads for CFA piles in granular soil ($N = 25$), $\phi = 35°$

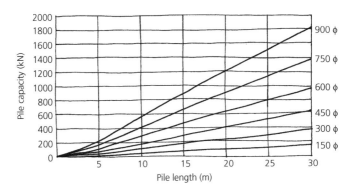

Working pile loads for CFA piles in cohesive soil ($C_u = 50$)

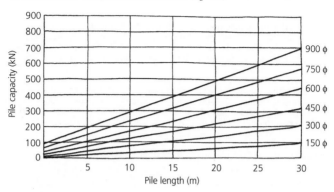

Working pile loads for CFA piles in cohesive soil ($C_u = 100$)

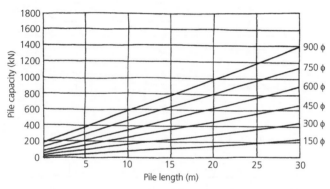

Single bored piles in clay

$$Q_{allow} = \frac{N_c A_b C_{base}}{\gamma_{fbase}} + \frac{\alpha \bar{c} A_s}{\gamma_{fshaft}}$$

where A_b is the area of the pile base, A_s is the surface area of the pile shaft in the clay, $\bar{c}$ is the average value of shear strength over the pile length and is derived from undrained triaxial tests, where $\alpha = 0.3$–0.6 depending on the time that the pile boring is left open. Typically, $\alpha = 0.3$ for heavily fissured clay and $\alpha = 0.45$–0.5 for firm to stiff clays (e.g. London clay). $N_c = 9$ where the embedment of the tip of the pile into the clay is more than five diameters. The factors of safety are generally taken as 2.5 for the base and 3.0 for the shaft.

Group action of bored piles in clay – The capacity of groups of piles can be as little as 25% of the collective capacity of the individual piles.

A quick estimate of group efficiency:

$$E = 1 - \left(\tan^{-1} \frac{D}{S} \right) \frac{[m(n-1) + n(m-1)]}{90mn}$$

where D is the pile diameter, S is the pile spacing and m and n represent the number of rows in two directions of the pile group.

Negative skin friction – Negative skin friction occurs when piles have been installed through a compressible material to reach firm strata. Cohesion in the soft soil will tend to drag down on the piles as the soft layer consolidates and compresses, causing an additional load on the pile. This additional load is due to the weight of the soil surrounding the pile. For a group of piles, a simplified method of assessing the additional load per pile can be based on the volume of soil which would need to be supported on the pile group. $Q_{skin\ friction} = AH\gamma/N_p$, where A is the area of the pile group, H is the thickness of the layer of consolidating soil or fill which has a bulk density of γ, and N_p is the number of piles in the group. The chosen area of the pile group will depend on the arrangement of the piles and could be the area of the building or part of the building. This calculation can be applied to individual piles, although it can be difficult to assess how much soil could be considered to contribute to the negative skin friction forces.

Piles in granular soil – Although most methods of determining driven pile capacities require information on the resistance of the pile during driving, capacities for both driven and bored piles can be estimated by the same equation. The skin friction and end bearing capacity of bored piles will be considerably less than driven piles in the same soil as a result of loosening caused by the boring, and design values of γ, N and $k_s \tan \delta$ should be selected for loose conditions.

$$Q_{allow} = \frac{N_q^* A_b q_o' + A_s q_{o\,mean}' k_s \tan\delta}{\gamma_f}$$

where N_q^* is the pile bearing capacity factor based on the work of Berezantsev, A_b is the area of the pile base, A_s is the surface area of the pile shaft in the soil, q_o' is the effective overburden pressure, k_s is the horizontal coefficient of earth pressure, K_0 is the coefficient of earth pressure at rest, δ is the angle of friction between the soil and the pile face, ϕ' is the effective internal angle of shearing resistance and the factor of safety, $\gamma_f = 2.5$–3.

Typical values of N_q[a]	Pile length / Pile diameter		
ϕ	5	20	70
25	16	11	7
30	29	24	20
35	69	53	45
40	175	148	130

[a] Berezantsev (1961) values from charts for N_q based on ϕ calculated from uncorrected N values.

Typical values of δ and k_s for sandy soils can therefore be determined based on the work by Kulhawy (1984) as follows:

Pile face/soil type	Angle of pile/soil friction δ/ϕ
Smooth (coated) steel/sand	0.5–0.9
Rough (corrugated) steel/sand	0.7–0.9
Cast in place concrete/sand	1.0
Precast concrete/sand	0.8–1.0
Timber/sand	0.8–0.9

Installation and pile type	Coefficients of horizontal soil stress/earth pressure at rest k_s/k_o
Driven piles large displacement	1.00–2.00
Driven piles small displacement	0.75–1.25
Bored cast in place piles	0.70–1.00
Jetted piles	0.50–0.70

Although pile capacities improve with depth, it has been found that at about 20 pile diameters, the skin friction and base resistances stop increasing and 'peak' for granular soils. Generally, the peak value for base bearing capacity is 110,000 kN/m² for a pile length of 10–20 pile diameters and the peak values for skin friction are 10 kN/m² for loose granular soil, 10–25 kN/m² for medium dense granular soil, 25–70 kN/m² for dense granular soil and 70–110 kN/m² for very dense granular soil.

Source: Kulhawy, F.H. 1984 – reproduced by permission of the ASCE.

Pile caps – Pile caps transfer the load from the superstructure into the piles and take up tolerances on the pile position (typically ±75 mm). The pile cap normally projects 150 mm in plan beyond the pile face and if possible, only one depth of pile cap should be used on a project to minimise cost and labour. The Federation of Piling Specialists suggests the following pile cap thicknesses which generally mean that the critical design case will be for the sum of all the pile forces to one side of the cap centre line, rather than the punching shear:

Pile diameter (mm)	300	350	400	450	500	550	600	750
Pile cap depth (mm)	700	800	900	1000	1100	1200	1400	1600

Retaining walls

Rankine's theory on lateral earth pressure is most commonly used for retaining wall design, but Coulomb's theory is easier to apply for complex loading conditions. The most difficult part of Rankine's theory is the appropriate selection of the coefficient of lateral earth pressure, which depends on whether the wall is able to move. Typically, where sufficient movement of a retaining wall is likely and acceptable, 'active' and 'passive' pressures can be assumed, but where movement is unlikely or unacceptable, the earth pressures should be considered 'at rest'. Active pressure will be mobilised if the wall moves 0.25–1% of the wall height, while passive pressure will require movements of 2–4% in dense sand or 10–15% in loose sand. As it is normally difficult to assume that passive pressure will be mobilised, unless it is absolutely necessary for stability (e.g. embedded walls), the restraining effects of passive pressures are often ignored in analysis. The main implications of Rankine's theory are that the engineer must predict the deflected shape, to be able to predict the forces which will be applied to the wall.

Rankine's theory assumes that movement occurs, that the wall has a smooth back, that the retained ground surface is horizontal and that the soil is cohesionless, so that $\sigma_h = K_a v$.

For soil at rest, $k = k_o$; for active pressure, $k = k_a$; and for passive pressure, $k = k_p$.

$$k_o \approx 1 - \sin\phi \quad k_a = \frac{(1 - \sin\phi)}{(1 + \sin\phi)} \quad k_p = \frac{1}{k_a} = \frac{(1 + \sin\phi)}{(1 - \sin\phi)}$$

For cohesive soil, k_o should be factored by the overconsolidation ratio:

$$OCR = \sqrt{\frac{\text{Pre-consolidation pressure}}{\text{Effective overburden pressure}}}$$

Typical k_o values are 0.35 for dense sand, 0.6 for loose sand, 0.5–0.6 for normally consolidated clay and 1.0–2.8 for overconsolidated clays such as London clay. The value of k_o depends on the geological history of the soil and should be obtained from a geotechnical engineer.

Rankine's theory can be adapted for cohesive soils, which can shrink away from the wall and reduce active pressures at the top of the wall as a tension 'crack' forms. Theoretically, the soil pressures over the height of the tension crack can be omitted from the design, but in practice, the crack is likely to fill with water, rehydrate the clay and remobilise the lateral pressure of the soil. The height of the crack is $h_c = 2c'/(\gamma\sqrt{k_a})$ for drained conditions and $h_c = 2C_u/\gamma$ for undrained conditions.

Preliminary sizing of retaining walls

Gravity retaining walls: Typically, they have a base width of about 60–80% of the retained height.

Propped embedded retaining walls: There are 16 methods for the design of these walls depending on whether they are considered flexible (sheet piling) or rigid (concrete diaphragm). A reasonable approach is to use the Free Earth Support method which takes moments about the prop position, followed by the Burland and Potts method as a check. Any tension crack height is limited to the position of the prop.

Embedded retaining walls: They must be designed for fixed earth support where passive pressures are generated on the rear of the wall, at the toe. An approximate design method is to design the wall with free earth support by the same method as the propped wall but with moments taken at the foot of the embedded wall, before adding 20% extra depth as an estimate of the extra depth required for the fixed earth condition.

Introduction to geotechnical design to BS EN 1997

In previous editions of the *Structural Engineer's Pocket Book*, the geotechnical chapter did not cover the codified aspects of geotechnical design, as most practising engineers only need enough detail to prepare preliminary designs, or to broadly understand what their geotechnical consultants and advisors are doing. During the transition from British Standards to Eurocodes, some readers might benefit from a short introduction to BS EN 1997 Geotechnical Design.

Like the rest of the Eurocode suite, BS EN 1997 uses partial factors and ultimate and serviceability limit states. This is a radical departure from traditional UK practice (as embodied in British Standards such as BS 8004) which historically used global factors of safety. The advantage of this is that foundation design and loadings are directly compatible with the codes of practice for superstructure design. The processes for obtaining derived values are essentially the same as current good practice and BS EN 1997 could be considered simply as attempting to codify these processes.

A raft of British Standards have been withdrawn and replaced with just the two parts of BS EN 1997:

BS EN 1997-1 Geotechnical Design (General Rules) which contains the following sections:

- Section 1 General
- Section 2 Basis of geotechnical design
- Section 3 Geotechnical data
- Section 4 Supervision of construction, monitoring and maintenance
- Section 5 Fill, dewatering, ground improvement and reinforcement
- Section 6 Spread foundations
- Section 7 Pile foundations
- Section 8 Anchorages
- Section 9 Retaining structures
- Section 10 Hydraulic failure
- Section 11 Site stability
- Section 12 Embankments
- Annexes A to J

BS EN 1997-2 Geotechnical Design (Ground Investigation and Testing) which contains the following sections:

- Section 1 General
- Section 2 Planning of ground investigations
- Section 3 Soil and rock sampling and groundwater measurements
- Section 4 Field tests in soil and rock
- Section 5 Laboratory tests on soil and rock
- Section 6 Ground investigation report
- Annexes A to X

BS EN 1997-1 describes the general Principles and Application Rules for geotechnical design, to ensure 'safety' (adequate strength and stability), 'serviceability' (acceptable movement and deformation) and 'durability' of structures founded on soil or rock.

Together, the key features of BS EN 1997 provide a single set of guiding principles for all geotechnical designs that is absent in the previous, diverse set of British Standard geotechnical design codes:

- Eurocodes state that Principles shall be honoured, while British Standards provided guidance which was more flexible.
- BS EN 1997-1 uses a design methodology that makes substructure and superstructure design fully compatible when using the Eurocode suite.
- BS EN 1997-1 explicitly identifies design limit states making a clear distinction between ultimate limit state (failure of the ground or collapse of a supported structure) and serviceability limit state (the consequences of deflections and deformations) where traditional practice has been less clear.
- Eurocodes use partial factors applied to characteristic values of parameters in calculations. This allows more sophistication when allowing for uncertainty in parameters than the use of the global factors of safety previously used in British Standards. However, it also requires much more skill and knowledge on the part of the designer.
- The formal adoption of the four commonly used alternative methods for achieving a geotechnical design: use of prescriptive measures, tests on models or full-scale tests, the observational method or calculations.
- BS EN 1997-1 makes it compulsory to provide a geotechnical design report to the client, while BS EN 1997-2 requires this report to include a ground investigation report.

BS EN 1997 applies to the design of new projects, and works to existing geotechnical structures. It does not, however, specifically deal with the re-use of existing foundations nor does it apply to the assessment of existing structures. The main guidance applies to greenfield sites and while 'clean' fill is covered, contaminated land is not. There is also no guidance on the design of reinforced earth structures and soil nailing, which must still be designed using BS 8006, BS EN 14475 and BS EN 14490.

BS EN 1997-1 is not a detailed geotechnical design manual but is intended to provide a framework for design and for checking that a design will not reach a 'limiting condition' in prescribed 'design situations'. The code therefore provides all the general requirements for conducting and checking design, but only gives limited assistance or information on how to perform design calculations and further detail is required from other non-contradictory complementary information (NCCI).

The following is a list of references that contain NCCI for use with BS EN 1997-1:2004: BS 1377, BS 5930, BS 6031, BS 8002, BS 8004, BS 8008, BS 8081, PD 6694-11, CIRIA C580 and the UK Design Manual for Roads and Bridges. Although some of the aspects of these NCCI will be in conflict with BS EN 1997, the Eurocode will take precedence until 'residual' documents are prepared to remove these conflicts.

So unlike other codes in the Eurocode suite, many of the British Standards remain current and applicable at present. Therefore, the transition period from British Standards to Eurocodes is even more complicated than other Eurocodes and there is ongoing debate about how and where traditional UK good practice guidance might be published.

Trees and shallow foundations

Trees absorb water from the soil which can cause consolidation and settlements in fine-grained soils. Shallow foundations in these conditions may be affected by these settlements and the National House Building Council (NHBC) publishes guidelines on the depth of shallow foundations on silt and clay soils to take the foundation to a depth beyond the zone of influence of tree roots. The information reproduced here is current in 2012 but the information may change over time and amendments should be checked with NHBC.

The effect depends on the plasticity index of the soil, the proximity of the tree to the foundation, the mature height of the tree and its water demand. Where the plasticity index of the soil is not known, assume high plasticity.

Plasticity index PI = Liquid limit – plastic limit		Zone of tree influence (D is mature height of tree) (m)	Minimum foundation depth outside zone of influence (m)
Low	10–20%	0.50 D	0.75
Medium	20–40%	0.75 D	0.9
High	>40%	1.25 D	1.0

Source: NHBC. 2013. The information may change at any time and revisions should be checked with NHBC.

Water demand and mature height of selected UK trees

The following common British trees are classified as having high, moderate or low water demand. Where the tree cannot be identified, assume high water demand.

Water demand	Broad-leaved trees				Conifers	
	Species	Mature height[a] (m)	Species	Mature height[a] (m)	Species	Mature height[a] (m)
High	Elm	18–24	Poplar	15–28	Cypress	18–20
	Eucalyptus	18	Willow	16–24		
	Oak	16–24	Hawthorn	10		
Moderate	Acacia false	18	Lime	22	Cedar	20
	Alder	18	Maple	8–18	Douglas fir	20
	Apple	10	Mountain ash	11	Larch	20
	Ash	23	Pear	12	Monkey puzzle	18
	Bay laurel	10	Plane	26	Pine	20
	Beech	20	Plum	10	Spruce	18
	Blackthorn	8	Sycamore	22	Wellingtonia	30
	Cherry	8–17	Tree of heaven	20	Yew	12
	Chestnut	20–24	Walnut	18		
			Whitebeam	12		
Low	Birch	14	Hornbeam	17		
	Elder	10	Laburnum	12		
	Fig	8	Magnolia	9		
	Hazel	8	Mulberry	9		
	Holly	12	Tulip tree	20		
	Honey locust	14				

Notes:
[a] For range of heights within species, see the full NHBC source table for complete details.
[b] Where species is known, but the subspecies is not, the greatest height should be assumed.
[c] Further information regarding trees and water demand is available from the Arboricultural Association or the Arboricultural Advisory and Help Service.

Source: NHBC. 2013. The information may change at any time and revisions should be checked with NHBC.

Suggested depths for foundations on cohesive soil

If D is the distance between the tree and the foundation, and H is the mature height of the tree, the following three charts (based on soil shrinkability) will estimate the required foundation depth for different water demand classifications. The full NHBC document allows for a reduction in the foundation depth for climatic reasons, for every 50 miles from the south-east of England.

Suggested depths for foundations on highly shrinkable soil

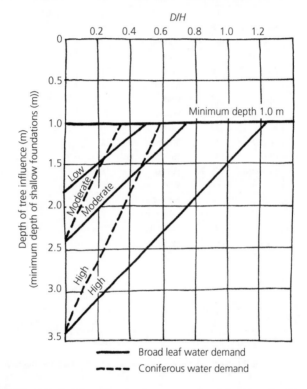

Broad leaf water demand
Coniferous water demand

Source: NHBC. 2013. The information may change at any time and revisions should be checked with NHBC.

Suggested depths for foundations on medium shrinkable soil

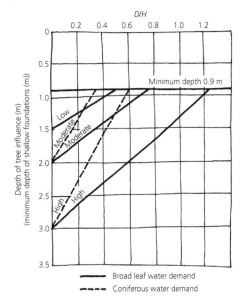

Suggested depths for foundations on low shrinkable soil

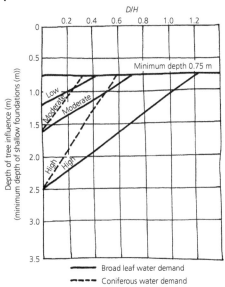

Source: NHBC. 2013. The information may change at any time and revisions should be checked with NHBC.

Contaminated land

Contamination can be present as a result of pollution from previous land usage or movement of pollutants from neighbouring sites by air or groundwater. The main categories of contamination are chemical, biological (pathological bacteria) and physical (radioactive, flammable materials etc.).

The Environmental Protection Act 1990 (in particular, Part IIA) is the primary legislation covering the identification and remediation of contaminated land. The Act defines contamination as solid, liquid, gas or vapour which might cause harm to 'targets'. This can mean harm to the health of living organisms or property, or other interference with ecological systems. The contamination can be on, in or under the land. The Act applies if the contamination is causing, or will cause, significant harm or results in the pollution of controlled waters, including coastal, river and groundwater. In order to cause harm, the pollution must have some way (called a 'pathway') of reaching the 'target'. The amount of harm which can be caused by contamination will depend on the proposed use for the land. Remediation of contaminated land can remove the contamination, reduce its concentrations below acceptable levels, or remove the 'pathway'.

The 1990 Act set up a scientific framework for assessing the risks to human health from land contamination. This has resulted in Contaminated Land Exposure Assessment (CLEA) and development of Soil Guideline Values for residential, allotment or industrial/commercial land use. Where contaminant concentration levels exceed the Soil Guideline Values, further investigation and/or remediation is required. Reports are planned for a total of 55 contaminants and some are available on the Environment Agency website. Without the full set, assessment is frequently made using Guideline Values from the Netherlands. Other frequently mentioned publications are Kelly and the now superseded ICRCL list. Zero Environment has details of the ICRCL, Kelly and Dutch lists on its website.

Before developing a 'brownfield site' (i.e. a site which has previously been used) a desk study on the history of the site should be carried out to establish its previous uses and therefore likely contaminants. Sampling should then be used to establish the nature and concentration of any contaminants. Remedial action may be dictated by law, but should be feasible and economical on the basis of the end use of the land.

Common sources of contamination

Specific industries can be associated with particular contaminants and the site history is invaluable in considering which soil tests to specify. The following list is a summary of some of the most common sources of contamination:

Common contaminants	Possible sources of contaminants
'Toxic or heavy metals' (cadmium, lead, arsenic, mercury etc.)	Metal mines; iron and steelworks; foundries and electroplating
'Safe' metals (copper, nickel, zinc etc.)	Anodising and galvanising; engineering/ship/scrap yards
Combustible materials such as coal and coke dust	Gas works; railways; power stations; landfill sites
Sulphides, chlorides, acids and alkalis	Made-up ground
Oily or tarry deposits and phenols	Chemical refineries; chemical plants; tar works
Asbestos	Twentieth century buildings

Effects of contaminants

Effect of contaminant	Typical contaminants
Toxic/narcotic gases and vapours	Carbon monoxide or dioxide, hydrogen sulphide, hydrogen cyanide, toluene, benzene
Flammable and explosive gases	Acetylene, butane, hydrogen sulphide, hydrogen, methane, petroleum hydrocarbons
Flammable liquids and solids	Fuel oils, solvents; process feedstocks, intermediates and products
Combustible materials	Coal residues, ash timber, variety of domestic commercial and industrial wastes
Possible self-igniting materials	Paper, grain, sawdust – microbial degradation of large volumes if sufficiently damp
Corrosive substances	Acids and alkalis; reactive feedstocks, intermediates and products
Zootoxic metals and their salts	Cadmium, lead, mercury, arsenic, beryllium and copper
Other zootoxic metals	Pesticides, herbicides
Carcinogenic substances	Asbestos, arsenic, benzene, benzo(a)pyrene
Substances resulting in skin damage	Acids, alkalis, phenols, solvents
Phytotoxic metals	Copper, zinc, nickel, boron
Reactive inorganic salts	Sulphate, cyanide, ammonium, sulphide
Pathogenic agents	Anthrax, polio, tetanus, Weils
Radioactive substances	Waste materials from hospitals, mine workings, power stations and so on
Physically hazardous materials	Glass, blades, hospital wastes – needles, and so on
Vermin and associated pests	Rats, mice and cockroaches (contribute to pathogenic agents)

Note: Zootoxic means toxic to animals and phytotoxic means toxic to plants.

Site investigation and sampling

Once a desk study has been carried out and the most likely contaminants are known, an assessment must be carried out to establish the risks associated with the contaminants and the proposed land use. These two factors will determine the maximum concentrations of contaminants which will be acceptable. These maximum concentrations are the Soil Guideline Values published by the Department for the Environment, Food and Rural Affairs as part of the CLEA range of documents.

Once the soil guideline or trigger values have been selected, laboratory tests can be commissioned to discover if the selected soil contaminants exist, as well as their concentration and their distribution over the site. Reasonably accurate information can be gathered about the site using a first stage of sampling and testing to get a broad picture and a second stage to define the extents of localised areas of contamination.

Sampling on a rectangular grid with cores of 100 mm diameter, it is difficult to assess how many samples might be required to get a representative picture of the site. British Standards propose 25 samples per 10,000 m^2 which is only 0.002% of the site area. This would only give a 30% confidence of finding a 100 m^2 area of contamination on the site, while 110 samples would give 99% confidence. It is not easy to balance the cost and complexity of the site investigations and the cost of any potential remedial work, without an appreciation of the extent of the contamination on the site!

Remediation techniques

There are a variety of techniques available depending on the contaminant and target user. The chosen method of treatment will not necessarily remove all of a particular contaminant from a site as in most circumstances it may be sufficient to reduce the risk to below the predetermined trigger level. In some instances, it may be possible to change the proposed layout of a building to reduce the risk involved. However, if the site report indicates that the levels of contaminant present in the soil are too high, four main remediation methods are available.

Excavation

Excavation of contaminated soil for specialist disposal or treatment (possibly in a specialist land-fill site) and reconstruction of the site with clean fill material. This is expensive and the amount of excavated material can sometimes be reduced, by excavating down to a limited 'cut-off' level, before covering the remaining soil with a barrier and thick granular layer to avoid seepage/upward migration. Removal of soil on restricted sites might affect existing, adjacent structures.

Blending

Clean material is mixed into the bulk of the contaminated land to reduce the overall concentrations taking the test samples below trigger values. This method can be cost-effective if some contaminated soil is removed and replaced by clean imported fill, but it is difficult to implement and the effects on adjacent surfaces and structures must be taken into account.

Isolation

Isolation of the proposed development from the contaminants can be attempted by displacement sheet piling, capping, horizontal/vertical barriers, clay barriers, slurry trenches or jet grouting. Techniques should prevent contaminated soil from being brought out of the ground to contaminate other areas.

Physical treatment

Chemical or biological treatment of the soil so that the additives bond with and reduce the toxicity of, or consume, the contaminants.

13
Structural Glass

Structural glass assemblies are those in which the self-weight of the glass, wind and other imposed loads is carried by the glass itself rather than by glazing bars, mullions or transoms, and the glass elements are connected together by mechanical connections or by adhesives.

Despite the increasing use of glass as a structural material over the last 25 years, there is no single code of practice that covers all of the issues relating to structural glass assemblies. Therefore, values for structural design must be based on first principles, research, experience and load testing. The design values given in this chapter should be used very carefully with this in mind.

The following issues should be considered:

- Glass is classed as a rigid liquid as its intermolecular bonds are randomly arranged, rather than the crystalline arrangement normally associated with solids. Glass will behave elastically until the applied stresses reach the level where the interatomic bonds break. If sufficient stress is applied, these cracks will propagate and catastrophic failure will occur. The random arrangement of the interatomic bonds means that glass is not ductile and therefore failure is sudden.
- Cracks in glass propagate faster as temperature increases.
- Without the ability to yield, or behave plastically, glass can fail due to local over-stressing. Steel can redistribute high local stresses by local yielding and small plastic deformations, but glass cannot behave like this and high local stresses will result in brittle failure.
- Modern glass is not thought to deform or creep under long-term stresses. It behaves perfectly elastically and will return to its original shape when applied loads are removed. However, some old glass has been found to creep.
- Glass will generally fail as a result of the build-up of tensile stresses. Generally, it is the outer surfaces of the glass that are subject to these stresses. Small flaws on glass surfaces encourage crack propagation, which can lead to failure. Structural glass should be carefully checked for flaws.
- Annealed glass can also fail as a result of 'static fatigue'. There are various theories on why this occurs, but in simple terms, microcracks form and propagate under sustained loads resulting in the failure of the glass. This means that the strength of glass is time dependent; in the short term, glass can carry about twice the load that it can carry in the long term. Long-term stresses are kept low in design to prevent propagation of cracks. There is a finite time for static fatigue failure to occur for each type of glass and although this is beyond the scope of this book, this period can be calculated. It is about 15 days for borosilicate glass (better known as Pyrex), but is generally much longer for annealed glass.
- Thermal shock must also be considered for annealed glass. Temperature differences across a single sheet of glass can result in internal stresses. Glass elements that are partly in direct sunlight and partly shaded are at most risk of failure. Thermal shock cracks tend to start at the edge of the glass, travelling inwards at about 90°, but this type of failure can depend on many things including edge restraint, and manufacturers should be consulted for each situation. If thermal shock is expected to be an issue, toughened or tempered glass should be specified in place of annealed.
- Glass must come from a known and reliable source to provide reliable strength and minimal impurities.

Types of glass products

Annealed/float glass – Glass typically consists of frit: sand (silica 72%), soda ash (sodium carbonate 13%), limestone (calcium carbonate 10%) and dolomite (calcium magnesium carbonate 4%). This mixture is combined with broken glass (called cullet) at about 80% frit to 20% cullet, and is heated to 1500°C to melt it. It leaves the first furnace at about 1050°C and goes on to the forming process.

There are a number of forming processes, but structural glass is generally produced by the float glass method, which was developed in 1959 by Pilkington. The molten glass flows out of the furnace on to a bed of molten tin in a controlled atmosphere of nitrogen and oxygen and is kept at high temperature. This means that defects and distortions are melted out of the upper and lower surfaces without grinding and polishing. The glass is progressively cooled as it is moved along the bath by rollers until it reaches about 600°C when the glass sheet becomes viscous enough to come into contact with the rollers without causing damage to the bottom surface. The speed at which the ribbon of glass moves along the tin bath determines the thickness of the glass sheet. Finally, the glass is cooled in a gradual and controlled manner to 200°C in the annealing bay. The term 'annealed' means that the glass has been cooled carefully to prevent the build-up of residual stresses. The surfaces of float glass can be described by using the descriptions 'tin side' and 'air side' depending on which way the glass was lying in the float bath. For use as structural glass, the material should be free of impurities and discoloration. At failure, annealed glass typically breaks into large pieces with sharp edges. Annealed glass must therefore be specified carefully so that on failure it will not cause injury.

Toughened/fully tempered glass – Sheets of annealed glass are reheated to their softening point at about 700°C before being rapidly cooled. This can be done by hanging the glass vertically gripped by tongs with cooling applied by air jets, or by rolling the glass through the furnace and cooling areas. The rapid cooling of the glass causes the outer surfaces to contract quicker than the inner core. This means that a permanent precompression is applied to the glass, which can make its capacity for tensile stress three to four times better than annealed glass. The strength of toughened glass can depend on its country of origin. The values quoted in this book relate to typical UK strengths. The 'tin side' of toughened glass can be examined using polarising filters to determine the residual stresses and hence the strength of the glass. Toughened glass cannot be cut or drilled after toughening; therefore, glass is generally toughened for specific projects rather than being kept as a stock item. Toughened glass is more resistant to temperature differentials within elements than annealed glass and therefore it tends to be used externally in elements such as floors where annealed glass would normally be used in internal situations. Toughened glass can fail as a result of nickel sulphide impurities as described in the section 'Heat-Soaked Glass'. If specified to BS 6206, toughened glass is regarded as a safety glass because it fractures into small cubes (40 particles per 50 mm square) without sharp edges when broken. However, these cubes have the same density as crushed concrete and the design should prevent broken glass falling out of place to avoid injury to the public.

Partly toughened/heat-tempered glass – Sheets of annealed glass are heated and then cooled in the same way as the toughening process; however, the cooling is not as rapid. This means that slightly less permanent precompression is applied to the glass, which will make its capacity for tensile stress 1.5–2 times better than annealed glass. The residual strength can be specified depending on the proposed use. Heat-tempered glass will not fail as a result of nickel sulphide impurities as described for toughened glass in the section 'Heat-Soaked Glass'. The fracture pattern is very similar to that of annealed glass, as the residual stresses are not quite enough to shatter the glass into the small cubes normally associated with toughened glass. Tempered glass can be laminated to the top of toughened glass to produce units resistant to thermal shock, which can remain in place with a curved shape even when both sheets of glass have been broken.

Heat-soaked glass – The float glass process leaves invisible nickel sulphide (NS) impurities in the glass called inclusions. When the glass is toughened, the heat causes these inclusions to become smaller and unstable. After toughening, and often after installation, thermal movements and humidity changes can cause the NS inclusions in the glass to revert to their original form by expanding. This expansion causes the glass to shatter and failure can be quite explosive. Heat soaking can be specified to reduce the risk of NS failure of toughened glass by accelerating the natural phenomenon. This accelerated fatigue will tend to break flawed glass during a period of prolonged heating at about 300°C. Heat-soaking periods are the subject of international debate and range between 2 and 8 h. The German DIN standard is considered the most reliable code of practice. Glass manufacturers indicate that one incidence of NS failure is expected in every 4 tonnes of toughened glass, but after heat soaking this is thought to reduce to about 1 in 400 tonnes.

Laminated glass – Two or more sheets of glass are bonded, or laminated, together using plastic sheet or liquid resin interlayers. The interlayer is normally polyvinyl butyral built up in sheets of 0.38 mm and can be clear or tinted. It is normal to use four of these layers (about 12 mm) in order to allow for the glass surface ripple which is produced by the rollers used in the float glass manufacturing process. Plastic sheets are used for larger numbers and sizes of panels. Liquid resins are more suited to curved glass and to small-scale manufacturing, as the glass sheets have to be kept spaced apart in order to obtain a uniform thickness of resin between the sheets. The bonding is achieved by applying heat and pressure in an autoclave. If the glass breaks in service, the interlayer tends to hold the fragmented glass to the remaining sheet until the panel can be replaced. Laminates can be used for safety, bullet proofing, solar control and acoustic control glazing. Toughened, tempered, heat-soaked and annealed glass sheets can be incorporated and combined in laminated panels. Ideally, the glass should be specified so that toughened or tempered glass is laminated with the 'tin side' of the glass outermost, so that the glass strength can be inspected if necessary. Laminated panels tend to behave monolithically for short time loading at temperatures below 70°C, but interlayer creep means that the layers act separately under long-term loads.

Summary of material properties

Density	25–26 kN/m³
Compressive strength	$F_{cu} = 1000$ N/mm²

Tensile strength: Strength depends on many factors, including duration of loading, rate of loading, country of manufacture, residual stresses, temperature, size of cross section, surface finish and microcracks. Fine glass fibres have tensile strengths of up to 1500 N/mm² but for the sections used in structural glazing, typical characteristic tensile strengths are 45 N/mm² for annealed, 80 N/mm² for tempered glass and 120 N/mm² for toughened. Patterned or wired glass can carry less load.

Modulus of elasticity	70–74 kN/mm²
Poisson's ratio	0.22–0.25
Linear coefficient of thermal expansion	8×10^{-6}/°C

Typical glass section sizes and thicknesses

The range of available glass section sizes changes as regularly as the plant and facilities in the glass factories are updated or renewed. There are no standard sizes, only maximum sizes. Manufacturers should be contacted for up-to-date information about the sheet sizes available. The contact details for Pilkington, Solaglass Saint Gobain, Firman, Hansen Glass, QTG, European, Bishoff Glastechnik and Eckelt are listed in the chapter on useful addresses. Always check that the required sheet size can be obtained and installed economically.

Annealed/float glass

The typical maximum size is 3210 mm × 6000 mm although sheets up to 3210 mm × 8000 mm can be obtained on special order or from continental glass manufacturers.

Typical float glass thicknesses

Thickness (mm)	3[a]	4	5[a]	6	8[a]	10	12	15	19	25
Weight (kg/m²)	7.5	10.0	12.5	15.0	20.0	25.0	30.0	37.5	47.5	62.5

Note:
[a] Generally used in structural glazing laminated units.

Toughened/fully tempered glass

The sizes of toughened sheets are generally smaller than the sizes of float glass available. Toughened glass in 25 mm is currently still only experimental and is generally not available.

Thickness mm	4	5	6	8	10	12	15	19
Sheet size[a] mm × mm	1500 × 2200	2000 × 4200	2000 × 4200	2000 × 4200	2000 × 4200	2000 × 4200	1700 × 4200	1500 × 4200

Note:
[a] Larger sizes are available from certain UK and European suppliers.

Heat-tempered/partly toughened glass

They are normally only produced in 8-mm-thick sheets for laminated units. Manufacturers should be consulted about the availability of 10 and 12 mm sheets. Sheet sizes are the same as those for fully toughened glass.

Heat-soaked glass

Sheet sizes are limited to the size of the heat-soaking oven, typically about 2000 mm × 6000 mm.

Laminated glass

They are limited only by the size of sheets available for the different types of glass and the size of autoclave used to cure the interlayers.

Curved glass

Curved glass can be obtained in the United Kingdom from Pilkington with a minimum radius of 750 mm for 12 mm glass; a minimum radius of 1000 mm for 15 mm glass and a minimum radius of 1500 mm for 19 mm glass. However, Sunglass in Italy and Cricursa in Spain are specialist providers who can provide a minimum radius of 300 mm for 10 mm glass down to 100 mm for 4–6 mm glass.

Durability and fire resistance

Durability

Glass and stainless steel components are inherently durable if they are properly specified and kept clean. Glass is corrosion resistant to most substances apart from strong alkalis. The torque of fixing bolts and the adhesives used to secure them should be checked approximately every 5 years and silicone joints may have to be replaced after 25–30 years depending on the exposure conditions. Deflection limits might need to be increased to prevent water ingress caused by rotations at the framing and sealing to the glass.

Fire resistance

Fire-resistant glasses are capable of achieving 60 min of stability and integrity when specially framed using intumescent seals and so on. There are several types of fire-resistant glass which all have differing amounts of fire resistance. The wire interlayer in Georgian wired glass maintains the integrity of the pane by holding the glass in place as it is softened by the heat of a fire. Intumescent interlayers in laminated glass expand to form an opaque rigid barrier to contain heat and smoke. Prestressed borosilicate glass (better known as Pyrex) can resist heat without cracking but must be specially made to order and is limited to $1.2 \text{ m} \times 2 \text{ m}$ panels.

Typical glass sizes for common applications

The following are typical sizes from Pilkington for standard glass applications. The normal design principles of determinacy and redundancy should also be considered when using these typical sizes. These designs are for internal use only. External use requires more careful consideration of thermal effects, where it may be more appropriate to specify toughened glass instead of annealed glass.

Toughened glass barriers

Horizontal line load (kN/m)	Toughened glass thickness[a] (mm)
0.36	12
0.74	15
1.50	19
3.00	25

Note:
[a]   For 1.1 m high barrier, clamped at foot.

Toughened glass infill to barriers bolted between uprights

Loading		Limiting glass span for glass thickness (m)			
UDL (kN/m²)	Point load (kN)	6 mm[a]	8 mm[a]	10 mm	12 mm
0.5	0.25	1.40	1.75	2.10	2.40
1.0	0.50	0.90	1.45	1.75	2.05
1.5	1.50	–	–	1.20	1.60

Note:
[a]   Not suitable if free path beside barrier is >1.5 as it will not contain impact loads as class A to BS 6206.

Laminated glass floors and stair treads

UDL (kN/m²)	Point load (kN)	Glass thickness (top + bottom annealed)[a] (mm + mm)	Typical use
1.5	1.4	19 + 10	Domestic floor or stair
5.0	3.6	25 + 15	Dance floor
4.0	4.5	25 + 25	Corridors
4.0	4.0	25 + 10	Stair

Note:
[a]   Based on a floor sheet size of 1 m² or a stair tread of 0.3 × 1.5 m supported on four edges with a minimum bearing length equal to the thickness of the glass unit. The 1 m² is normally considered to be the maximum size/weight can be practically handled on site.

Glass mullions or fins in toughened safety glass

Mullion height (m)	Mullion thickness/depth for wind loading (mm)[a]			
	1.00 kN/m²	1.25 kN/m²	1.50 kN/m²	1.75 kN/m²
<2.0	19/120	19/130	25/120	25/130
2.0–5	19/160	25/160	25/170	–
2.5–3.0	25/180	25/200	–	–
3.0–3.5	25/230	–	–	–
3.5–4.0	25/280	–	–	–

Note:
[a] Assuming restraint at head and foot plus sealant to main panels. Normal maximum spacing is approximately 2 m.

Glass walls and planar glazing

Suspended structural glass walls can typically be up to 23 m high and of unlimited length, while ground supported walls are usually limited to a maximum height of 9 m.

Planar glazing is limited to a height of 18 m with glass sheet sizes of less than 2 m² so that the weights do not exceed the shear capacity of the planar bolts and fixings.

In a sheltered urban area, 2×2 m square panels will typically need a bolt at each of the four corners; 2×3.5 m panels will need six bolts and panels taller than 3.5 m will need eight bolts.

Source: Pilkington. 2002.

Structural glass design

Summary of design principles

- Provide alternative routes within a building so that users can choose to avoid crossing glass structures.
- Glass is perfectly elastic, but failure is sudden.
- Deflection and buckling normally govern the design. Deflections of vertical panes are thought to be acceptable to span/150, while deflections of horizontal elements should be limited to span/360.
- Glass works best in compression, although bearing often determines the thickness of beams and fins.
- For designs in pure tension, the supports should be designed to distribute the stresses uniformly across the whole glass area.
- Glass can carry bending both in and out of its own plane.
- Use glass in combination with steel or other metals to carry tensile and bending stresses.
- The sizes of glass elements in external walls can be dictated by energy efficiency regulations as much as the required strength.
- Keep the arrangement of supports simple, ensuring that the glass only carries predictable loads to avoid failure as a result of stress concentrations. Isolate glass from shock and fluctuating loads with spring and damped connections.
- Sudden failure of the glass elements must be allowed for in the design by provision of redundancy, alternative load paths, and by using the higher short-term load capacity of glass.
- Glass failure should not result in a disproportionate collapse of the structure.
- Generally, long-term stresses in annealed glass are kept low to prevent failure as a result of static fatigue, that is, time-dependent failure. For complex structures with simple loading conditions, it is possible to stress glass elements for a calculated failure period in order to promote failure by static fatigue.
- The effects of failure and the method of repair or replacement must be considered in the design, as well as maintenance and access issues.
- The impact resistance of each element should be considered to establish an appropriate behaviour as a result of damage by accident or vandalism.
- It is good practice to laminate glass sheets used overhead. Sand blasting, etching and fritting can be used to provide slip resistance and modesty for glass to be used underfoot.
- Toughened glass elements should generally be heat soaked to avoid nickel sulphide failure if they are to be used to carry load as single or unlaminated sheets.
- Consider proof testing elements/components if the design is new or unusual, or where critical elements rely on the additional strength of single ply toughened glass.
- Glass sheet sizes are limited to the standard sizes produced by the manufacturers and the size of sheets which cutting equipment can handle.
- When considering large sheet sizes, thought must be given to the practicalities of weight, method of delivery and installation and possible future replacement.
- Inspect glass delivered to site for damage or flaws which might cause failure.
- Check that the glass can be obtained economically, in the time available.

Codes of practice and design standards

There is no single code of practice to cover structural glass, although the draft Eurocode pr EN 13474 'Glass in Building' is the nearest to an appropriate code of practice for glass design. It is thought to be slightly conservative to account for the varying quality of glass manufacture coming from different European countries. Other useful references are BS 6262, Building Regulations Part N, Glass and Glazing Federation Data Sheets, Pilkington Design Guidance Sheets; the IStructE Guide to Structural Use of Glass in Buildings and the Australian standard AS 1288.

Glass can carry load in compression, tension, bending, torsion and shear, but the engineer must decide how the stresses in the glass are to be calculated, what levels of stress are acceptable, what factor of safety is appropriate and how can unexpected or changeable loads be avoided. Overdesign will not guarantee safety.

Although some design methods use fracture mechanics or Weibull probabilities, the simplest and most commonly used design approach is elastic analysis.

Guideline allowable stresses

The following values are for preliminary design using elastic analysis with unfactored loads and are based on the values available in pr EN 13474.

Glass type	Characteristic bending strength (N/mm²)	Loading condition	Typical factor of safety	Typical allowable bending stress (N/mm²)
Annealed	45	Long term	6.5	7
		Short term	2.5	20
Heat tempered	70	Long term	3.5	20
		Short term	2.4	30
Toughened	120	Long term	3.0	40
		Short term	2.4	50

Connections

Connections must transfer the load in and out of glass elements in a predictable way avoiding any stress concentrations. Clamped and friction grip connections are the most commonly used for single sheets. Glass surfaces are never perfectly smooth and connections should be designed to account for differences of up to 1 mm in the glass thickness. Cut edges can have tolerances of 0.2–0.3 mm if cut with a CNC laser, otherwise dimension tolerances can be 1.0–1.5 mm.

Simple supports

The sheets of glass should sit perfectly on to the supports, either in the plane, or perpendicular to the plane, of the glass. Gaskets can cause stress concentrations and should not be used to compensate for excessive deviation between the glass and the supports. The allowable bearing stress is generally limited to about 0.42–1.5 N/mm^2 depending on the glass and setting blocks used.

Friction grip connections

Friction connections use patch plates to clamp the glass in place and are commonly used for single ply sheets of toughened glass. More complex clamped connections can use galvanized fibre gaskets and holes lined with nylon bushes to prevent stress concentrations. Friction grip bolt torques should be designed to generate a frictional clamping force of $N = F/\mu$, where the coefficient of friction is generally $\mu = 0.2$.

Holes

Annealed glass can be drilled. Toughened or tempered glass must be machined before toughening. The Glass and Glazing Federation suggests that the minimum clear edge distance should be the greater of 30 mm or 1.5 times the glass thickness (t). The minimum clear corner distance and minimum clear bolt spacing should be $4t$. Holes should be positioned in low stress areas, should be accurately drilled and the hole diameter must not be less than the glass thickness.

Bolted connections

Bolted connections can be designed to resist loads, both in and out of the plane of the glass. Pure bolted connections need to be designed for strength, tolerance, deflection, thermal and blast effects. They can be affected by minor details (such as drilling accuracy or the hole lining/bush) and this is why proprietary bolted systems are most commonly used. Extensive testing should be carried out where bolted connections are to be specially developed for a project.

Non-silicon adhesives

The use of adhesives (other than silicon) is still fairly experimental and as yet is generally limited to small glass elements. Epoxies and UV cure adhesives are among those which have been tried. It is thought that failure strengths might be about 10 times those of silicon, but suitable factors of safety have not been made widely available. Loctite and 3M have some adhesive products which might be worth investigating/testing.

Structural silicones

Sealant manufacturers should be contacted for assistance with specifying their silicon products. This assistance can include information on product selection, adhesion, compatibility, thermal/creep effects and calculation of joint sizes. Data from one project cannot automatically be used for other applications.

Structural silicon sealant joints should normally be a minimum of 6 mm × 6 mm, with a maximum width-to-depth ratio of 3:1. If this maximum width-to-depth ratio is exceeded, the glass sheets will be able to rotate, causing additional stresses in the silicon. A simplified design approach to joint rotation can be used (if the glass deflection is less than $L/100$) where reduced design stresses are used to allow additional capacity in the joint to cover any rotational stresses. If joint rotation is specifically considered in the joint design calculations, higher values of allowable design stresses can be used.

Dow Corning manufactures two silicon adhesives for structural applications. Dow Corning 895 is one part adhesive, site-applied silicon used for small-scale remedial applications or where a two-sided structurally bonded system has to be bonded on site. Dow Corning 993 is a two part adhesive, and normally factory applied. The range of colours is limited and availability should be checked for each product and application.

Technical data on the Dow Corning silicones is set out below:

Dow Corning silicon	Young's modulus (kN/mm²)	Type of stress	Failure stress (N/mm²)	Loading condition	Typical allowable design stress (N/mm²)
933 (2 part)	0.0014	Tension/compression	0.95	Short-term/live loads. Design stress for comparison with simplified calculations not allowing for stresses due to joint rotation	0.140
				Short-term/live loads. Design stress where the stresses due to joint rotation for a particular case have been specifically calculated	0.210
				Long-term/dead loads	n/a
		Shear	0.68	Short-term/live loads	0.105
				Long-term/dead loads	0.011
895 (1 part)	0.0009	Tension/compression	1.40	Short-term/live loads. Design stress for comparison with simplified calculations not allowing for stresses due to joint rotation	0.140
				Short-term/live loads. Design stress where the stresses due to joint rotation for a particular case have been specifically calculated	0.210
				Long-term/dead loads	n/a
		Shear	0.07	Short-term/live loads	0.140
				Long-term/dead loads	0.007

Source: Dow Corning. 2002.

14
Building Elements, Materials, Fixings and Fastenings

Waterproofing

Although normally detailed and specified by an architect, the waterproofing must coordinate with the structure and the engineer must understand the implications of the waterproofing on the structural design.

Damp proof course

A damp proof course (DPC) is normally installed at the top and bottom of external walls to prevent the vertical passage of moisture through the wall. Cavity trays and weep holes are required above the position of elements which bridge the cavity, such as windows or doors, in order to direct any moisture in the cavity to the outside. The inclusion of a DPC will normally reduce the flexural strength of the wall.

DPCs should

- Be bedded both sides in mortar to prevent damage.
- Be lapped with damp proof membranes (DPMs) in the floor or roof.
- Be lapped in order to ensure that moisture will flow over and not into the laps.
- Not project into cavities where they might collect mortar and bridge the cavity.

Different materials are available to suit different situations:

- Flexible plastic sheets or bitumen impregnated fabric can be used for most DPC locations but can be torn if not well protected and the bituminous types can sometimes be extruded under high loads or temperatures.
- Semi-rigid sheets of copper or lead are expensive but are most effective for intricate junctions.
- Rigid DPCs are layers of slate or engineering brick in Class I mortar and are only used in the base of retaining walls or freestanding walls. These combat rising damp and (unlike the other DPC materials) can transfer tension through the DPC position.

Damp proof membrane

DPMs are sheet or liquid membranes that are typically installed at roof and ground floor levels. In roofs they are intended to prevent the ingress of rain and at ground floor level they are intended to prevent the passage of moisture from below by capillary action. Sheet membranes can be polyethylene, bituminous or rubber sheets, while liquid systems can be hot or cold bitumen or epoxy resin.

Basement waterproofing

Basement waterproofing is problematic as leaks are only normally discovered once the structure has been occupied. The opportunity for remedial work is normally limited, quite apart from the difficulty of reaching externally applied tanking systems. Although an architect details the waterproofing for the rest of the building, sometimes the engineer is asked to specify the waterproofing for the basement. In this case very careful co-ordination of the lapping of the waterproofing above and below ground must be achieved to ensure that there are no weak points. Basement waterproofing should always be considered as a three-dimensional problem.

It is important to establish whether the system will be required to provide basic resistance to water pressure, or whether special additional controls on water vapour will also be required.

Basement waterproofing to BS 8102

BS 8102 sets out guidance for the waterproofing of basement structures according to their use. The following table has been adapted from Table 1 in BS 8102: 1990 to include some of the increased requirements suggested in CIRIA Report 139.

Methods of basement waterproofing

The following types of basement waterproofing systems can be used individually or together depending on the building requirements:

Tanked: This can be used internally or externally using painted or sheet membranes. Externally it is difficult to apply and protect under building site conditions, while internally water pressures can blow the waterproofing off the wall; however, it is often selected as it is relatively cheap and takes up very little space.

Integral: Concrete retaining walls can resist the ingress of water in differing amounts depending on the thickness of the section, the applied stresses, the amount of reinforcement and the density of the concrete. The density of the concrete is directly related to how well the concrete is compacted during construction. Integral structural waterproofing systems require a highly skilled workforce and strict site control. However, moisture and water vapour can still pass through a plain wall and additional protection should be added if this moisture will not be acceptable for the proposed basement use. BS 8007 provides guidance on the design of concrete to resist the passage of water, but this still does not stop water vapour. Alternatively, Caltite or Pudlo additives can be used with a BS 8110 structure to create 'waterproof concrete'. This is more expensive than standard concrete but this can be offset against any saving on the labour and installation costs of traditional forms of waterproofing.

Drained: Drained cavity and floor systems allow moisture to penetrate the retaining wall. The moisture is collected in a sump to be pumped away. Drained cavity systems tend to be expensive to install and can take up quite a lot of basement floor area, but they are thought to be much more reliable than other waterproofing systems. Draining ground water to the public sewers may require a special licence from the local water authority. Access hatches for the inspection and maintenance of internal gulleys should be provided where possible.

Basement waterproofing to BS 8102

Grade of basement to BS 8102	Basement use	Performance of water proofing	Form of construction	Comments
1	Car parking, plant rooms (excluding electrical equipment) and workshops	Some water seepage and damp patches tolerable (typical relative humidity > 65%)	Type B – RC to BS 8110 (with crack widths limited to 0.3 mm)	Provides integral protection and needs waterstops at construction joints. Medium risk. Consider ground chemicals for durability and effect on finishes. The BS 8102 description of a workshop is not as good as the workshop environment described in the Building Regulations
2	Workshops and plantrooms requiring drier environment. Retail storage areas	No water penetration but moisture Vapour tolerable (typical relative humidity = 35–50%)	Type A	Requires drainage to external basement perimeter below the level of the wall/floor membrane lap. Medium risk with multiple membrane layers and strict site control
			Type B – RC to BS 8007	Provides integral protection and needs waterstops at construction joints. Medium risk. Consider ground chemicals for durability and effect on finishes. Additional tanking is likely to be needed to meet retail storage requirements
3	Ventilated residential and working areas, offices, restaurants and leisure centres	Dry environment, but no specific control on moisture vapour (typical relative humidity 40–60%)	Type A	Not recommended unless drainage is provided above the wall/floor membrane lap position and the site is relatively free draining. High risk
			Type B – RC to BS 8007	Provides integral protection and needs waterstops at construction joints. Medium risk. Consider ground chemicals for durability and effect on finishes. Additional tanking is recommended
			Type C – wall and floor cavity system	A drained cavity allows the wall to leak and it is therefore foolproof. Sumps may need back-up pumps. High safety factor

continued

(continued)

Grade of basement to BS 8102	Basement use	Performance of water proofing	Form of construction	Comments
4	Archives and computer stores	Totally dry environment with strict control of moisture vapour (typical relative humidity = 35% for books – 50% for art storage)	*Type A*	Unlikely to be able to provide the controlled conditions required. Very high risk
			Type B – RC to BS 8007 plus vapour barrier	High risk. Medium risk with addition of a drainage cavity to reduce water penetration
			Type C – wall and floor cavity system with vapour barrier to inner skin and floor cavity with DPM	Medium risk. Addition of a water resistant concrete wall would provide the maximum possible safety for sensitive environments

Notes:

a *Type A* = tanked construction, *Type B* = integral structural waterproofing and *Type C* = drained protection.

b Relative humidity indicates the amount of water vapour in the air as a percentage of the maximum amount of water vapour which would be possible for air at a given temperature and pressure. Typical values of relative humidity for the UK are about 40–50% for heated indoor conditions and 85% for unheated external conditions.

Source: BS 8102:1990.

Remedial work

Failed basement systems require remedial work. Application of internal tanking in this situation is not normally successful. The junction of the wall and floor is normally the position where water leaks are most noticeable.

An economical remedial method is to turn the existing floor construction into a drained floor by chasing channels in the existing floor finishes around the perimeter. Additional channels may cross the floor where there are large areas of open space. Proprietary plastic trays with perforated sides and bases can be set into the chases, connected up and drained to a sump and pump. New floor finishes can then be applied over the original floor and its new drainage channels, to provide ground water protection with only a small thickness of additional floor construction.

The best way to avoid disrupting, distressing and expensive remedial work is to design and detail a good drained cavity system in the first place!

Screeds

Screeds are generally specified by an architect as a finish to structural floors in order to provide a level surface, to conceal service routes and/or as a preparation for application of floor finishes. Historically screeds fail due to inadequate soundness, cracking and curling and therefore, like waterproofing, it is useful for the engineer to have some background knowledge. Structural toppings generally act as part of a precast structural floor to resist vertical load or to enhance diaphragm action. The structural issues affecting the choice of screed are type of floor construction, deflection, thermal or moisture movements, surface accuracy and moisture condition.

Deflection

Directly bonded screeds can be successfully applied to solid reinforced concrete slabs as they are generally sufficiently rigid, while floating screeds are more suitable for flexible floors (such as precast planks or composite metal decking) to avoid reflective cracking of the screed. Floating screeds must be thicker than bonded screeds to withstand the applied floor loadings and are laid on a slip membrane to ensure free movement and avoid reflective cracking.

Thermal/moisture effects

Drying shrinkage and temperature changes will result in movement in the structure, which could lead to the cracking of an overlying bonded screed. It is general practice to leave concrete slabs to cure for 6 weeks before laying screed or applying rigid finishes such as tiles, stone or terrazzo. For other finishes the required floor slab drying times vary. If movement is likely to be problematic, joints should be made in the screed at predetermined points to allow expansion/contraction/stress relief. *Sand*: cement screeds must be cured by close covering with polythene sheet for 7 days while foot traffic is prevented and the screed is protected from frost. After this the remaining free moisture in the screed needs time to escape before application of finishes. This is especially true if the substructure and finish are both vapour proof as this can result in moisture being trapped in the screed. Accurate prediction of screed drying times is difficult, but a rough rule is 4 weeks per 25 mm of screed thickness (to reach about 75% relative humidity). Accelerated heating to speed the drying process can cause the screed to crack or curl, but dehumidifiers can be useful.

Surface accuracy

The accuracy of surface level and flatness of a laying surface is related to the type of base, accuracy of the setting out and the quality of workmanship. These issues should be considered when selecting the overall thickness of the floor finishes to avoid problems with the finish and/or costly remedial measures.

Resin anchors

Resin anchor is the generic name for bolts or studding that are fixed using a chemical adhesive. Fixings can be made to concrete, brickwork, solid/hollow blockwork and stone. Where the hole is drilled cleanly and the dust removed, the chemical bond can often be stronger than the base material, without the local crushing problems which can be caused by expansion type anchors. As a result high loads, closely spaced anchors and low edge distances can be achieved. Installation methods include bulk mix, capsule or injection cartridge, but all use the same basic principle of a base resin which mixes with a second component to begin the chemical curing process. Cartridge delivery makes for an easier installation but a higher cost and more waste packaging than a capsule. Manufacturers generally offer anchor design advice and load testing services. Load values for general purpose resin anchors set out below are for Grade 8.8 bolts in RawlCFS RM50 resin in capsule form. Details of Rawlplug's high strength and fast cure resin anchor products (including epoxy and vinylester products for fixing threaded studs and rebar) are available from Rawl Technical Services.

C20/25 concrete										
	Characteristic resistance (kN)		Design resistance (ULS) (kN)		Recommended load (working) (kN)		Characteristic edge distance (mm)		Characteristic spacing (mm)	Standard embedment depth (mm)
Bolt size	Tension N_{Rk}	Shear V_{Rk}	Tension N_{Rd}	Shear V_{Rd}	Tension N_{Rec}	Shear V_{Rec}	Tension	Shear	Tension & shear	
M8	11.5	9.9	5.1	7.9	3.6	5.7	80	100	100	80
M10	20.0	15.7	8.3	12.6	5.9	9.0	90	130	130	90
M12	28.7	22.9	11.5	18.3	8.2	13.1	110	150	150	110
M16	45.7	42.5	16.6	34.0	11.9	24.3	130	170	170	125
M20	62.0	66.8	20.9	53.4	15.0	38.2	150	190	210	145
M20 deep	69.7	66.8	23.2	53.4	16.6	38.2	180	190	230	180
M24	79.3	95.7	26.4	76.6	18.9	54.7	190	240	240	180
M24 deep	87.9	95.7	29.3	76.6	20.9	54.7	220	240	270	210

	Recommended load (SLS)[4] kN			
Bolt size	Brickwork 20.5 N/mm²	Blockwork 7 N/mm²	Blockwork 3.5 N/mm²	Blockwork 2.8 N/mm²
M8	1.4	0.6	0.5	0.4
M10	2.9	1.3	0.9	0.7
M12	4.0	2.0	1.1	0.9
M16	5.0	3.0	–	–

Notes:
[a] Reductions should be made for reduced edge distances and spacings. For quick approximation, an M12 has 50% shear capacity and 78% tension capacity at 70 mm edge distance, with 100% capacity at 170 mm for shear and 110 mm for tension. For combined tension and shear, an M12 resin anchor would have 80% capacity at 70 mm spacing, increasing to 100% capacity at 150 mm spacing. Refer to the manufacturer's literature for more accurate guidance and values.
[b] Characteristic resistance is the ultimate load with a statistical probability factor applied (95% Fractile). This factor is dependent on the variance of a set of results and also the number of anchors tested.
[c] Design resistance is based upon the characteristic resistance divided by partial safety factors as determined from testing. Design resistance always be greater than the value of any design action – a factored load taking into account variable actions.
[d] Those who prefer to use the global safety factor method can continue to do so, using the recommended load (previously called 'safe working load').
[e] Minimum spacing 100 mm and minimum edge distance 200 mm.

Source: Rawl Fixings UK. 2012.

Precast concrete hollowcore slabs

The values for the hollowcore slabs set out below are for precast prestressed concrete slabs by Bison Concrete Products. The prestressing wires are stretched across long shutter beds before the concrete is extruded or slip formed along beds up to 130 m long. The prestress in the units induces a precamber. The overall camber of associated units should not normally exceed $L/300$. Some planks may need a concrete topping (not screed) to develop their full bending capacity or to contribute to diaphragm action. Minimum bearing lengths of 100 mm are required for masonry supports, while 75 mm is acceptable for supports on steelwork or concrete. Planks are normally 1200 mm wide at their underside and are butted up tight together on site. The units are only 1180 to 1190 mm wide at the top surface and the joints between the planks are grouted up on site. Narrower planks are normally available on special order in a few specific widths. Special details, notches, holes and fixings should be discussed with the plank manufacturer early in the design.

Typical hollowcore working load capacities

Nominal hollowcore plank depth mm	Fire resistance hours	Clear span for imposed loads[a] m				
		1.5 kN/m²	3.0 kN/m²	5.0 kN/m²	10.0 kN/m²	15.0 kN/m²
150[b]	Up to 2	7.5	7.5	6.7	5.2	4.4
200	2	10.0	9.1	8.0	6.4	5.5
250	2	11.7	10.5	9.3	7.5	6.4
300	2	14.6	13.2	11.9	9.8	8.5
350	2	16.1	14.7	13.3	11.0	9.5
400	2	17.2	15.8	14.3	11.9	10.4
450	2	18.0	17.1	15.5	13.0	11.4

Notes:
[a] 1.5 kN/m² for finishes included in addition to self-weight of plank.
[b] 35 mm screed required for 2 h fire resistance.
[c] The reinforcement pattern within a Bison section will vary according to the design loading specified.
[d] Ask for 'Sound Slab' where floor mass must be greater than 300 kg/m³.

Source: Bison Manufacturing. 2012. Note that this information is subject to change at any time. Consult the latest Bison literature for up-to-date information.

Bi-metallic corrosion

When two dissimilar metals are put together with an 'electrolyte' (normally water), an electrical current passes between them. The further apart the metals are on the galvanic series, the more pronounced this effect becomes.

The current consists of a flow of electrons from the anode (the metal higher in the galvanic series) to the cathode, resulting in the 'wearing away' of the anode. This effect is advantageous in galvanising where the zinc coating slowly erodes, sacrificially protecting the steelwork. Alloys of combined metals can produce mixed effects and should be chosen with care for wet or corrosive situations in combination with other metals.

The amount of corrosion is dictated by the relative contact surface (or areas) and the nature of the electrolyte. The effect is more pronounced in immersed and buried objects. The larger the cathode, the more aggressive the attack on the anode. Where the presence of electrolyte is limited, the effect on mild steel sections is minimal and for most practical building applications where moisture is controlled, no special precautions are needed. For greater risk areas where moisture will be present, gaskets, bushes, sleeves or paint systems can be used to separate the metal surfaces.

The galvanic series

Anode	Magnesium
	Zinc
	Aluminium
	Carbon and low alloy steels (structural steel)
	Cast iron
	Lead
	Tin
	Copper, brass, bronze
	Nickel (passive)
	Titanium
Cathode	Stainless steels (passive)

Structural adhesives

There is little definite guidance on the use of adhesives in structural applications which can be considered if factory-controlled conditions are available. Construction sites rarely have the quality control which is required. Adhesive manufacturers should be consulted to ensure that a suitable adhesive is selected and that it will have appropriate strength, durability, fire resistance, effect on speed of fabrication, creep, surface preparation, maintenance requirements, design life and cost. Data for specific products should be obtained from manufacturers.

Adhesive families

Epoxy resins	Good gap-filling properties for wide joints, with good strength and durability; low cure shrinkage and creep tendency and good operating temperature range. The resins can be cold or hot cure, in liquid or in paste form but generally available as two part formulations. Relatively high cost limits their use to special applications
Polyurethanes	Very versatile, but slightly weaker than epoxies. Good durability properties (resistance to water, oils and chemicals but generally not alkalis) with operating temperatures of up to 60°C. Moisture is generally required as a catalyst to curing, but moisture in the parent material can adversely affect the adhesive. Application includes timber and stone, but concrete should generally be avoided due to its alkalinity
Acrylics	Toughened acrylics are typically used for structural applications which generally need little surface preparation of the parent material to enhance bond. They can exhibit significant creep, especially at higher operating temperatures and are best suited to tight fitting (thin) joints for metals and plastics
Polyesters	Polyesters exhibit rapid strength gain (even in extremely low temperatures) and are often used for resin anchor fixings, etc. However, they can exhibit high cure shrinkage and creep, and have poor resistance to moisture
Resorcinol-formaldehydes (RF) and phenol-formaldehydes (PRF)	Intended for use primarily with timber. Curing can be achieved at room temperature and above. These adhesives are expensive but strong, durable, water and boil proof and will withstand exposure to salt water. They can be used for internal and external applications, and are generally used in thin layers, for example, finger joints in glulam beams
Phenol-formaldehydes (PF)	Typically used in factory 'hot press' fabrication of structural plywood. Cold curing types use strong acids as catalysts which can cause staining of the wood. The adhesives have similar properties to RF and PRF adhesives
Melamine-urea – formaldehydes (MUF) and urea-formaldehydes (UF)	Another adhesive typically used for timber, but these need protection from moisture. These are best used in thin joints (of less than 0.1 mm) and cure above 10°C
Caesins	Derived from milk proteins, these adhesives are less water resistant than MUF and UF adhesives and are susceptible to fungal attack
Polyvinyl acetates and elastomerics	Limited to non-loadbearing applications indoors as they have limited moisture resistance
Adhesive tapes	Double sided adhesive tapes are typically contact adhesives and are suitable for bonding smooth surfaces where rapid assembly is required. The tapes have a good operating temperature range and can accommodate a significant amount of strain. Adhesive tapes are typically used for metals and/or glass in structural applications

Surface preparation of selected materials in glued joints

Surface preparation is essential for the long-term performance of a glued joint and the following table describes the typical steps for different materials. Specific requirements should normally be obtained from the manufacturer of the adhesive.

Material	Surface preparation	Typical adhesive
Concrete	1. Test parent material for integrity 2. Grit blast or water jet to remove the cement-rich surface, curing agents and shutter oil, etc. 3. Vacuum dust and clean surface with solvent approved by the glue manufacturer 4. Apply a levelling layer to the roughened concrete surface before priming for the adhesive	Epoxies are commonly used with concrete, while polyesters are used in resin fixings and anchors. Polyurethanes are not suitable for general use due to the alkalinity of the concrete
Steel and cast iron	1. Degrease the surface 2. Mechanically wire brush, grit blast or water jet to remove mill scale and surface coatings 3. Vacuum dust then prime surface before application of the adhesive	Epoxies are the most common for use with structural iron/steel. Where high strength is not required acrylic or polyurethane may be appropriate, but only where humidity can be controlled or creep effects will not be problematic
Zinc coated steel	1. Test the steel/zinc interface for integrity 2. Degrease the surface 3. Lightly abrade the surface and avoid rupturing the zinc surface 4. Vacuum dust and then apply an etch primer 5. Thoroughly clean off the etch primer and prime the surface for the adhesive	Epoxies are suitable for structural applications. Acrylics are not generally compatible with the zinc surface
Stainless steel	Factory method: 1. Acid etch the surface and clean thoroughly 2. Apply primer Site method: 1. Degrease the surface with solvent 2. Grit blast 3. Apply chemical bonding agent, e.g. silane	Toughened epoxies are normally used for structural applications
Aluminium	Factory method: 1. Degrease with solvent 2. Use alkaline cleaning solution 3. Acid etch, then neutralise 4. Prime surface before application of the adhesive Site method (as 1 and 2): 3. Grit blast 4. Apply a silane primer/bonding agent	Epoxies and acrylics are most commonly used. Anodised components are very difficult to bond

continued

(continued)

Material	Surface preparation	Typical adhesive
Timber	1. Remove damaged parent material 2. Dry off contact surfaces and ensure both surfaces have a similar moisture content (which is also less than 20) 3. Plane to create a clean flat surface or lightly abrade for sheet materials 4. Vacuum dust then apply adhesive promptly	Epoxies are normally limited to special repairs. RF and PRF adhesives have long been used with timber. Durability of the adhesive must be carefully considered. They are classified: WBP – weather proof and boil proof; BR – boil resistant; MR – moisture resistant INT – Interior
Plastic and fibre composites	1. Dust and degrease surface 2. Abrade surface to remove loose fibres and resin-rich outer layers 3. Remove traces of solvent and dust	Epoxies usual for normal applications. In dry conditions polyurethanes can be used, and acrylics if creep effects are not critical
Glass	1. Degreasing should be the only surface treatment. Abrading or etching the surface will weaken the parent material 2. Silane primer is occasionally used	Structural bonding tape or modified epoxies. The use of silicon sealant adhesives if curing times are not critical

Fixings and fastenings

Although there are a great number of fixings available, the engineer will generally specify nails, screws or bolts. Within these categories there are variations depending on the materials to be fixed. The fixings included here are standard gauges generally available in the United Kingdom.

Selected round wire nails to BS 1202

	Diameter (standard wire gauge (swg) and mm)						
	11 swg	10	9	8	7	6	5
Length (mm)	3.0 mm	3.35	3.65	4.0	4.5	5.0	5.6
50	•	•					
75		•	•				
100			•	•	•	•	
125						•	
150							•

Selected wood screws to BS 1210

	Diameter (standard gauge (sg) and mm)						
	6 sg	7	8	10	12	14	16
Length (mm)	3.48 mm	3.50	4.17	4.88	5.59	6.30	6.94
25	•	•	•	•	•		
50	•	•	•	•	•	•	•
75	•		•	•	•	•	•
100				•	•	•	•
125				•	•	•	•

Selected self-tapping screws to BS 4174

Self-tapping screws can be used in metal or plastics, while thread cutting screws are generally used in plastics or timber.

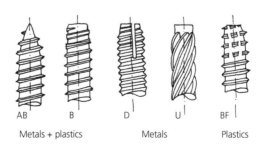

AB   B   D   U   BF

Metals + plastics   Metals   Plastics

Selected ISO metric black bolts to BS 4190 and BS 3692

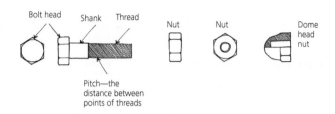

Bolt head   Shank   Thread   Nut   Nut   Dome head nut

Pitch—the distance between points of threads

Nominal diameter (mm)	Coarse pitch (mm)	Maximum width of head and nut (mm) Across flats	Across corners	Maximum height of head (mm)	Maximum thickness of nut (black) (mm)	Minimum distance between centres (mm)	Tensile stress area (mm²)	Normal sizeª (form E) round washers to BS 4320 Inside diameter (mm)	Outside diameter (mm)	Nominal thickness (mm)
M6	1.00	10	11.5	4.375	5.375	15	20.1	6.6	12.5	1.6
M8	1.25	13	15.0	5.875	6.875	20	36.6	9.0	17.0	1.6
M10	1.50	17	19.6	7.450	8.450	25	58.0	11.0	21.0	2.0
M12	1.75	19	21.9	8.450	10.450	30	84.3	14.0	24.0	2.5
M16	2.00	24	27.7	10.450	13.550	40	157.0	18.0	30.0	3.0
M20	2.50	30	34.5	13.900	16.550	50	245.0	22.0	37.0	3.0
M24	3.00	36	41.6	15.900	19.650	60	353.0	26.0	44.0	4.0
M30	3.50	46	53.1	20.050	24.850	75	561.0	33.0	56.0	4.0

ª Larger diameter washers as Form F and Form G are also available to BS 4320.

Lengthª (mm)	Bolt size						
	M66	M88	M10	M12	M16	M20	M24
30	•	•					
50	•	•	•	•	•		
70		•	•	•	•	•	•
100			•	•	•	•	•
120			•	•	•	•	•
140					•	•	•
150				•	•	•	
180				•			

Note:
M6, M8, M10 and M12 threaded bar (called studding) is also available in long lengths.
ª Intermediate lengths are available.

Spanner and podger dimensions

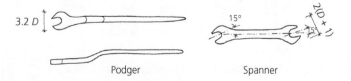

3.2 D

15°

2D + 1

Podger   Spanner

Selected metric machine screws to BS 4183

Available in M3 to M20, machine screws have the same dimensions as black bolts but they are threaded full length and do not have a plain shank. Machine screws are often used in place of bolts and have a variety of screw heads:

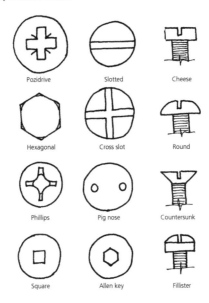

Pozidrive	Slotted	Cheese
Hexagonal	Cross slot	Round
Phillips	Pig nose	Countersunk
Square	Allen key	Fillister

Selected metric countersunk allen key machine screws

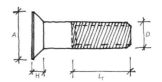

Nominal diameter (mm)	Coarse pitch (mm)	Maximum width of head (mm)	Maximum depth of tapered heat (mm)	Minimum threaded length (mm)
M3	0.50	6.72	1.7	18
M4	0.70	8.96	2.3	20
M5	0.80	11.20	2.8	22
M6	1.00	13.44	3.3	24
M8	1.25	17.92	4.4	28
M10	1.50	22.40	5.5	32
M12	1.75	26.88	6.5	36
M16	2.00	33.60	7.5	44
M20	2.50	40.32	8.5	52
M24	3.00	40.42	14.0	60

Selected coach screws to BS 1210

Typically used in timber construction. The square head allows the screw to be tightened by a spanner.

Length (mm)	Diameter			
	6.25	7.93	9.52	12.3
25	•	•		
37.5	•	•	•	
50	•	•	•	
75	•	•	•	•
87.5	•	•		
100	•	•	•	•
112	•			
125	•	•	•	•
150		•	•	•
200				•

Selected welding symbols to BS 449

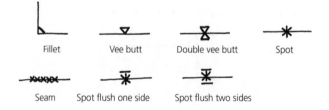

Fillet   Vee butt   Double vee butt   Spot

Seam   Spot flush one side   Spot flush two sides

Cold weather working

Cold weather and frosts can badly affect wet trades such as masonry and concrete; however, rain and snow may also have an effect on ground conditions, make access to the site and scaffolds difficult, and cause newly excavated trenches to collapse. Site staff should monitor weather forecasts to plan ahead for cold weather.

Concreting

Frost and rain can damage newly laid concrete which will not set or hydrate in temperatures below 1°C. At lower temperatures, the water in the mixture will freeze, expand and cause the concrete to break up. Heavy rain can dilute the top surface of a concrete slab and can also cause it to crumble and break up.

- Concrete should not be poured below an air temperature of 2°C or if the temperature is due to fall in the next few hours. Local conditions, frost hollows or wind chill may reduce temperatures further.
- If work cannot be delayed, concrete should be delivered at a minimum temperature of 5°C and preferably at least 10°C, so that the concrete can be kept above 5°C during the pour.
- Concrete should not be poured in more than the lightest of rain or snow showers and poured concrete should be protected if rain or snow is forecast. Formwork should be left in place longer to allow for the slower gain in strength. Concrete that has achieved 5 N/mm² is generally considered frost safe.
- Mixers, handling plant, subgrade/shuttering, aggregates and materials should be free from frost and be heated if necessary. If materials and plant are to be heated, the mixing water should be heated to 60°C. The concrete should be poured quickly and in extreme cases, the shuttering and concrete can be insulated or heated.

Bricklaying

Frost can easily attack brickwork as it is usually exposed on both sides and has little bulk to retain heat. Mortar will not achieve the required strength in temperatures below 2°C. Work exposed to temperature below 2°C should be taken down and rebuilt. If work must continue and a reduced mortar strength is acceptable, a mortar mix of 1 part cement to 5 to 6 parts sand with an air entraining agent can be used. Accelerators are not recommended and additives containing calcium chloride can hold moisture in the masonry resulting in corrosion of any metalwork in the construction.

- Bricks should not be laid at air temperatures below 2°C or if the temperature is due to fall in the next few hours. Bricklaying should not be carried out in winds of force 6 or above, and walls without adequate returns to prevent instability in high winds should be propped.
- Packs, working stacks and tops of working sections should be covered to avoid soaking, which might lead to efflorescence and/or frost attack. An airspace between any polythene and the brickwork will help to prevent condensation. Hessian and bubble wrap can be used to insulate. The protection should remain in place for about 7 days after the frost has passed. In heavy rain, scaffold boards nearest the brickwork can be turned back to avoid splashing, which is difficult to clean off.
- If bricks have not been dipped, a little extra water in the mortar mix will allow the bricks to absorb excess moisture from the mortar and reduce the risk of expansion of the mortar due to freezing.

Effect of fire on construction materials

This section is a brief summary of the effect of fire on structural materials to permit a quick assessment of how a fire may affect the overall strength and stability of a structure.

It is necessary to get an accurate history of the fire and an indication of the temperatures achieved. If this is not available via the fire brigade, clues must be gathered from the site on the basis of the amount of damage to the structure and finishes. At 150°C paint will be burnt away, at 240°C wood will ignite, at 400–500°C PVC cable coverings will be charred, zinc will melt and run off and aluminium will soften. At 600–800°C aluminium will run off and glass will soften and melt. At 900–1000°C most metals will be melting and above this, temperatures will be near the point where a metal fire might start.

The effect of heat on structure generally depends on the temperature, the rate and duration of heating, and the rate of cooling. Rapid cooling by dousing with water normally results in the cracking of most structural materials.

Reinforced concrete

Concrete is likely to blacken and spall, leaving the reinforcement exposed. The heat will reduce the compressive strength and elastic modulus of the section, resulting in cracking and creep/permanent deflections. For preliminary assessment, reinforced concrete heated to 100–300°C will have about 85% of its original strength, by 300–500°C it will have about 40% of its original strength and above 500°C it will have little strength left. As it is a poor conductor of heat only the outer 30–50 mm will have been exposed to the highest temperatures and therefore there will be temperature contours within the section, which may indicate that any loss of strength reduces towards the centre of the section. At about 300°C concrete will tend to turn pink and at about 450–500°C it will tend to become a dirty yellow colour. Bond strengths can normally be assumed to be about 70% of pre-fire values.

Prestressed concrete

The concrete will be affected by fire as listed for reinforced concrete. More critical is the behaviour of the steel tendons, as non-recoverable extension of the tendons will result in loss of prestressing forces. For fires with temperature of 350–400°C, the tendons may have about half of their original capacity.

Timber

Timber browns at 120–150°C, blackens at 200–250°C and will ignite and char at temperatures about 400°C. Charring may not affect the whole section and there may be sufficient section left intact that can be used in calculations of residual strength. Charring can be removed by sandblasting or planing. Large timber sections have often been found to perform better in fire than similarly sized steel or concrete sections.

Brickwork

Bricks are manufactured at temperatures above 1000°C, therefore they are only likely to be superficially or aesthetically damaged by fire. It is the mortar which can lose its strength as a result of high temperatures. Cementitious mortar will react very similarly to reinforced concrete, except without the reinforcement and section mass, it is more likely to be badly affected. Hollow blocks tend to suffer from internal cracking and separation of internal webs from the main block faces.

Steelwork

The yield strength of steel at 20°C is reduced by about 50% at 550°C and at 1000°C it is 10% or less of its original value. Being a good conductor of heat, the steel will reach the same temperature as the fire surrounding it and transfer the heat away from the area to affect other remote areas of the structure. Steelwork heated up to about 600°C can generally be reused if its hardness is checked. Cold worked steel members are more affected by increased temperature. Connections should be checked for thread stripping and general soundness. An approximate guide is that connections heated to 450°C will retain full strength, to 600°C will retain about 80% of their strength and to 800°C will retain only about 60% of their strength.

15
Sustainability

Sustainability is steadily moving into mainstream building projects through the Building Regulations and other legislative controls. It is a particularly difficult area to cover as it is a relatively new topic, involving changes in public opinion as well as traditional construction industry practices. It is therefore difficult for the engineer to find good practice guidance beyond the minimum standards found in legislation and there are often no 'right answers'.

Context

Sustainability was defined by the 1987 Brundtland Report as meeting the needs of the present without compromising the ability of future generations to meet their own needs. This is also often described as the Triple Bottom Line, which aims to balance environmental, social and economic factors. As this balance depends on each individual's moral framework, it is generally very difficult to come to an agreement about what constitutes sustainable design.

Currently, the environmental sustainability debate is focussed on climate change and there is consensus within the scientific community that this is happening due to carbon emissions (although there is less agreement about whether these changes are due to human activity (anthropogenic) or natural cycles). While the debate about causes and action continues, *The Precautionary Principle* states that the effects of climate change are potentially so bad, that action should be taken now to reduce carbon emissions. The 2006 Stern Review supported this approach by concluding that it was likely to be more economic to pay for sustainable design now (in an attempt to mitigate climate change) rather than wait and potentially pay more to deal with its consequences. These are the principles which form the basis of the UK government's drive to reduce carbon emissions through policy on transport and energy, and legislative controls such as Planning and Building Regulations.

Environmental indicators

As the built environment is a huge consumer of resources and source of pollution, the construction industry has a significant role to play. It is estimated that building construction's use and demolition account for nearly half of the UK's carbon emissions, impacting on the environment as follows:

- Climate change (CO_2)
- Ozone depletion (CFC, HCFC)
- Ecological loss
- Fossil fuel depletion
- Land and materials depletion
- Water depletion
- Waste generation
- Acid rain (SO_2, NO_x)
- Toxicity and health (VOCs).

Although there are many indicators and targets in relation to environmental impact, carbon emissions are generally used as a simplified, or key, indicator of environmental sustainability. Other, more sophisticated, methods include the *Ecopoint* system developed by the UK's Building Research Establishment (BRE) or *Eco-Footprinting* as supported by the World Wildlife Fund. Although there is significant controversy surrounding the Eco–Footprinting methodology, it does provide some simplified concepts to help to understand the scale of the climate change problem:

1. The earth's renewable resources are currently being consumed faster than they can be regenerated.
2. Three planets worth of resources would be required if all of the world's population had a westernised lifestyle.

Climate change predictions for the United Kingdom

The following scenarios are predicted for the United Kingdom:

- The climate will become warmer – by possibly 2–3.5°C by the 2080s.
- Hot summers will be more frequent and extreme cold winters less common.
- In 2004 one day per summer was expected to reach 31°C, but by the 2080s this could be nearer 10 days, with one day per summer reaching 38.5°C.
- Summers will become drier and winters wetter.
- There will be less snowfall.
- Heavy winter precipitation events will become more frequent.
- Sea levels will continue to rise by an estimated 26–86 cm by the 2080s.
- Extreme sea levels will return more frequently.

These changes to the climate are likely to have the following implications for building design:

- Increased flooding events (coastal, river and urban/flash).
- Buildings less weather-tight in face of more inclement weather.
- Increased foundation movement on clay soils due to drier summers.
- Increased summer overheating.
- Disruption of site activities due to inclement weather.
- Potential modifications to design loadings (e.g. wind) and durability predictions for building materials (in particular sealants, jointing materials, plastics, coatings and composites).

Source: UKCIP02 Climate Change Scenarios (funded by DEFRA, produced by the Tyndall & Hadley Centres for UKCIP). Copyright BRE, reproduced from Good Building Guide 63 with permission.

Sustainability scenarios and targets

In addition to dealing with the effects of climate change as it happens, sustainable design must also consider how buildings will help to achieve governmental carbon emissions reductions targets. In broad terms, environmental impact can be reduced by applying the following principles in order: *Reduce – Reuse – Recycle – Specify Green*. However, this sentiment is meaningless without specific targets and this is where the individual's view affects what action (if any) is taken.

As climate change involves complex interactions between human and natural systems over the long term, scenario planning is a frequently used assessment tool. Scenarios can be qualitative, narrative or mathematical predictions of the future based on different actions and outcomes. In 2001, the Intergovernmental Panel on Climate Change (IPCC) identified more than 500 mathematical scenarios and over 120 narrative scenarios, which fall into four simplified categories:

1. *Pessimist* – climate change is happening; little can be done to prevent or mitigate this.
2. *Economy paramount* – business should continue as usual, unless the environment affects the economy.
3. *High-tech optimist* – technological developments and a shift to renewable energy will provide the required efficiencies.
4. *Sustainable development* – self-imposed restrictions and lifestyle change with energy efficiencies in developed countries, to leave capacity for improved quality of life in developing countries.

UK policy is broadly based on the findings of the IPCC study and at the time of writing, the draft UK Climate Change Bill 2008 proposes the following carbon reduction targets based on a *Sustainable Development* agenda (although there are many organisations campaigning for higher targets):

- 26–32% reduction by 2020 compared to 1990 emissions
- 60% reduction by 2050 compared to 1990 emissions

These targets are likely to have a significant impact on the average UK lifestyle, and it is unlikely that the required carbon emissions reductions will be achieved by voluntary lifestyle changes:

> In 1990; carbon emissions were about 10.9 tonnes of CO_2 per person in the UK and we can assume that this represents the emissions generated by the average UK lifestyle. Government policy aims to reduce emissions by 60% by 2050 which equates to 4.4 tonnes of CO_2 per person per year. However a UK citizen who holidays in the UK, only travels on foot or by bike, likes a cool house, conserves energy and has lower than average fuel bills, while buying energy from renewable sources, eating locally grown fresh food and producing less than the average amount of household waste (most of which would be recycled), might generate about 5.6 tonnes of CO_2 per year.

As carbon emission targets are non-negotiable – if less is done in one area, more will need to be achieved in another area. For some time, the UK government policy has concentrated on making alterations to electricity generation to reduce carbon emissions (hence the public consultation on nuclear power in 2007). This has not been entirely successful as the proliferation of computers, personal electronics and air conditioning have offset most of the savings made by altering electricity supply.

Therefore, as the built environment is thought to be the second largest consumer of raw materials after food production, the construction industry is considered to be a sector where disproportionate reductions can be made to offset against other areas. Increasing controls are being applied via Planning, Building Regulations, Energy Performance of Buildings Directive and other initiatives such as the Code for Sustainable Homes. This legislation aims to achieve 'zero carbon' housing by 2016 and it is likely that similar controls are to be applied to commercial buildings by 2019. It is highly likely that, until the nuclear debate is settled, construction professionals will be expected to deliver increasingly more efficient buildings to minimise the impact on other areas of the average UK lifestyle.

It is worth noting that buildings can vary their energy in use by up to a factor of 3 depending on how the people inside choose to operate them, thus carbon emission reductions from energy-efficient buildings are not guaranteed. Carbon reductions from the built environment will only be delivered if we all choose to use and operate our buildings more efficiently.

For climate change sceptics, increasing energy costs and security of energy supply are two alternative reasons why energy efficiency measures are being pursued.

Sustainable building design priorities

In addition to the wider issues regarding targets, sustainable building design is difficult to grasp because of the many conflicting design constraints, as well as the need to involve the client, design team, contractors and building users in decision making. If substantial carbon emission savings are to be achieved, it is likely that similar substantial changes will also be required to broader industry practices, such as changes to contract structures, standard specifications and professional appointments being extended to include post-occupation reviews. However, while the industry infrastructure develops to support sustainable design in mainstream practice, there are still many issues for designers to tackle.

Order of priorities

To ensure that design efforts to improve environmental sustainability are efficiently targeted, a broad hierarchy should be applied by the design team:

- **Building location:** Transportation of building users to and from a rural location can produce more carbon emissions over the building's life than those produced by the building services. Encourage clients to choose a location which encourages the use of public transport.
- **Energy in use:** The energy used by building services can account for 60–80% of the total carbon emissions produced in the life of a building. Heavily serviced buildings can produce almost double the lifetime emissions of naturally ventilated buildings. Selection of natural ventilation and simple services can substantially reduce carbon emissions – as well as protecting clients against future energy costs or shortages.
- **Embodied energy:** The relative importance of embodied energy increases as energy in use is reduced, if an efficient location and services strategy has been selected, embodied energy can account for about 40% of a building's lifetime emissions. This should leave plenty of scope for some emissions reductions, although higher embodied energy materials can be justified if they can contribute to the thermal performance of the building and therefore reduce the energy in use emissions.
- **Renewables:** After efficiencies have been made in all other areas, renewable energy sources can be harnessed to reduce carbon emissions – assuming that the lifetime energy savings justify the embodied energy used in the technology!

Design team actions

In the early stages of design, design teams should aim to:

- Choose an efficient building shape to minimise energy in use and details to minimise air leakage and heat loss.
- Design simple buildings with reduced interfaces.
- Assess the implications of design life – lightweight short-lived construction, versus heavier, flexible, durable buildings.
- Anticipate change and make those changes easy to achieve. Consider the effect of future adaptions based on fashion and the expected service life of different building layers: 5–15 years for the fixtures and fittings, 5–20 years for the space plan, 5–30 years for building services, 30–60 years for the facade and 60–200 years for the structure.
- Establish whether the structure should contribute to the thermal performance of the building.
- Agree reduction targets for pollution, waste and embodied and operational energy.
- Consider a specification catchment area (or radius around the site) from which all the materials for the project might come, to minimise transport emissions and benefit the local economy and society.

Actions for the structural engineer

Structural engineers are most likely to have direct control of, or influence over:

- Assessment of structures and foundations for reuse.
- On-site reuse of materials from demolitions or excavations.
- Minimum soil movements around and off site.
- Selection of a simple structural grid and efficient structural forms.
- Detailing structures with thermal mass to meet visual/aesthetic requirements.
- Balancing selection of design loadings to minimise material use, versus provision of future flexibility/adaptability/deconstruction.
- Use of reclaimed, recycled, 'A-rated' or 'green' building materials.
- Use of specifications to ensure material suppliers use environmental management systems (e.g. ISO 14001 or EMAS).
- Avoidance of synthetic chemicals, polyvinyl chloride, etc.
- Limiting numbers of building materials to reduce waste.
- Design to material dimensions to limit off-cuts and waste.
- Assessment of embodied energy and potential reductions.
- Assessment of prefabrication to minimise waste, if the carbon emissions resulting from transport do not outweigh the benefits.
- Specifications to reduce construction and packaging waste.
- Drainage systems to minimise run-off.
- Use of flood protection measures and flood-resistant materials.
- Keep good records to help enable future reuse or refurbishment.

Exposed slabs and thermal mass

Thermal mass is the name given to materials which (when exposed to air flows) regulate temperature by slowly absorbing, retaining and releasing heat; preventing rapid temperature fluctuations. This property is quantified by the specific heat capacity (kJ/kg K). There are three situations where a service engineer might use exposed thermal mass to minimise or eliminate the need for mechanical heating and cooling, and therefore make considerable energy and carbon emissions savings:

- **Night-time cooling:** Air is let into buildings overnight to pre-cool walls and slabs, to increase their ability to absorb heat during the day. In the United Kingdom, this system is particularly suited to offices, but can be difficult to achieve in urban areas where heat tends to be retained, reducing the temperature differentials required to make passive cooling systems work.
- **Passive solar heating:** Heat from the sun is collected during daylight hours, stored in the building fabric and then slowly released overnight. In the United Kingdom, this system is particularly suited to domestic houses which require heating throughout most of the year.
- **Temperature stabilisation:** Where day and night-time temperatures vary significantly above and below the average temperature. Not typically required in the United Kingdom.

Although the decision to use thermal mass is generally driven by the desire for an energy-efficient building services strategy, the need for exposed thermal mass means that the architect and building users will be more concerned about how the structural elements look. Extra care has to be taken when selecting and specifying exposed elements in terms of finish, details, erection and protection from damage and weather. Thermal mass is likely to become increasingly important in the light of increasingly hot summers and the UK government policy to reduce carbon emissions.

Typical specific heat capacity of different building materials

Material	Typical specific heat capacity (kJ/m³ K)
Water	4184
Granite	2419
Concrete	2016
Sandstone	1806
Clay tiles	1428
Compressed earth block	1740
Rammed earth	1675
Brick	1612
Earth wall (adobe)	1300
Wood	806
Rockwool insulation	25
Fibreglass insulation	10

Source: Adapted from data in Guide A, CIBSE. 2006.

Embodied energy

The embodied energy of a material is the energy used to extract, process, refine and transport it for use. Typically the more processing steps, or distance travelled, the higher the embodied energy – which is often reflected in its price. The higher the embodied energy, the higher the carbon emissions generated by production.

In many cases it is possible to justify higher embodied energy if there is some other benefit, for example increased design loadings resulting in a more flexible building or concrete slabs providing thermal mass to regulate temperature. Although energy in use is more significant, it is still worth reducing embodied energy when this can be achieved without compromising performance standards or incurring other adverse environmental impacts.

Typical embodied energy contribution of building elements

Volume and service life for different materials affect each building element's environmental impact.

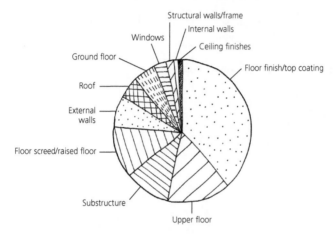

Although all specification choices are important, designers might concentrate on the building elements which have the greatest environmental impact:

- Floor construction, floor finishes
- External walls and windows
- Roofs

General strategies for reduction of embodied energy

It is a complex area and each case should be studied independently using the best method available at the time. Therefore, as the available figures for the embodied energy of typical building materials can vary by up to a factor of 10, it is better to follow general guidelines if no site-specific data are available for a particular project. The general themes are:

- Do not build more than you need – optimise rather than maximise space.
- Design long life, durable, simple and adaptable buildings.
- Modify or refurbish, rather than demolish or extend.
- 'High-tech' normally means higher levels of embodied energy.
- Consider higher design loadings to maximise the building life.
- Reuse material found on, or excavated from, site.
- Source materials locally.
- Use salvaged materials in preference to recycled materials.
- Use recycled materials in preference to new materials.
- Use high grade salvaged or recycled materials, not just as bulk fill, etc.
- Select low embodied energy materials.
- Give preference to materials produced using renewable energy.
- Specify standard sizes and avoid energy-intensive fillers.
- Avoid wasteful material use and recycle off-cuts and leftovers.
- Design for demountability for reuse or recycling.

Source: Anderson et al. (2002).

Typical embodied energy values for the UK building materials

Materials		Typical embodied energy[a] (MJ/kg)
Concrete	1:2:4 Mass concrete	0.99
	1% Reinforced section	1.81
	2% Reinforced section	2.36
	3% Reinforced section	2.88
	Precast	2.00
Steel	General	24.00
	Bar and rod	19.7
	Stainless	51.50
Facing bricks		8.20
Lightweight concrete blocks		3.50
Aggregate (general)		0.15
Timber	Sawn softwood	7.40
	Sawn hardwood	7.80
	Plywood	15.00
	Glue-laminated	11.00
Polycarbonate		112.90
Glass	Annealed	13.50 ± 5.00
	Toughened	16.20–20.70
Insulation	Mineral wool	16.60
	Polyurethane	72.10
	Sheeps' wool	3.00
Wallpaper		36.40
Plaster	Gypsum	1.8
	Plasterboard	2.7
Aluminium		154.30
Stone	Granite	5.90
	Imported granite	13.90
	Limestone	0.24
	Marble	2.00
	Slate	0.1–1.0
Drainage	Clay pipe	6.19–7.86
	Polyvinyl chloride pipe	67.50

Note:

[a] These figures are for illustrative purposes only. Embodied energy figures can vary up to a factor of 10 from site-specific data. The Green Guide to Specification might be a more useful guide for lay persons.

Source: University of Bath. 2006.

Construction waste

It is estimated that a minimum of 53 million tonnes of construction and demolition waste are produced annually in the United Kingdom. Of this, 24 million tonnes are recycled and 3 million tonnes is reclaimed – leaving 26 million tonnes being dumped in landfill, with associated air and water pollution. However, with waste disposal costs rising and the UK government's plan to halve the amount of construction waste going to landfill by 2012, waste reduction strategies are likely to become more widespread. Some of the first steps towards this are the creation of WRAP (Waste and Resources Action Programme) and introduction of compulsory Site Waste Management Plans for all sites in England from April 2008.

The benefits of reducing construction waste are threefold:

- Reduction of carbon emissions associated with less transport and processing.
- Reduction of waste going to landfill.
- Reduction of raw material use.

Using the *Reduce – Reuse – Recycle – Specify Green* hierarchy, construction waste can be minimised by using the specification and contract documents to:

- Encourage use of reusable protection and packaging systems for deliveries.
- Reduce soil movements by reusing material on site.
- Increase recovery and segregation of waste for reuse on site or recycling elsewhere (preferably locally).
- Use of efficient installation and temporary works systems.
- Design simple, efficient buildings with minimum materials and interfaces to reduce waste and off-cuts.
- Design buildings for flexibility, adaption and demountability.

Main potential for waste recovery and reuse after demolition

Steel, masonry, concrete and timber comprise the vast bulk of construction materials and all offer possibilities for reuse where fixings have been designed to facilitate this. Timber tends to be susceptible to poor practice and is not reused as often as steel and masonry. Glass and plastics tend to have limited reuse potential, and are generally more suited to recycling.

Material	Value	Potential for reuse	Potential for recycling
Concrete	High	• Precast concrete elements	• Crushed for use as aggregate in concrete mixes
		• Large concrete pieces to form thermal store for passive heating system	• Crushed for use as unbound fill
Masonry	Medium	• Bricks and blocks if used with soft mortar	• Crushed for use as aggregate in low strength concrete mixes
		• Stone and slate	• Crushed for use as unbound fill
Metals	High	• Steel beams and columns if dismantled rather than cut with thermal lance	• Aluminium and copper
			• Steel beams and reinforcement
Timber	Medium	• Generally reused in non-structural applications for indemnity reasons	• Chipped for use in landscaping, engineered timber products, etc., if not prevented by remains of old fixing and preservatives (treated timber is considered hazardous waste)
		• High value joinery	• Composting
			• Energy production
Glass	High	• Re-glazing	• Crushed for use as sand or fine aggregate in unbound or cement-bound applications
			• Crushed for use in shot blasting, water filtration, etc.

Reclaimed materials

Reclaimed materials are considered to be any materials that have been used before either in buildings, temporary works or other uses and are reused as construction materials without reprocessing. Reclaimed materials may be adapted and cut to size, cleaned up and refinished, but they fundamentally are being reused in their original form.

This is the purest and most environmental friendly form of recycling and therefore, where possible, should be investigated before the use of recycled or reprocessed materials in line with the *Reduce – Reuse – Recycle – Specify Green* hierarchy.

Although recycled content is generally resolved by manufacturers and their quality control processes, the use of reclaimed materials is generally more difficult as it must be resolved by the design team, within the limits of the site and project programme. The following should be considered if substantial amounts of reclaimed materials are proposed:

- Early discussions with reclaimed material dealers will help to identify materials that are easily available at the right quality and quantity.
- Basic modern salvage direct from demolition is often cheap or free, whereas older antique or reclaimed materials from salvage yards and stockholders (particularly in large quantities) may be quite costly.
- Material specifications need to be flexible to allow for the normal variations in reclaimed materials. Specifications should outline the essential performance properties required of a material, avoiding specification of particular products.
- Early design information helps in the sourcing of reclaimed materials, which might have considerably longer lead times than for off-the-shelf materials.
- It can be helpful to use agreed samples as part of the specification process – indicating acceptable quality, colour and state of wear and tear, etc.
- Identify nearby demolition projects and negotiate for reclaimed materials which might be useful.
- The building contractor will often need to set up relationships with new suppliers in the salvage trade.
- Additional storage space on, or near, the site is essential. Reclaimed materials do not fall into the 'just in time' purchasing process normally used by contractors.
- Use of experience, provenance, visual inspections, testing and/or clear audit trails to confirm material standards and satisfy indemnity requirements. Reclaimed materials inspectors are available in some areas if certain aspects are outside the design team's expertise.

Recycled materials

Recycled materials are considered to be any materials that have been taken from the waste stream and reprocessed or remanufactured to form part of a new product. A further refinement of this is:

- *Recycling* is where materials can be reclaimed with broadly the equivalent value to their original, for example, steel, paper, aluminium, glass and so on.
- *Downcycling* is where materials are reclaimed but can only be used in a lesser form than previously, for example, crushed concrete frame used as hardcore.

Ideally downcycling should be limited, but different materials are particularly suited to specific reprocessing techniques; metals being easily recycled, concrete most easily downcycled and timber relatively easily reused.

Most building designs achieve reasonable amounts of recycled content without explicitly trying as many manufacturers have traditionally used high levels of recycling. For example, a typical steel framed, masonry clad building might achieve 15–20% and a timber framed building about 10% (by value). WRAP (the Waste and Resources Action Plan) predict that most building projects could achieve a further 5–10% recycled content by specifying similar products (with higher recycled contents) without affecting the cost or affecting the proposed building design. This means that typical buildings should easily be able to achieve 20–25% (by value) recycled content without any radical action.

Although this sort of target reduces the amount of waste going to landfill, it does very little in the context of carbon emissions reductions. On the basis of the 30–60% reductions stated in the UK's Climate Change Bill, recycled content targets might be more meaningful in the 25–40% (by value) range.

Recycled content is normally calculated as a percentage of the *material value* to:

- Maximise carbon emissions savings – high value items generally have higher embodied energy values.
- Encourage action across the whole specification – rather than allowing design teams to concentrate on small savings on high volume materials.

WRAP have an online tool for calculating the recycled content of projects for most construction materials.

Finally, after deciding on a target for the project, adequate time must be set aside to ensure that appropriate materials and/or recycled contents are specified, as well as monitoring on site whether the materials specified are actually those being used.

Design for demountability

Although the use of reclaimed and recycled materials deals with waste from past building operations, building designers should perhaps consider the role of demountability in limiting the amount of waste produced in the future, in line with the *Reduce – Reuse – Recycle – Spec ify Green* hierarchy.

Basic principles of demountability

Specific considerations for demountability are as follows:

- Anticipate change and design/detail the building to allow changes for fit-out, replannig, major refurbishment and demolition to be made easily.
- Pay particular attention to the differential weathering/wearing of surfaces and allow for those areas to be maintained or replaced separately from other areas.
- Detail the different layers of the building so that they can be easily separated.
- Try to use durable components which can be reused and avoid composite elements, wet/ applied finishes, adhesives, resins and coatings as these tend to result in contamination of reclaimed materials on demolition.
- Try to specify elements which can be overhauled, renovated or redecorated as a 'second hand' appearance is unlikely to be acceptable.
- Selection of small elements (e.g. bricks) allows design flexibility and therefore allows more scope for reuse.
- Adopt a fixing regime which allows all components to be easily and safely removed, and replaced through the use of simple/removable fixings.
- Ensure the client, design team, contractors and subcontractors are briefed.
- Develop a detailed Deconstruction Plan with each design stage and ensure this forms part of the final CDM Building Manual submitted to the client.
- Carefully plan services to be easily identified, accessed and upgraded or maintained with minimum disruption. Design the services to allow long-term servicing rather than replacement.
- There is unlikely to be a direct cost benefit to most clients and design for complete demountability; however, it should be possible to make some degree of provision on most projects without cost penalty.

Demountable structure

For structure, design for demountability should probably be the last design consideration after flexible design loadings and layout, as the latter aim to keep the structural materials out of the waste stream for as long as possible. However, with rising landfill costs, demountability of structures may become more interesting in the future.

Connections are probably the single most important aspect of designing for deconstruction. The best fixings are durable and easily removable without destroying the structural integrity and finish of the joined construction elements. Dry fixing techniques are preferable and recessed or rebated connections which involve mixed materials should be avoided.

Implications of connections on deconstruction

Type of connection	Advantages	Disadvantages
Nail fixing	• Quick construction • Cheap	• Difficult to remove and seriously limits potential for timber reuse • Removal usually destroys a key area of element
Screw fixing	• Relatively easy to remove with minimum damage	• Limited reuse • Breakage during removal very problematic
Rivets	• Quick installation	• Difficult to remove without destroying a key area of element
Bolt fixing	• Good strength • Easily reused	• Can seize up, making removal difficult • Fairly expensive
Clamped	• Easily reused • Reduced fabrication	• Limited choice of fixings • Expensive
Traditional/ tenon	• Quick installation • Easily reused	• Additional fabrication • Expensive labour
Mortar	• Strength can be varied • Soft lime mortar allows easy material reclamation	• Cement mortars difficult to reuse and prevent reclamation of individual units • Lime mortar structures require more mass to resist tension compared to cement mortars
Adhesives	• Strong and efficient • Durable • Strength can be varied • Good for difficult geometry	• Likely to prevent effective reuse of parent materials • Relatively few solvents available for separation of bonded layers at the end of life • Adhesive not easily recycled or reused

Green materials specification

Assessment and specification of environmentally friendly materials is incredibly complex as research is ongoing and good practice is under constant review. There are a number of companies who provide advice on this area, but one of the best and simplest sources of information is *The Green Guide to Specification*. The most important issue is to feed this into the design early to inform decision making, rather than trying to justify a design once complete.

Environmental rating for selected suspended floors

Structural element	Summary rating
Beam and block floor with screed	A
Hollowcore slabs with screed	A
Hollowcore slabs with structural topping	B
In situ reinforced concrete slab	C
In situ reinforced concrete ribbed slab	B
In situ reinforced concrete waffle slab	B
Omni-deck-type precast lattice and structural topping	B
Omni-deck-type precast lattice with polystyrene void formers and structural topping	B
Holorib-type in situ concrete slab with mesh	B
Solid prestressed composite planks and structural topping	C

Notes:
[a] Suspended floor assessment based on 7.5 m grid, 2.5 kN/m^2 design load and 60-year building life.
[b] Timber joisted floors are typically not viable for this sort of arrangement, but perform significantly better in environmental terms than the flooring systems listed.
[c] Weight reductions in profiled slabs outweigh the environmental impact of increased amounts of shuttering.

Source: Green Guide to Specification. 2002.

Toxicity, health and air quality

Issues regarding toxicity generate considerable debate. Industrial chemicals are very much part of our lives and are permitted by the UK and European laws based on risk assessment analysis. Chemicals present in everyday objects, such as paints, flooring and plastics, leach out into air and water. However, many chemicals have not undergone risk assessment and assessment techniques are still developing. Environmental organisations, such as Greenpeace and the World Health Organization, suggest that we should substitute less or non-hazardous materials wherever possible to protect human health and the environment on the basis of the *Precautionary Principle*. With such large quantities specified, small changes to building materials could make considerable environmental improvements.

Summary of toxins, associated problems and substitutes

Material or product	Typical uses	Associated problem	Possible substitutes
Any containing VOCs	Petrochemical-based paints, plastics, flooring, etc.	See note	Water or vegetable oil-based products (e.g. linoleum, ceramic tile, wood, etc.)
Polyvinyl chloride (PVC or 'vinyl')	About 50% of all PVC is used in construction: pipes, conduit, waterproofing, roof membranes, door and window frames; flooring and carpet backing, wall coverings, furniture and cable sheathing	Lifetime VOC emissions; regular combination with heavy metals and release of hydrochloric acid if burnt	Aim for bio-based plastics such as polyethylene terephthalate (PET), polyolefins (PE, PP, etc.) or second choice PET, polyolefins (PE, PP, etc.) but try to avoid polyurethane, polystyrene, acrylonitrile butadiene styrene, polycarbonate
Phthalates	Used to make PVC flexible. Typically in vinyl flooring, carpet backing and PVC wall or ceiling coverings	Bronchial irritants; potential asthma triggers and have been linked to developmental problems	Subject to ongoing research and risk assessment, but options being considered include: adipates, citrates and cyclohexyl-based plasticisers
Polychloroprene (neoprene)	Geotextile, weather stripping, water seals, expansion joint filler, gaskets and adhesives	Same as PVC	
Composite wood products and insulation (using urea or phenol formaldehyde)	Panelling, furniture, plywood, chipboard, MDF, adhesives and glues	Formaldehyde is a potent eye, upper respiratory and skin irritant and is a carcinogen	
Preserved wood	Chromium copper arsenic (CCA), creosote and pentachlorophenol, PCP, lindane, tributyl tin oxide, dichlofluanid, permethrin	Carcinogenic	No treatment, boron-based compounds, Cu and Zn naphthanates or acypectas zinc
Heavy metals	Flashings, roofing, solder, switches, thermostats, thermometers, fluorescent lamps, paints and PVC products as stabilisers	Lead, mercury and organotins are particularly damaging to the brains of children. Cadmium can cause kidney and lung damage	
Halogenated flame retardants	Flame retardants used on polyurethane and polyisocyanurate (PIR) insulation in buildings	PBDEs and other brominated flame retardants disrupt thyroid and oestrogen hormones, causing problems with the brain and reproductive system	Area of ongoing research looking at halogen-free phosphorous compounds for PIR products. Rarely available

Notes:

[a] Volatile organic compounds (VOCs) are thousands of different chemicals (e.g. formaldehyde and benzene) which evaporate readily in air. VOCs are associated with dizziness, headaches, eye, nose and throat irritation or asthma, but some can also cause cancer, provoke longer-term damage to the liver, kidney and the nervous system.

[b] Greenpeace publishes a list of Chemicals for Priority Action (after OSPAR, 1998), which they are lobbying to have controlled.

Sustainable timber

Timber is generally considered a renewable resource as harvested trees can be replaced by new saplings. However, this is not always the case, with deforestation and illegal logging devastating ancient forests around the world. Since 1996 a number of certification schemes have been set up to help specifiers select 'legal' and 'sustainable' timber. The *Chain of Custody Certification* standards address management planning, harvesting, conservation of biodiversity, pest and disease management and social impacts of the forestry operations. At present only 7% of the world's forests are certified; are mainly located in the Northern Hemisphere and are relatively free from controversy. Sustainable forest management has great significance for the world climate and if all building specifications insist on certified timber, industry practices worldwide will be forced to improve.

The first action is to consider where timber is to be used on a project: structure, temporary works, shuttering, joinery, finishes and so forth and avoid selection of materials that are highly likely to come from illegal or unsustainable sources. At specification stage, options which should be included for all types of timber to be used are

- Certified timber from an official scheme.
- Timber from independently certified and reliable suppliers, with documentary evidence that supplies are from legal and well-managed forests.
- Timber from suppliers that have adopted a formal Environmental Purchasing Policy (such as those freely available from Forests Forever, VWVF 95+) for those products and that can provide evidence of commitment to that policy.
- Timber for illegal sources must not be used and any timber must be shown to come from legal sources. Environmental statements alone are not to be used as demonstration that materials are from a sustainable source.
- Where possible provide a list of FSC-accredited suppliers and request evidence of certified timber purchase. This might include requests for custody certificate numbers, copies of invoices and delivery notes.

Despite some weaknesses in some certification schemes, they are a significant step towards sourcing of sustainable timber. The main certification schemes are as follows:

- **COC: Chain of Custody**

Independent audit trail to prevent timber substitution and ensure an unbroken chain from well-managed forest to user.

- **FSC: Forest Stewardship Council**

Independent, non-profit organisation implementing a COC system. Greenpeace considers this to be the only scheme which is truly effective.

- **PEFC: Programme for the Endorsement of Forest Certification Schemes**

Global umbrella organisation for forestry industry bodies covering about 35 certification schemes including FSC, although with weaker social and environmental criteria.

- MTCC: Malaysian Timber Certification Council
- CSA: Canada Standards Authority
- SFI: Sustainable Forests Initiative

Timber preservatives

Although many hardwoods can be left untreated, softwood (whether internal or external) is routinely treated to provide durability via resistance to rot and insect infestation. This practice has only really developed since about 1940 and may be because the fast-grown timber used today does not have the same natural resistance to decay as the close-ringed, slow-grown timber which used to be standard (and can still be sourced at a premium today). Alternatively it may be because clients want the warranties and guarantees which preservative companies provide. However, the reasons for considering avoiding timber preservatives are as follows:

• Many preservatives release toxins to air, surface water and soil, to which workers and consumers are exposed.
• Impregnating wood hampers the sustainable reuse of wood.
• Treated timber is classed as 'Hazardous Waste' and should not be burned or sent to landfill to avoid air and water pollution.

Specifying timber preservatives

Timber will generally only deteriorate when its moisture content is higher than 20%. Therefore, with careful detailing, preservative treatment might be reduced or avoided, in accordance with BS 5589 and BS 5268: Part 5.

If preservatives are required, try to use non-toxic boron-based compounds such as borate oxide. However, the treatment can only be carried out on green timber with a moisture content of over 50%. As this is well above the desired moisture content at installation, sufficient time will need to be allowed in the programme for suppliers to be sourced and the timber to be seasoned (preferably by air, rather than kiln, drying).

If time is an issue, copper or zinc naphthanates, acypectas zinc, ammoniacal copper quaternary, copper azole and copper citrate can be considered – as a last resort. However, creosote, arsenic, chromium salts, dieldrin (banned in the United Kingdom), PCP, lindane, tributyl tin oxide, dichlofluanid, permethrin and copper chrome arsenate should be completely avoided.

Cement substitutes

Old Portland cement production (firing limestone and clay in kilns at high temperature) produces about 1 tonne of CO_2 for every tonne of cement produced. Cement production accounts for 5% of all European carbon emissions. In addition to improving energy efficiency at cement production plants, carbon emissions savings can be made by using cement substitutes, which also improve concrete durability. One drawback is that construction programmes need to allow for the slower curing times of the substitutes that work in two ways:

1. Hydration and curing like Portland cement, although slightly slower.
2. 'Pozzolans' providing silica that reacts with the hydrated lime which is an unwanted by-product of concrete curing. While stronger and more durable in the end, pozzolans take longer to set, although this can be mitigated slightly by reducing water content.

Ground-granulated blast furnace slag aggregate cement

Ground-granulated blast furnace slag aggregate cement (GGBS) is a by-product of iron and steel production. Molten slag is removed from the blast furnaces, rapidly quenched in water and then ground into a fine cementitious powder. GGBS tends to act more like Portland cement than a pozzolan and can replace Portland cement at rates of 30–70%, up to a possible maximum of 90%. As the recovery and production of 1 tonne of GGBS produces about 0.1 tonne of CO_2, considerable carbon emissions savings can be made. It is common practice in the United Kingdom for ready mixed concrete companies to produce concrete with a cementitious component of 50% GGBS and 50% Portland cement. Concrete using GGBS tends to be lighter in colour than those with Portland cements and can be considered as an alternative to white cement (which results in higher carbon emissions than Portland cement) for aesthetics or integration with daylighting strategies.

Pulverised fuel ash

Pulverised fuel ash (PFA) is a by-product of burning coal in power stations and is also known as 'Fly ash'. The ash is removed from flue gases using electrostatic precipitators and is routinely divided into two classes: 'Type C' and 'Type F' according to the lime (calcium) content. Type F has a higher calcium content and acts more like a pozzolan than Type C, which has pozzolanic and Portland cement qualities. Both types can be used in concrete production, replacing Portland cement at rates of 10–30%, though there have been examples of over 50% replacement. Available from ready mix suppliers, concrete mixed with PFA cement substitute tends to be darker in colour than Portland cement mixes.

Non-hydraulic and naturally hydraulic lime

Fired at lower temperatures than Portland cement and with the ability to reabsorb CO_2 while curing (as long as the volume of lime material itself does not prevent this) has led some lime manufacturers to claim that lime products are responsible for 50% less carbon emissions than similar cement products. Lime products are not commonly used in new-build projects, but are generating increasing interest. Traditionally, lime is used for masonry bedding and lime-ash floors. Being softer than cement, lime allows more movement and reduces the need for masonry movement joints (as long as the structure has sufficient mass to resist tensile stresses), as well as allowing easier recycling of both the masonry units and the lime itself.

Magnesite

The idea of replacement of the calcium carbonate in Portland cement with magnesium carbonate (magnesite or dolomite) dates back to the nineteenth century. Less alkaline than Portland cement mixes, they were not pursued due to durability problems. However, MgO cement uses 'reactive' magnesia that is manufactured at much lower temperatures than Portland cement (reducing emissions by about 50%), is more recyclable than polycarbonate, is expected to provide improved durability and to have a high propensity for binding with waste materials. It is claimed that magnesite can be used in conjunction with other cement replacements without such problems as slow curing times. The main barrier to use seems to be that although magnesite is an abundant mineral, it is expensive to mine. Research is currently being carried out and further information is available from the University of Cambridge, TecEco and the BRE.

Sustainable aggregates

Sustainable aggregates fall into two categories:

1. *Recycled aggregates (RAs)*: derived from reprocessing materials previously used in construction.
2. *Secondary aggregates*: usually by-products of other industrial processes not previously used in construction. Secondary aggregates can be further sub-divided into manufactured and natural, depending on their source.

Although the United Kingdom is a leading user of sustainable aggregates, with about 25% of the total UK aggregate demand being met with sustainable products, there is scope for this to be expanded further. RAs can be used in unbound, cement bound and resin bound applications subject to various controls.

RAs can be purchased directly from demolition sites or from suitably equipped processing centres, and the quality of the product depends on the selection, separation and processing techniques used. RA can be produced on site, at source or off site in a central processing plant, with economic and environmental benefits maximised with on-site processing.

Many materials have a strong regional character, with China clay sand from South West England, slate waste from North Wales and metallurgical slag from South Wales, Yorkshire and Humberside. Clearly, the biggest economic and environmental benefits will be gained when materials are used locally.

Typical uses for the UK recycled and secondary aggregates

Aggregate type	Potential for reuse				Notes
	Unbound aggregate	Concrete aggregate	Lightweight aggregate	Building components aggregate	
Recycled aggregate					
Crushed concrete (recycled concrete aggregate)	High	High	None	Some	Alkali silica reaction (ASR), frost resistance and weathering should be considered
Crushed masonry (RA)	High	High	High	Some	
Ceramic waste	Some	Some	None	Some	
Recycled glass	Some	High	Some	Some	
Spent rail ballast	High	Low	None	High	
Mixed plastic	None	Low	Some	Some	
Scrap tyres	None	None	Low	Some	
Secondary aggregate – manufactured					
Blast furnace slag (Lytag)	High	High	High	High	Regular use. BS EN and BRE IP18/01:2001 guidance
Steel slag	None	None	None	None	See BRE Reports.
Non-ferrous slags	Low	Some	None	Low	Unbound use to comply with BRE Digest SD1:2001
Pulverised fuel ash	Some	Some	High	Some	ASR, sulphates, frost resistance and weathering should be considered
Incinerator bottom ash from municipal waste incinerators	Some	Some	Some	Some	
Furnace bottom ash	Some	Some	High	High	Fully utilised in concrete blocks
Used foundry sand	Some	High	None	Some	
Sewage sludge as synthetic aggregate	None	None	High	None	Use as a substitute for natural gypsum
FGD gypsum (desulphogypsum results from desulphurisation of coal-fired power station flue gases)	None	None	None	High	
Secondary aggregate – natural					
Slate waste	High	Some	High	High	
China clay sand	Some	High	None	High	Sulphate content can be high. Unbound use to comply with BRE Digest SD1:2001
Burnt colliery spoil	Some	Low	None	Low	
Unburnt colliery spoil	High	None	Some	None	
Clay waste	None	None	High	None	

Detailed descriptions of these materials can be found within the Recycled Content Specifier Tool available on the WRAP website.

Unbound use

RAs are highly suitable for use under floor slabs and for pipe bedding as well as for general fill materials in building construction. Contaminants such as metals, plastic and wood should normally be kept below 2% and the grading should be suitable for full compaction where this is required.

The only other consideration when using them is that the fine fractions of both recycled concrete aggregate (RCA) and RA could be contaminated with sulphate salts (e.g. from some types of gypsum plaster) to a degree sufficient to cause sulphate attack on concrete in contact with it. Therefore, the soluble sulphate content of material, containing fine RA, should be tested and appropriate precautions taken.

RAs in concrete mixes

Based on the *Reduce – Reuse – Recycle – Specify Green* hierarchy, and the desire to avoid *Downcycling*, engineers should be aiming to use sustainable aggregates in concrete mixes, in addition to lower grade unbound applications such as fill and pipe bedding and so on.

Most ready mix concrete suppliers can offer concrete containing RAs, but mixes are of limited availability and may not be available at the right time, in the right place or in the required quantities. It is therefore not practical to insist on their use on every project at present, but expressing a preference in specifications should encourage their use where possible.

BS 8500 gives limits on the permitted composition of recycled coarse aggregates as well as guidance on where and how their use in concrete is permitted, but the use of fine aggregates is not covered at present as their increased water demand generally leads to low strength mixes. The use of concrete containing recycled coarse aggregates is restricted to the least severe exposure classes and is not yet practical for use in site batching. Designated concrete mixes, which require strength tests to be carried out, are the easiest way to specify recycled content while maintaining quality control.

BS 8500: Part 2 Clause 4.3 defines two categories of coarse RA, that is, RCA consisting primarily of crushed concrete (i.e. where less than 5% is crushed masonry) and RA that may include a higher proportion of masonry and must meet a default value for aggregate drying shrinkage of 0.075%.

RA is limited to use in concrete with a maximum strength class of C16/20 and in only the mildest exposure conditions, whereas RCA can be used up to strength class C40/50 and in a wider range of exposure conditions, but is generally restricted to use in non-aggressive soils (DC-1 conditions).

Although it is generally accepted that the use of coarse RCA to replace up to 30% of the natural coarse aggregate will have an insignificant effect on the properties of concrete, for BS 8500 designated concretes RC25–RC50, the amount of RCA or RA is restricted to 20% by weight of the total coarse aggregate fraction unless the specifier gives permission to relax this requirement.

Therefore, where appropriate for exposure conditions, specification clauses should include the following in order to promote the use of RAs:

- The use of recycled materials (RCA or RA), if available, as coarse aggregate is the preferred option.
- The proportion of RA or RCA (as a mass fraction of the total coarse aggregate) is permitted to exceed 20%.

16
Useful Mathematics

Trigonometric relationships

Addition formulae

$\sin(A \pm B) = \sin A \cos B \pm \cos A \sin B$

$\cos(A \pm B) = \cos A \cos B \mp \sin A \sin B$

$\tan(A \pm B) = \dfrac{\tan A \pm \tan B}{1 \mp \tan A \tan B}$

Sum and difference formulae

$\sin A + \sin B = 2\sin\dfrac{1}{2}(A + B) \cos\dfrac{1}{2}(A + B)$

$\sin A - \sin B = 2\cos\dfrac{1}{2}(A + B) \sin\dfrac{1}{2}(A - B)$

$\cos A + \cos B = 2\cos\dfrac{1}{2}(A + B) \cos\dfrac{1}{2}(A - B)$

$\cos A - \cos B = -2\sin\dfrac{1}{2}(A + B) \sin\dfrac{1}{2}(A - B)$

$\tan A + \tan B = \dfrac{\sin(A + B)}{\cos A \cos B}$

$\tan A - \tan B = \dfrac{\sin(A - B)}{\cos A \cos B}$

Product formulae

$2\sin A \cos B = \sin(A - B) + \sin(A + B)$

$2\sin A \sin B = \cos(A - B) - \cos(A - B)$

$2\cos A \cos B = \cos(A - B) + \cos(A + B)$

Multiple angle and powers formulae

$\sin 2A = 2 \sin A \cos A$

$\cos 2A = 2 \cos^2 A - \sin^2 A$

$\cos 2A = 2 \cos^2 A - 1$

$\cos 2A = 1 - 2\sin^2 A$

$\tan 2A = \dfrac{2 \tan A}{1 - \tan^2 A}$

$\sin^2 A + \cos^2 A = 1$

$\sin^2 A = \tan^2 A + 1$

Relationships for plane triangles

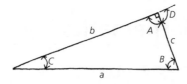

Pythagoras for right-angled triangles: $b^2 + c^2 = a^2$

Sin rule:

$$\frac{a}{\sin A} = \frac{b}{\sin B} = \frac{c}{\sin C}$$

$$\sin A = \frac{2}{bc}\sqrt{s(s-a)(s-b)(s-c)},$$

where $s = (a + b + c)/2$

Cosine rule:

$$a^2 = b^2 + c^2 - 2bc\cos A$$
$$a^2 = b^2 + c^2 + 2bc\cos D$$

$$\cos A = \frac{b^2 + c^2 - a^2}{2bc}$$

Special triangles

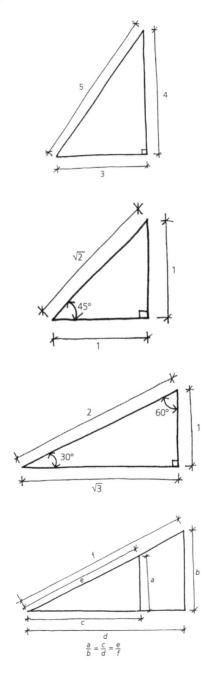

$$\frac{a}{b} = \frac{c}{d} = \frac{e}{f}$$

Algebraic relationships

Quadratics

$$ax^2 + bx + c = 0 \qquad x = \frac{-b \pm \sqrt{b^2 - 4ac}}{2a}$$

$$x^2 + 2xy + y^2 = (x+y)^2$$

$$x^2 - y^2 = (x+y)(x-y)$$

$$x^3 - y^3 = (x-y)(x^2 + xy + y^2)$$

Powers

$$a^x a^y = a^{x+y} \qquad \frac{a^x}{a^y} = a^{x-y} \qquad (a^x)^y = a^{xy}$$

Logarithms

$$x \equiv e^{\log_e x} \equiv e^{\ln x}$$

$$x \equiv \log_{10}(10^x) \equiv \log_{10}(\text{antilog}_{10} x) \equiv 10^{\log_{10} x}$$

$$e = 2.71828$$

$$\ln x = \frac{\log_{10} x}{\log_{10} e} = 2.30259 \log_{10} x$$

Equations of curves

Circle

$$x^2 + y^2 = a^2$$

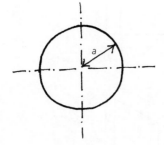

Ellipse

$$\frac{x^2}{a^2} + \frac{y^2}{b^2} = 1$$

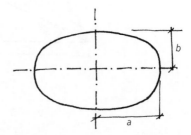

Hyperbola

$$\frac{x^2}{a^2} - \frac{y^2}{b^2} = 1$$

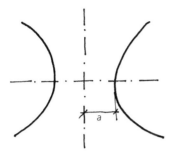

Parabola

$$y^2 = ax$$

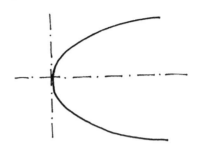

Circular arc

$$R = \left(d^2 + \frac{L^2}{4} \right) \frac{1}{2d}$$

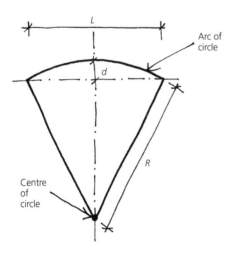

Rules for differentiation and integration

$$\frac{d}{dx}(uv) = u\frac{dv}{dx} + v\frac{du}{dx}$$

$$\frac{d}{dx}\left(\frac{u}{v}\right) = \frac{1}{v^2}\left(v\frac{du}{dx} - u\frac{dv}{dx}\right)$$

$$\frac{d}{dx}(uvw) = uv\frac{dw}{dx} + uw\frac{dv}{dx} = vw\frac{du}{dx}$$

$$\int(uv)dx = u\int(v)dx - \int\frac{du}{dx}\int(v)dx$$

Standard differentials and integrals

$$\frac{d}{dx}x^n = nx^{n-1}$$

$$\int x^n dx = \frac{x^{n+1}}{n+1} \quad n \neq 1$$

$$\frac{d}{dx}\ln x = \frac{1}{x}$$

$$\int\frac{1}{x}dx = \ln x$$

$$\frac{d}{dx}e^{ax} = ae^{ax}$$

$$\int e^{ax}dx = \frac{e^{ax}}{a} \quad a \neq 0$$

$$\frac{d}{dx}a^x = a^x\ln a$$

$$\int a^x dx = \frac{a^x}{\ln a} \quad a > 0, a \neq 0$$

$$\frac{d}{dx}x^x = x^x(1 + \ln x)$$

$$\int \ln x\, dx = x(\ln x - 1)$$

$$\frac{d}{dx}\sin x = \cos x$$

$$\int \sin x\, dx = -\cos x$$

$$\frac{d}{dx}\cos x = -\sin x$$

$$\int \cos x\, dx = \sin x$$

$$\frac{d}{dx}\tan x = \sec^2 x$$

$$\int \tan x\, dx = -\ln(\cos x)$$

$$\frac{d}{dx}\cot x = -\csc^2 x$$

$$\int \cot x\, dx = \ln(\sin x)$$

$$\frac{d}{dx}\sin^{-1} x = \frac{1}{\sqrt{1 - x^2}}$$

$$\int \sec^2 x\, dx = \tan x$$

$$\frac{d}{dx}\cos^{-1} x = \frac{1}{\sqrt{1 + x^2}}$$

$$\int \csc^2 x\, dx = -\cot x$$

$$\frac{d}{dx}\tan^{-1} x = \frac{1}{1 + x^2}$$

$$\int \frac{1}{\sqrt{1 - x^2}}dx = \sin^{-1} x \quad |x| < 1$$

$$\frac{d}{dx}\cot^{-1} x = \frac{-1}{1 + x^2}$$

$$\int \frac{1}{\sqrt{1 + x^2}}dx = \tan^{-1} x$$

Useful Addresses

Advisory organisations

Aluminium Federation Ltd

National Metal Forming Centre, 47 Birmingham Road,
W Bromwich B70 6PY
www.alfed.org.uk

Tel: 0121 601 6363
Fax: 0870 138 9714

Ancient Monuments Society

St Anne's Vestry Hall, 2 Church Entry, London
EC4V 5HB
www.ancientmonumentssociety.org.uk

Tel: 020 7236 3934

Arboricultural Advisory & Information Service

Alice Holt Lodge, Wreclesham, Farnham GU10 4LH
www.treehelp.info

Tel: 01420 22022
Fax: 01420 22000

Arboricultural Association

The Malthouse, Stroud Green, Standish, Stonehouse, Gloucestershire
GL10 3DL
www.trees.org.uk

Tel: 01242 522152
Fax: 01242 577766

Architects Registration Board (ARB)

8 Weymouth Street, London W1W 5BU
www.arb.org.uk

Tel: 020 7580 5861
Fax: 020 7436 5269

Asbestos Removal Contractors Association (ARCA)

Unit 1 Stretton Business Park 2, Brunel Drive Stretton, Burton upon
Trent, DE13 0BY
www.arca.org.uk

Tel: 01283 566467
Fax: 01283 505770

Association for Project Safety

5 New Mart Place, Edinburgh EH14 1RW
www.aps.org.uk

Tel: 0131 442 6600
Fax: 0131 442 6601

Association for the Conservation of Energy

Westgate House, 2a Prebend Street, London N1 8PT
www.ukace.org

Tel: 020 7359 8000
Fax: 020 7359 0863

Brick Development Association Ltd (BDA)

The Building Centre, 26 Store Street, London WC1E 7BT
www.brick.org.uk

Tel: 020 7323 7030
Fax: 020 7580 3795

British Adhesives & Sealants Association (BASA)

5 Alderson Road, Worksop, Notts S80 1UZ
www.basa.uk.com

Tel: 01909 480888
Fax: 01909 473834

British Architectural Library

RIBA, 66 Portland Place, London W1N 4AD   Tel: 020 7307 3882
www.architecture.com/LibraryDrawingsAndPhotographs

British Board of Agrément (BBA)

Bucknalls Lane, Garston, Watford, Herts WD25 9BA   Tel: 01923 665300
www.bbacerts.co.uk   Fax: 01923 665301

British Cement Association (BCA)

Riverside House, 4 Meadows Business Park, Blackwater, Camberley   Tel: 01276 608700
GU17 9AB   Fax: 01276 608701
www.cementindustry.co.uk

British Constructional Steelwork Association Ltd (BCSA)

4 Whitehall Court, London SW1A 2ES   Tel: 020 7839 8566
www.steelconstruction.org   Fax: 020 7976 1634

British Library

96 Euston Road, London NW1 2DB   Tel: 0843 2081144
www.bl.uk   Fax: 020 7412 7954

British Precast Concrete Federation (BPCF)

60 Charles Street, Leicester LE1 1FB   Tel: 0116 253 6161
www.britishprecast.org   Fax: 0116 251 4568

British Rubber and Polyurethane Products Association Ltd (BRPPA)

5 Berewyk Hall Court, White Colne, Colchester, Essex   Tel: 01787 226995
www.brppa.co.uk   Fax: 0845 301 6853

British Safety Council (BSC)

70 Chancellor's Road, London W6 9RS   Tel: 020 8741 1231
www.britsafe.org   Fax: 020 8741 4555

British Stainless Steel Association

Broomgrove, 59 Clarkehouse Road, Sheffield S10 2LE   Tel: 0114 267 1260
www.bssa.org.uk   Fax: 0114 266 1252

British Standards Institution (BSI)

389 Chiswick High Road, London W4 4AL   Tel: 020 8996 9001
www.bsigroup.com   Fax: 020 8996 7001

Building Centre

26 Store Street, London WC1E 7BT   Tel: 020 7692 4000
www.buildingcentre.co.uk   Fax: 020 7580 9641

Building Research Establishment (BRE)

Bucknalls Lane, Garston, Watford WD25 9XX   Tel: 01923 664000
www.bre.co.uk   Fax: 01923 664787

Building Services Research and Information Association (BSRIA)

Old Bracknell Lane West, Bracknell, Berks RG12 7AH   Tel: 01344 465600
www.bsria.co.uk   Fax: 01344 465626

CADW – Welsh Historic Monuments

Plas Carew, Unit 5/7, Cefn Coed, Parc Nantgarw, Cardiff CF15 7QQ
www.cadw.wales.gov.uk

Tel: 01443 336 000
Fax: 01443 336 001

Canal & River Trust (formerly British Waterways)

First Floor North, Station House, 500 Elder Gate, Milton Keynes
MK9 1BB
www.canalrivertrust.org.uk

Tel: 0303 040 4040

Cares (UK Certification Authority for Reinforcing Steels)

Pembroke Mews, Pembroke Road, Sevenoaks, Kent TN13 1XR
www.ukcares.com

Tel: 01732 450000
Fax: 01732 455917

Cast Metal Federation

47 Birmingham Road, West Bromwich, West Midlands B70 6PY
www.castmetalsfederation.com

Tel: 0121 601 6397
Fax: 0121 601 6391

Castings Development Centre

Castings Technology International, Advanced Manufacturing Park,
Brunel Way, Rotherham S60 5WG
www.castingstechnology.com

Tel: 0114 254 1144
Fax: 0114 254 1155

Commission for Architecture and the Built Environment (CABE) at The Design Council

Angel Building, 407 St John Street, London EC1V 4AB
http://www.designcouncil.org.uk/our-work/cabe

Tel: 020 7420 5200
Fax: 020 7420 5300

Concrete Repair Association (CRA)

Kingsley House, Ganders Business Park, Kingsley, Bordon, Hampshire
GU35 9LU
www.cra.org.uk

Tel: 01420 471615

Concrete Society

Riverside House, 4 Meadows Business Park, Station Approach,
Blackwater, Camberley GU17 9AB
www.concrete.org.uk

Tel: 01276 607140
Fax: 01276 607141

Construction Fixings Association

65 Dean Street, Oakham LE15 6AF
www.fixingscfa.co.uk

Tel/fax: 01664 823687

Construction Industry Research & Information Association (CIRIA)

Classic House, 174–180 Old Street, London EC1V 9BP
www.ciria.org

Tel: 020 7549 3300
Fax: 020 7253 0523

Copper Development Association (CDA)

5 Grovelands Business Centre, Boundary Way, Hemel Hempstead HP2
7TE
www.copperinfo.co.uk

Tel: 01442 275705
Fax: 01442 275716

Council for Aluminium in Building

Bank House, Bond's Mill, Stonehouse, Glos GL10 3RF
www.c-a-b.org.uk

Tel: 01453 828 851
Fax: 01453 828 861

Design Council

Angel Building, 407 St John Street, London EC1V 4AB
http://www.designcouncil.org.uk

Tel: 020 7420 5200
Fax: 020 7420 5300

English Heritage

The Engine House, Fire Fly Avenue, Swindon SN2 2EH
www.english-heritage.org.uk

Tel: 0870 333 1181
Fax: 01793 414926

European Glaziers Association (UEMV)

PO Box 41, NL-1483 ZG De Rijp, the Netherlands
www.uemv.com

Tel: +31 299 68 26 14
Fax: +31 299 68 26 19

European Stainless Steel Advisory Body (Euro-Inox)

21st Century Building, 19 rue de Bitbourg, L-1273
Luxembourg
www.euro-inox.org

Tel: +35 226 10 30 50
Fax: +35 226 10 30 51

Federation of Manufacturers of Construction Equipment & Cranes

Airport House, Purley Way, Croydon CR0 0XZ
www.coneq.org.uk

Tel: 020 8253 4502
Fax: 020 8253 4510

Federation of Master Builders

Gordon Fisher House, 14-15 Great James Street, London WC1N 3DP
www.fmb.org.uk

Tel: 020 7242 7583
Fax: 020 7404 0296

Federation of Piling Specialists

Forum Court, 83 Copers Cope Road, Beckenham, Kent BR3 1NR
www.fps.org.uk

Tel: 020 8663 0947
Fax: 020 8663 0949

Fire Protection Association (FPA)

London Road, Moreton in Marsh, Glos GL56 0RH
www.thefpa.co.uk

Tel: 01608 812500

Forest Stewardship Council (FSC)

11–13 Great Oak Street, Llanidloes, Powys SY18 6BU
www.fsc-uk.org

Tel: 01686 413916
Fax: 01686 412176

Friends of the Earth

26–28 Underwood Street, London N1 7JQ
www.foe.co.uk

Tel: 020 7490 1555

Galvanizers' Association

Wren's Court, 56 Victoria Road, Sutton Coldfield, W. Midlands B72 1SY
www.galvanizing.org.uk

Tel: 0121 355 8838

Georgian Group

6 Fitzroy Square, London W1P 6DX
www.georgiangroup.org.uk

Tel: 0871 750 2936

Glass and Glazing Federation

54 Ayres Street, London SE1 1EU
www.ggf.org.uk

Tel: 020 7939 9101

Glue Laminated Timber Association

Chiltern House, Stocking Lane, High Wycombe HP14 4ND
www.glulam.co.uk

Tel: 01494 565180
Fax: 01494 565487

Health and Safety Executive (HSE)

Redgrave Court, Merton Road, Bootle, Merseyside L20 7HS
www.hse.gov.uk

Tel: 0151 951 4000

Historic Scotland

Longmore House, Salisbury Place, Edinburgh EH9 1SH
www.historic-scotland.gov.uk

Tel: 0131 668 8600

HM Land Registry

Ty Cwm Tawe, Phoenix Way, Llansamlet, Swansea SA7 9FQ
www.landreg.gov.uk

Tel: 0844 892 1111

Institution of Civil Engineers (ICE)

1–7 Great George Street, London SW1P 3AA
www.ice.org.uk

Tel: 020 7222 7722
Fax: 020 7222 7500

Institution of Structural Engineers (IStructE)

11 Upper Belgrave Street, London SW1X 8BH
www.istructe.org

Tel: 020 7235 4535
Fax: 020 7235 4294

London Metropolitan Archives

40 Northampton Road, London EC1R 0HB
www.lma.gov.uk

Tel: 020 7332 3820

Mineral Products Association

38–44 Gillingham Street, London SW1V 1HU
www.mineralproducts.org

Tel: 020 7963 8000
Fax: 020 7963 8001

Meteorological Office

Fitzroy Road, Exeter, Devon EX1 3PB
www.metoffice.gov.uk

Tel: 0870 900 0100
Fax: 0870 900 5050

National Building Specification Ltd (NBS)

The Old Post Office, St Nicholas Street, Newcastle upon Tyne NE1 1RH
www.thenbs.co.uk

Tel: 0191 244 5500
Fax: 0191 232 5714

National House-Building Council (NHBC)

NHBC House, Davy Anveue, Knowhill, Milton Keynes
MK5 8FP
www.nhbc.co.uk

Tel: 0844 633 1000
Fax: 01908 747255

Network Rail

Kings Place, 90 York Way, London N1 9AG
www.networkrail.co.uk

Tel: 020 7557 8000

Nickel Development Institute (NIDI)

8th Floor, Avenue des Arts 14, Brussels, Belgium 1210
www.nickelinstitute.org

Tel: +32 2290 3200

Northern Ireland Environment Agency Built Heritage

Waterman House, 5–33 Hill Street, Belfast BT1 2LA   Tel: 028 9054 3095
www.doeni.gov.uk/niea   Fax: 028 9054 3150

Ordnance Survey

Adanac Drive, Southampton SO16 0AS   Tel: 08456 050505
www.ordnancesurvey.co.uk

Paint Research Association (PRA)

14 Castle Mews, High Street, Hampton, Middx TW12 2NP   Tel: 020 8487 0800
www.pra-world.com   Fax: 020 8487 0801

Plastics and Rubber Advisory Service, British Plastics Federation (BPF)

6 Bath Place, Rivington Street, London EC2A 3JE   Tel: 020 7457 5000
www.bpf.co.uk   Fax: 020 7457 5020

The Property Care Association (formerly BWPDA)

Lakeview Court, Ermine Business Park Huntingdon, Cambridgeshire   Tel: 0844 3754301
PE29 6XR   Fax: 01480 417587
www.property-care.org

Pyramus and Thisbe Club

Administration Office, Rathdale House, 30 Back Road, Rathfriland,   Tel: 028 4063 2082
Belfast BT34 5QF   Fax: 028 4063 2083
www.partywalls.org.uk

Royal Incorporation of Architects in Scotland (RIAS)

15 Rutland Square, Edinburgh EH1 2BE   Tel: 0131 229 7545
www.rias.org.uk   Fax: 0131 228 2188

Royal Institute of British Architects (RIBA)

66 Portland Place, London W1B 1AD   Tel: 020 7580 5533
www.architecture.com   Fax: 020 7255 1541

Royal Institution of Chartered Surveyors (RICS)

Parliament Square, London SW1P 3AD   Tel: 0870 333 1600
www.rics.org   Fax: 020 7334 3811

Royal Society of Architects in Wales (RSAW)

4 Cathedral Road, Cardiff CF11 9LJ   Tel: 029 2022 8987
www.architecture.com   Fax: 029 2023 0030

Royal Society of Ulster Architects (RSUA)

2 Mount Charles, Belfast BT7 1NZ   Tel: 028 9032 3760
www.rsua.org.uk   Fax: 028 9023 7313

Scottish Canals (formerly British Waterways)

Caledonian Canal Office, Seaport Marina, Muirtown Wharf, Inverness   Tel: 01463 725500
IV3 5LE
www.canalrivertrust.org.uk

Scottish Building Standards Agency

Scottish Government, Denholm House, Almondvale Business Park,   Tel: 01506 600400
Livingston EH54 6GA   Fax: 01506 600401
www.sbsa.gov.uk

Society for the Protection of Ancient Buildings

37 Spital Square, London E1 6DY
www.spab.org.uk

Tel: 020 7377 1644
Fax: 020 7247 5296

Stainless Steel Advisory Service

Broomgrove, 59 Clarkehouse Street, Sheffield S10 2LE
www.bssa.org.uk

Tel: 0114 267 1260
Fax: 0114 266 1252

The Stationery Office (TSO previously HMSO)

St Crispins, Duke Street, Norwich NR3 1PD
www.tso.co.uk

Tel: 01603 622211

Steel Construction Institute (SCI)

Silwood Park, Buckhurst Road, Ascot, Berks SL5 7QN
www.steel-sci.org.uk

Tel: 01344 636 525
Fax: 01344 636 570

Stone Federation Great Britain (SFGB)

Channel Business Centre, Ingles Manor, Castle Hill Ave, Folkestone
CT20 2RD
www.stone-federationgb.org.uk

Tel: 01303 856123
Fax: 01303 856117

TATA Steel Construction Centre

PO Box 1, Brigg Road, Scunthorpe DN16 1BP
www.corusconstruction.com

Tel: 01724 405060
Fax: 01724 404224

Thermal Spraying & Surface Engineering Association (TSSEA)

38 Lawford Lane, Bilton, Rugby, Warwickshire CV22 7JP
www.tssea.co.uk

Tel: 01788 522792
Fax: 01788 522905

Timber Trade Federation

Building Centre, 26 Store Street, London WC1E 7BT
www.ttf.co.uk

Tel: 020 3205 0067
Fax: 020 7291 5379

TRADA Technology Ltd

Stocking Lane, Hughenden Valley, High Wycombe HP14 4ND
www.tradatechnology.co.uk

Tel: 01494 569600
Fax: 01494 565487

UK Cast Stone Association

15 Stonehill Court, The Arbours, Northampton NN3 3RA
www.ukcsa.co.uk

Tel/fax: 01604 405666

UK Climate Impact Programme (UKCIP)

Oxford University Centre for the Environment,
South Parks Road, Oxford OX1 3QY
www.ukcip.org.uk

Tel: 01865 285717
Fax: 01865 285710

Victorian Society

1 Priory Gardens, Bedford Park, London W4 1TT
www.victorian-society.org.uk

Tel: 020 8994 1019

Waste & Resources Action Plan (WRAP)

The Old Academy, 21 Horse Fair, Banbury OX16 0AH
www.wrap.org.uk

Tel: 0808 100 2040

Water Authorities Association

1 Queen Anne's Gate, London SW1H 9BT
www.water.org.uk

Tel: 020 7344 1844
Fax: 020 7344 1853

Water Jetting Association

Thames Innovation Centre, Veridion Way, Erith,
Kent DA18 4AL
www.waterjetting.org.uk

Tel: 020 8320 1090

Wood Panel Industries Federation

28 Market Place, Grantham, Lincolnshire NG31 6LR
www.wpif.org.uk

Tel: 01476 563707
Fax: 01476 579314

The Wood Protection Association (formerly BWPDA)

5C Flemming Court, Castleford, West Yorks WF10 5HW.
www.wood-protection.org

Tel: 01977 558274

Manufacturers

3M Tapes & Adhesives UK Ltd

3M Centre, Cain Road, Bracknell RG12 8HT
www.solutions3 m.co.uk

Tel: 01344 858000
Fax: 01344 858278

Abraservice (special steels)

Arley Road, Saltley, Birmingham, West Midlands B8 1BB
www.abraservice.com

Tel: 0121 326 3100
Fax: 0121 326 3105

Angle Ring Company Ltd

Bloomfield Road, Tipton, West Midlands DY4 9EH
www.anglering.co.uk

Tel: 0121 557 7241
Fax: 0121 522 4555

Aplant Acrow/Ashtead Group plc

102 Dalton Ave, Birchwood Park, Warrington WA3 6YE
www.aplant.com

Tel: 01925 281000
Fax: 01925 281001

BGT Bischoff Glastechnik

Alexanderstraβe 2, 705015 Bretten, Germany
www.bgt-bretten.de

Tel: +49 7252 5030
Fax: +49 7252 503283

BRC Building Products

South Yorkshire Industrial Estate, Whaley Road, Barugh,
Barnsley, South Yorkshire S75 1HT
www.brc.ltd.uk

Tel: 01226 729793
Fax: 01226 248738

Caltite/Cementaid (UK) Ltd

1 Baird Close, Crawley, West Sussex RH10 9SY
www.cementaid.com

Tel: 0845 658 2000

Catnic

Corus UK Ltd, Pontypandy Industrial Estate, Caerphilly CF83 3GL
www.catnic.com

Tel: 029 2033 7900
Fax: 0870 0241809

Cricursa

Cami de Can Ferran s/n, Pol. Industrial Coll de la Manya, 08403
Granollers (Barcelona), Spain
www.cricursa.com

Tel: +34 93 840 4470
Fax: +34 93 840 1460

Dow Corning Ltd

Parc Industriel-Zone C, Rue Jules Bordet, 7180 Seneffe, Belgium
www.dowcorning.com

Tel: +32 64 888 000
Fax: +32 64 888 401

Eckelt Glass/Saint Gobain

Zentrale/Produktion, Resthofstaβe 18, 4400 Steyr, Austria
www.eckelt.at

Tel: +43 7252 8940
Fax: +43 7252 89424

European Glass Ltd

European House, Abbey Point, Abbey Road, London NW10 7DD
www.europeanglass.co.uk

Tel: 020 8961 6066
Fax: 020 8961 1411

F. A. Firman (Harold Wood) Ltd

19 Bates Road, Harold Wood, Romford, Essex RM3 0JH
www.firmanglass.com

Tel: 01708 374534
Fax: 01708 340511

Hanson

14 Castle Hill, Maidenhead SL6 4JJ
www.hanson.com/uk

Tel: 01628 774100

IG Lintels Ltd

Avondale Road, Cwmbran, Gwent NP44 1XY
www.iglintels.com

Tel: 01633 486486
Fax: 01633 486465

James Latham plc

Unit 3, Swallow Park, Finway Road, Hemel Hempstead HP2 7QU
www.lathamtimber.co.uk

Tel: 01442 849100
Fax: 01442 239287

James Latham plc

Unit 3, Swallow Park, Finway Road, Hemel Hempstead HP2 7QU
www.lathamtimber.co.uk

Tel: 01442 849100
Fax: 01442 239287

Loctite UK (Henkel Technologies)

Technologies House, Wood Lane End, Hemel Hempstead HP2 4RQ
www.loctite.co.uk

Tel: 01442 278 000
Fax: 01442 278 293

Metsawood

Mayne House, Fenton Way, Southfields Business Park, Basildon Essex
SS15 6RZ
www.metsawood.co.uk

Tel: 0845 601 2401
Fax: 01268 364 617

Perchcourt Stainless Ltd

Crompton Way, Bolton, Lancashire BL1 8TY
www.benteler-distribution.co.uk

Tel: 01204 301611
Fax: 01204 306907

Permasteelisa

7th Floor, Fountain House, 130 Fenchurch Street,
London EC3M 5DJ
www.permasteelisagroup.com

Tel: 020 7618 0461
Fax: 020 7618 0492

Pilkington UK Ltd

Alexandra Business Park, Prescot Road, St Helens WA10 3TT
www.pilkington.com

Tel: 01744 28882
Fax: 01744 692660

Pudlo/David Bell Group plc

Huntingdon Road, Bar Hill, Cambridge CB3 8HN
www.pudloconcrete.co.uk

Tel: 01954 780687

Quality Tempered Glass (QTG)

Concorde Way, Millennium Business Park, Mansfield, Notts NG19 7JZ
www.independentglass.co.uk

Tel: 01623 416300
Fax: 01623 416303

Rawlplug Ltd

Skibo Drive, Thornliebank Industrial Estate, Glasgow G46 8JR
www.rawlplug.co.uk

Tel: 0844 800 3320
Fax: 0844 300 3340

Richard Lees Steel Decking Ltd

Moor Farm Road West, The Airfield, Ashbourne, Derbyshire DE6 1HN
www.rlsd.com

Tel: 01335 300999
Fax: 01335 300888

RMD Kwikform

Brickyard Lane, Aldridge, Walsall WS9 8BW
www.rmdkwikform.com/uk

Tel: 01922 743743
Fax: 01922 743400

Solaglass Saint Gobain

Saint Gobain House, Binley Business Park, Coventry CV3 2TT
www.saint-gobain.co.uk

Tel: 02476 560700
Fax: 02476 560705

Staytite Self Tapping Fixings

Staytite House, Coronation Road, Cressex Business Park, High
Wycombe HP12 3RP
www.staytite.com

Tel: 01494 462322
Fax: 01494 464747

Sunglass

via Piazzola 13E, 35010 Villafranca, Padova, Italy
www.sunglass.it

Tel: 139 049 90500100
Fax: 139 049 9050964

Supreme Concrete Ltd

Coppingford Road, Sawtry, Huntingdon PE28 5GP
www.supremeconcrete.co.uk

Tel: 01487 833300
Fax: 01487 833348

Tarmac Building Products

Millfields Road, Ettingshall, Wolverhampton WV4 6JP
www.tarmacbuildingproducts.co.uk

Tel: 08000 324020

TATA Steel

PO Box 1, Brigg Road, Scunthorpe DN16 1BP
www.corusconstruction.com

Tel: 01724 405060
Fax: 01724 404224

TecEco Pty. Ltd

497 Main Road, Glenorchy, Tasmania 7010, Australia
www.tececo.com

Tel: +61 3 6249 7868
Fax: +61 3 6273 0010

Unbrako Deepak Fasteners

12-14 Tower Street, Newtown, Birmingham B19 3RR
www.unbrako.com

Tel: 0121 333 4610
fax: 0121 333 4525

Valbruna UK Ltd

Oldbury Road, West Bromwich, West Midlands B70 9BT
www.valbruna.co.uk

Tel: 0121 553 5384
Fax: 0121 500 5095

W. J. Leigh

Tower Works, Kestor Road, Bolton BL2 2AL
www.leighspaints.co.uk

Tel: 01204 521771
Fax: 01204 382115

Zero Environment Ltd

PO Box 1659, Warwick CV35 8ZD
www.zeroenvironment.co.uk

Tel: 01926 624966
Fax: 01926 624926

Further Reading

1 General Information

ACE. 1998. *Standard Conditions of Service Agreement B1*, 2nd Edition. Association of Consulting Engineers.
Blake, L. S. 1989. *Civil Engineer's Reference Book*, 4th Edition. Butterworth-Heinemann.
CPIC. 1998. *Selected CAWS Headings from Common Arrangement of Work Sections*, 2nd Edition. CPIC.
DD ENV. 1991. Eurocode 1. *Basis of Design and Actions on Structures*. BSI.
Hunt, T. 1999. *Tony Hunt's Sketchbook*. Architectural Press.

2 Statutory Authorities and Permissions

DETR. 1997. *The Party Wall etc. Act: Explanatory Booklet*. HMSO.
HSE. 2001. *Managing Health and Safety in Construction. CDM Regulations 1994. Approved Code of Practice*. HSE.
HSE. 2001. *Health and Safety in Construction*. HSE.
Information on UK regional policies: www.defra.gov.uk/www.dft.gov.uk/www.odpm.gov.uk/-www.wales.gov.uk www.scotland.gov.uk/www.nics.gov.uk
PTC. 1996. *Party Wall Act Explained. A Commentary on the Party Wall Act 1996*. Pyramus and Thisbe Club.

3 Design Data

CIRIA Report 111. 1986. *Structural Renovation of Traditional Buildings*. CIRIA.
Hunt, T. 1997. *Tony Hunt's Structures Notebook*. Architectural Press.
Information on the transportation of abnormal indivisible loads: www.dft.gov.uk
Lisborg, N. 1967. *Principles of Structural Design*. Batsford.
Lyons, A. R. 1997. *Materials for Architects and Builders – An Introduction*. Arnold.
Morgan, W. 1964. *The Elements of Structure*. Pitman.
Richardson, C. 2000. The Dating Game. *Architect's Journal*. **23/3/00**, 56–59, **30/3/00**, 36–39. **6/4/00**, 30–31.

4 Basic Shortcut Tools for Structural Analysis

Bolton, A. 1978. Natural Frequencies of Structures for Designers. *The Structural Engineer*. **Vol. 9, No. 56A**, 245–253.
Brohn, D. M. 2005. *Understanding Structural Analysis*. New Paradigm Solutions.
Calvert, J. R. and Farrer, R. A. 1999. *An Engineering Data Book*. Macmillan Press.
Carvill, J. 1993. *Mechanical Engineer's Data Book*. Butterworth-Heinemann.
Gere, J. M. and Timoshenko, S. P. 1990. *Mechanics of Materials*, 3rd SI Edition. Chapman & Hall.
Hambly, E. 1994. *Structural Analysis by Example*. Archimedes.
Heyman, J. 2005. Theoretical analysis and real-world design. *The Structural Engineer*. **Vol. 83, No. 8**. 19 Apr 2005, 14–17.
Johansen, K. W. 1972. *Yield Line Formulae for Slabs*. Cement and Concrete Association.
Megson, T. H. G. 1996. *Structural and Stress Analysis*. Butterworth-Heinemann.
Mosley, W. H. and Bungey, J. H. 1987. *Reinforced Concrete Design*, 3rd Edition. Macmillan.
Sharpe, C. 1995. *Kempe's Engineering Yearbook*, 100th Edition. M-G Information Services Ltd.
Wood, R. H. 1961. *Plastic and Elastic Design of Slabs and Plates*. Thames and Hudson.

5 Eurocodes

BS EN 1990:2005 and NA to BS EN 1990:2005. *Basis of Structural Design. BSI*.
DCLG. 2003. *Implementation of Structural Eurocodes in the UK*.
DCLG. 2006. *Guide to the Use of EN 1990 Basis of Structural Design*.

Draycott, T and Bullman, P. 2009. *Structural Elements Design Manual – Working with Eurocodes.* Butterworth Heinemann.

Gulvanessian, H. et al. 2002. *Designers' Guide to EN 1990 Eurocode: Basis of Structural Design.* Thomas Telford Books.

IStructE. 2010. *Manual for the Design of Building Structures to Eurocode 1 and Basis of Structural Design.*

Summary of legislation covering European Public Procurement: http://www.oecd.org/document/36/0,3343,en_33638100_34612958_35017124_1_1_1,00.html

6 Actions on Structures

BS EN 1991-1-1:2002 and NA to BS EN 1991-1-1:2002. *Actions on Structures: General Actions–Densities, Self-Weight and Imposed Loads.* BSI.

BS EN 1991-1-2:2002. *Actions on Structures: General Actions – Actions on Structures Exposed to Fire.* BSI.

BS EN 1991-1-3:2003. *Actions on Structures: General Actions – Snow Loads.* BSI.

BS EN 1991-1-4:2005. *Actions on Structures: General Actions – Wind Actions.* BSI.

BS EN 1991-1-5:2003. *Actions on Structures: General Actions – Thermal Actions.* BSI.

BS EN 1991-1-6:2005. *Actions on structures: General Actions – Actions during execution.* BSI.

BS EN 1991-1-7:2006. *Actions on Structures: General Actions – Accidental Actions.* BSI.

BS EN 1991-2:2003. *Actions on Structures: Traffic Loads on Bridges.* BSI

BS EN 1991-3:2006. *Actions on Structures: Actions Induced by Cranes and Machinery.* BSI.

BS EN 1991-4:2006. *Actions on Structures: Silos and Tanks.* BSI.

BS 648:1970. *Schedule of Weights of Building Materials.* BSI.

BS 5606:1990. *Guide to Accuracy in Building.* BSI

BS 6180:2011. *Code of Practice for Protective Barriers in and about Buildings.* BSI.

IStructE. 2010. *Manual for the Design of Building Structures to Eurocode 1 and Basis of Structural Design.*

Moore, D. et al. 2007. Designers' Guide to EN 1991-1-2, EN1992-1-2, EN 1993-1-2 and EN 1994-1-2 (Eurocode). Thomas Telford Books.

IStructE. 2010. *Practical Guide to Structural Robustness and Disproportionate Collapse in Buildings.*

7 Reinforced Concrete

Bennett, D. 2007. *Architectural In-situ Concrete.* RIBA Publishing.

BS 4483:1985. BS 4483. 1998. *Steel Fabric for the Reinforcement of Concrete.* BSI.

BS 8110: Part 1:1997. *Structural Use of Concrete. Code of Practice for Design and Construction.* BSI.

BS 8500-1:2006. Concrete. Complementary British Standard to BS EN 206-1. Method of specifying and guidance for the specifier.

BS 8666:2000. *Specification for Scheduling, Dimensioning, Bending and Cutting of Steel Reinforcement for Concrete.* BSI.

BS EN 1992-1-1:2004. *Eurocode 2. Design of Concrete Structures. General Rules and Rules for Buildings.* BSI.

BS EN 1992-1-2:2004. *Eurocode 2. Design of Concrete Structures. General Rules. Structural Fire Design.* BSI.

BS EN 1992-2:2005. *Eurocode 2. Design of Concrete Structures. Concrete Bridges. Design and Detailing Rules.* BSI.

BS EN 1992-3:2006. *Eurocode 2. Design of Concrete Structures. Liquid Retaining and Containing Structures.* BSI.

Concrete Society. 2009. *Designed and Detailed. Eurocode 2. Good Concrete Guide 9.*

Goodchild, C. H. 1997. *Economic Concrete Frame Elements.* BCA.

Goodchild, C. H. 2009. *Economic Concrete Frame Elements to Eurocode 2.* BCA.

ICE/IStructE. 2002. *Manual for the Design of Reinforced Concrete Building Structures.*

IStructE. 2006. *Manual for the Design of Concrete Building Structures to Eurocode 2.*

Mosley, W. H. and Bungey, J. H. 1987. *Reinforced Concrete Design,* 3rd Edition. Macmillan.

Neville, A. M. 1977. *Properties of Concrete.* Pitman.

8 Steel

Access Steel resources: www.access-steel.com

Baddoo, N. R. and Burgan, B. A. 2001. *Structural Design of Stainless Steel*. Steel Construction Institute, P291.

Brettle, M. 2009. *Steel Building Design: Introduction to the Eurocodes*. SCI.

Brettle, M. 2009. *Steel Building Design: Worked Examples in Accordance with Eurocodes and the UK National Annexes*. SCI.

BS 449: Part 2:1969. (as amended) *The Use of Structural Steel in Building*. BSI.

BS 5950 *Structural Use of Steelwork in Buildings*. Part 1: 2000. *Code of Practice for Design – Rolled and Welded Sections*. Part 5:1998. *Code of Practice for Design – Cold Formed Thin Gauge Sections*. BSI.

BS EN 1993-1-1:2005. and NA to BS EN 1993-1-1: 2005.

BS EN 1993-1-8:2005. and NA to BS EN 1993-1-8: 2005.

BS EN 1993-1-1:2005. *Eurocode 3. Design of Steel Structures. General Rules and Rules for Buildings*. BSI.

BS EN 1993-1-2:2005. *Eurocode 3. Design of Steel Structures. General Rules. Structural Fire Design*. BSI.

BS EN 1993-1-3:2006. *Eurocode 3. Design of Steel Structures. General Rules. Supplementary Rules for Cold-Formed Members and Sheeting*. BSI.

BS EN 1993-1-4:2006. *Eurocode 3. Design of Steel Structures. General Rules. Supplementary Rules for Stainless Steels*. BSI.

BS EN 1993-1-5:2006. *Eurocode 3. Design of Steel Structures. Plated Structural Elements*. BSI.

BS EN 1993-1-6:2007. *Eurocode 3. Design of Steel Structures. Strength and Stability of Shell Structures*. BSI.

BS EN 1993-1-7:2007. *Eurocode 3. Design of Steel Structures. Plated Structures Subject to Out of Plane Loading*. BSI.

BS EN 1993-1-8:2005. *Eurocode 3. Design of Steel Structures. Design of Joints*. BSI.

BS EN 1993-1-9:2005. *Eurocode 3. Design of Steel Structures. Fatigue*. BSI.

BS EN 1993-1-10:2005. *Eurocode 3. Design of Steel Structures. Material Toughness and through-Thickness Properties*. BSI.

BS EN 1993-1-11:2006. *Eurocode 3. Design of Steel Structures. Design of Structures with Tension Components*. BSI.

BS EN 1993-1-12:2007. *Eurocode 3. Design of Steel Structures. Additional Rules for the Extension of EN 1993 up to Steel Grades S 700*. BSI.

BS EN 1993-2:2006. *Eurocode 3. Design of Steel Structures. Steel Bridges*. BSI.

BS EN 1993-3-1:2006. *Eurocode 3. Design of Steel Structures. Towers, Masts and Chimneys. Towers and Masts*. BSI.

BS EN 1993-3-2:2006. *Eurocode 3. Design of Steel Structures. Towers, Masts and Chimneys. Chimneys*. BSI.

BS EN 1993-4-1:2007. *Eurocode 3. Design of Steel Structures. Silos*. BSI.

BS EN 1993-4-2:2007. *Eurocode 3. Design of Steel Structures. Tanks*. BSI.

BS EN 1993-4-3:2007. *Eurocode 3. Design of Steel Structures. Pipelines*. BSI.

BS EN 1993-5:2007. *Eurocode 3. Design of Steel Structures. Piling*. BSI.

BS EN 1993-6:2007. *Eurocode 3. Design of Steel Structures. Crane Supporting Structures*. BSI.

BS EN 10025-2:2004. *Hot Rolled Products of Structural Steels. Technical Delivery Conditions for Non-Alloy Structural Steels*. BSI.

ICE/IStructE. 2002. *Manual for the Design of Steelwork Building Structures*.

IStructE. 2010. *Manual for the Design of Steel Work Building Structures to Eurocode3*.

MacGinley, T. J. and Ang, T. C. 1992. *Structural Steelwork Design to Limit State Theory*, 2nd Edition. Butterworth-Heinemann.

Nickel Development Institute. 1994. *Design Manual for Structural Stainless Steel*. NIDI, 12011.

Owens, G. W. and Davison, B. 2012. *Steel Designer's Manual*, 7th Edition. Wiley-Blackwell.

SCI. 2001. *Steelwork Design Guide to BS 5950: Part 1:2000 Volume 1 Section Properties and Member Capacities*, 6th Edition. Steel Construction Institute, P202.

SCI. 1995. *Joints in Steel Construction: Moment Connections Volumes 1 1 2*. BCSA 207/95.

SCI. 2005. *Joints in Steel Construction: Simple Connections Volumes 1 1 2*. BCSA P212.

Simms, W. and Hughes, A. 2011. Composite Design of Steel Framed Buildings in accordance with Eurocodes and the UK National Annexes. SCI

9 Composite Steel and Concrete

BS 5950 Structural Use of Steelwork in Buildings. Part 3: 1990. Code of Practice for Design – Composite Construction. Part 4: 1994. Code of Practice for Design – Composite Slabs with Profiled Metal Sheeting. BSI.

BS EN 1994-1-1:2004. *Eurocode 4. Design of Composite Steel and Concrete Structures. General Rules and Rules for Buildings*. BSI.

BS EN 1994-1-2:2005. *Eurocode 4. Design of Composite Steel and Concrete Structures. General Rules. Structural Fire Design*. BSI.

BS EN 1994-2:2005. *Eurocode 4. Design of Composite Steel and Concrete Structures. General Rules and Rules for Bridges*. BSI.

Noble, P. W. and Leech, L. V. 1986. *Design Tables for Composite Steel and Concrete Beams*. Constrado.

SCI. 1990. *Commentary on BS 5950: Part 3: Section 3.1 Composite Beams*. Steel Construction Institute, P78.

10 Timber and Plywood

BS 5268: Part 2:2002. *Structural Use of Timber*. BSI.

BS EN 338:2009. *Structural Timber Strength Classes*. BSI.

BS EN 1995-1-1:2004 + A1:2008. *Eurocode 5. Design of Timber Structures. General. Common Rules and Rules for Buildings*. BSI.

BS EN 1995-1-2:2004. *Eurocode 5. Design of Timber Structures. General. Structural Fire Design*. BSI.

BS EN 1995-2:2004. *Eurocode 5. Design of Timber Structures. Bridges*. BSI.

BS EN 12369-1:2001. *Wood Based Panels. Characteristic Values for Structural Design: OSB, Particleboards and Fireboards*. BSI.

BS EN 12369-2:2004. *Wood Based Panels. Characteristic Values for Structural Design: Plywood*. BSI.

IStructE. 2007. *Manual for the Design of Timber Building Structures to Eurocode5*.

Ozelton, E. C. and Baird, J. A. 2002. *Timber Designers' Manual,* 3rd Edition. Blackwell.

TRADA. 2006. *Wood Information Sheet 37: Eurocode 5 – An Introduction*.

TRADA. 2006. *Engineering Guidance Document Series*.

11 Masonry

BS 5977: Part 1: 1981. *Lintels. Method for Assessment of Load*. BSI.

BS 5628 *Code of Practice for Masonry*. Part 1: 1992. *Structural use of Unreinforced Masonry*. Part 2:2001. *Materials and Components, Design and Workmanship*. BSI.

BS EN 1996-1-1:2005. *Eurocode 6. Design of Masonry Structures. General Rules for Reinforced and Unreinforced Masonry Structures*. BSI.

BS EN 1996-1-2:2005. *Eurocode 6. Design of Masonry Structures. General Rules. Structural Fire Design*. BSI.

BS EN 1996-2:2006. *Eurocode 6. Design of Masonry Structures. Design Considerations, Selection of Materials and Execution of Masonry*. BSI.

BS EN 1996-3:2006. *Eurocode 6. Design of Masonry Structures. Simplified Calculation Methods for Unreinforced Masonry Structures*. BSI.

CP111:1970. *Code of Practice for the Design of Masonry in Building Structures*. BSI.

Curtin, W. G., Shaw, G. and Beck, J. K. 1987. *Structural Masonry Designers' Manual*, 2nd Edition. BSP Professional Books.

Heyman, J. 1995. *The Stone Skeleton*. Cambridge University Press.

Howe, J. A. 1910. *The Geology of Building Stones*. Edward Arnold. Reprinted Donhead Publishing 2000.

ICE/IStructE. 1997. *Manual for the Design of Plain Masonry in Building Structures*

IStructE 2008. *Manual for the Design of Plain Masonry in Building Structures to Eurocode 6*.

12 Geotechnics

Berezantsev, V. G. 1961. Load bearing capacity and deformation of piled foundations. *Proc. of the 5th International Conference on Soil Mechanics. Paris.* **Vol. 2**, 11–12.

BS 5930:1981. *Code of Practice for Site Investigations*. BSI.

BS 8004:1986. *Code of Practice for Foundations*. BSI.

BS 8002:1994. *Code of Practice for Earth Retaining Structures*. BSI.

BS EN 1997-1:2004. *Eurocode 7. Geotechnical Design. General Rules*. BSI.

BS EN 1997-2:2007. *Eurocode 7. Geotechnical Design. Ground Investigation and Testing*. BSI.
BS EN 14688-2:2004. *Geotechnical Investigation*. BSI.
Craig, R. F. 1993. *Soil Mechanics*, 5th Edition. Chapman & Hall.
DCLG. 2007. *A Designer's Simple Guide to BS EN 1997.*
Environment Agency. 1997. *Interim Guidance on the Disposal of Contaminated Soils*, 2nd Edition. HMSO.
Environment Agency. 2002. *Contaminants in Soil: Collation of Toxological Data and Intake Values for Humans*. CLR Report 9, HMSO.
Hansen, J. B. 1961. A general formula for bearing capacity. *Danish Geotechnical Institute Bulletin.* **No. 11.** Also Hansen, J. B. 1968. A revised extended formula for bearing capacity. *Danish Geotechnical Institute Bulletin.* **No. 28.** Also Code of Practice for Foundation Engineering 1978, *Danish Geotechnical Institute Bulletin.* **No. 32.**
ICRCL. 1987. *Guidance on the Redevelopment of Contaminated Land*, 2nd Edition. Guidance Note 59/83, DoE.
Kelly, R. T. 1980. Site investigation and material problems. *Proc. of the Conference on the Reclamation of Contaminated Land*. Society of Chemical Industry. **B2**, 1–14.
NHBC. *National House-Building Council Standards*.
Terzaghi, K. and Peck, R. B. 1996. *Soli Mechanics in Engineering Practice*, 3rd Edition. Wiley.
Tomlinson, M. J. 2001. *Foundation Design and Construction*, 7th edition. Pearson.

13 Glass

Glass and Mechanical Strength Technical Bulletin. 2000. Pilkington.
pr EN. 13474. *Glass in Building – Design of Glass Panes*. Part 1, *Basis for Design*. Part 2, *Design for Uniformly Distributed Loads*. BSI.
Structural Use of Glass in Buildings. 1999. IStructE.

14 Building Elements, Materials, Fixings and Fastenings

BRE. 2002. *Thermal Insulation Avoiding Risks. A Good Practice Guide Supporting Building Regulation Requirements*, 3rd Edition. BR 262. CRC Ltd.
BS 8007:1987. *Code of Practice for Design of Concrete Structures for Retaining Aqueous Liquids*. BSI.
BS 8102:1990. *Protection of Structures against Water from the Ground*. BSI.
BS 8118: Part 1:1991. *Structural Use of Aluminium. Code of Practice for Design*. BSI.
CIRIA. 1995. *Water-Resisting Basement Construction – A Guide. Safeguarding New and Existing Basements against Water and Dampness*. CIRIA, Report 139.
CIRIA. 1998. *Screeds, Flooring and Finishes – Selection, Construction and Maintenance*. CIRIA, Report 184.
DD ENV. 1999. *Eurocode 9. Design of Aluminium Structures*. BSI.
General BRE Publications: BRE Digests, *Good Building Guides and Good Repair Guides*.
Guide to the Structural Use of Adhesives. 1999. IStructE.
Russell, J. R. and Ferry, R. L. 2002. *Aluminium Structures*. Wiley.

15 Sustainability

Addis, W. and Schouten, J. 2004. *Design for Deconstruction*. CIRIA.
Anderson, J. and Shiers, D. 2002. *The Green Guide to Specification*. Blackwell Science.
Berge, B. 2001. *The Ecology of Building Materials*. Architectural Press.
Bioregional. 2002. *Toolkit for Carbon Neutral Developments*. BedZed Construction Materials Report.
CIBSE. 2007. *Sustainability*. Guide L.
Friends of the Earth. 1996. *The Good Wood Guide*. Friends of the Earth.
FSC. 1996. *Forest Stewardship Council and the Construction Sector*. Factsheet.
IStructE. 1999. *Building for a Sustainable Future: Construction without Depletion.*
RCEP. 2000. *Energy – the Changing Climate*. Royal Commission on Environmental Pollution 22nd Report.
WCED. 1987. *Our Common Future* (the Brundtland Report). Oxford.
Woolley, T. and Kimmins, S. 2000. *The Green Building Handbook*, Volume 2. E&F Spon.
WRAP. 2007. *Reclaimed Building Products Guide*. Waste 1 Resources Action Programme.
WRAP. 2007. *Recycled Content Toolkit*. Online at www.wrap.org.uk.
WRAP. 2008. *Choosing Construction Products: Guide to Recycled Content of Mainstream Construction Products*. Waste 1 Resources Action Programme.
WWF. 2004. *The Living Planet Report*.

Sources

ACE. 2004. *Standard Conditions of Service Agreement B1*, 2nd Edition. Association of Consulting Engineers. General summary of normal conditions.

Access Steel. 2010. *SN003b NCCI: Elastic Critical Moment for Lateral Torsional Buckling*.

Access Steel. 2010. *SN0048 NCCI: Verification of Columns in Simple Construction – A Simplified Interaction Criterion*.

Anderson, J. and Shiers, D. 2002. *Green Guide to Specification*, 3rd Edition. P19 and 27 extracts.

Angle Ring Company Limited. 2002. *Typical Bend Radii for Selected Steel Sections*.

Bison Manufacturing. 2012. *Loading Data for Hollowcore Precast Planks*.

Berezantsev, V. G. 1961. Load bearing capacity and deformation of piled foundations. *Proc. of the 5th International Conference on Soil Mechanics. Paris*, **Vol. 2**, 11–12.

Bolton, A. 1978. Natural frequencies of structures for designers. *The Structural Engineer* 56A(9):245–253. Table 1.

BRE Digests 299, 307, 345. Extracts on durability of timber. Reproduced by permission of Building Research Establishment.

BRE. 2004. *Good Building Guide 63 – Climate Change: Impact on Building Design and Construction.* Box 4.

BS EN 338. 2009. *Structural Timber: Strength Classes.* Table 1.

BS 449: Part 2:1969. (as amended) *The Use of Structural Steel in Building.* BSI. Tables 2, 11 and 19.

BS EN 1194. 1999. Timber structures. Glued laminated timber. Strength classes and determination of characteristic values. Tables 1 and 2.

BS EN 1990. 2005. *Eurocode. Basis of Structural Design.* Table A.12 (A, B, C).

BS EN 1991. 2002. Part 1-1. Eurocode 1. Actions on structures. General actions. Densities, self-weight, imposed loads for buildings Tables A.1, A.2, A.3,.4, A.5, A.7 and 6.9.

BS EN 1991. 2006. Part 1-7. *Eurocode 1. Actions on Structures. Accidental actions.* Tables A.1 and Appendixes A&E

BS EN 1992. 2004. Part 1-2. *Eurocode 2. Design of Concrete Structures. General Rules. Structural Fire Design* Figure 3.5 modified, Extracts from Tables 5.2a, 5.4, 5.5, 5.6, 5.8 and 5.9.

BS EN 1993. 2005. Part 1-1. *Eurocode 3. Design of Steel Structures. General Rules and Rules for Buildings.* Tables 3.1, 6.2, 6.3 and Figure 6.4

BS EN 1993. 2005. Part 1-8. *Eurocode 3. Design of Steel Structures. Design of Joints.* Tables 4.1.

BS EN 1995. 2008. Part 1-1. *Eurocode 5. Design of Timber Structures. General. Common Rules and Rules for Buildings.* Tables 2.3, 3.1, 3.2, 6.1.

BS EN 1996. 2005. Part 1-1. *Eurocode 6. Design of Masonry Structures. General Rules for Reinforced and Unreinforced Masonry Structures.* Table 5.1 and Figure G.1.

BS 4483:1985. BS 4483:1998. *Steel Fabric for the Reinforcement of Concrete.* BSI. Table 1.

BS 5268: Part 2:1991. *Structural Use of Timber.* BSI. Appendix D.

BS 5268: Part 2:2002. *Structural Use of Timber.* BSI. Tables 8, 17, 19, 20, 21, 24 and extracts from Tables 28, 31, 33 and 34.

BS 5606:1990. *Guide to Accuracy in Building.* Figure 4.

BS 5628: Part 1:2005. *Structural use of Unreinforced Masonry.* BSI. Tables 1, 5 and 9, extracts from Tables 2 and 3, Figure 3.

BS 5628: Part 3:2005. *Materials and Components, Design and Workmanship.* BSI. Table 9 and Figure 6.

BS 5950 Part 1:2000. *Structural Use of Steelwork in Buildings. Code of Practice for Design – Rolled and Welded Sections.* BSI. Tables 2, 9, 14 and 22.

BS 5977: Part 1:1981. *Lintels. Methods for Assessment of Load.* BSI. Figures 1 and 4.

BS 6180:1990. *Code of Practice for Protective Barriers in and about Buildings.* BSI. Table 1.

BS 6399 *Loading for Buildings.* Part 1:1996. *Code of Practice for Dead and Imposed Loads.* BSI. Tables 1 and 4.

BS 8004:1986. *Code of Practice for Foundations.* BSI. Table 1 adapted.

BS 8102:1990. *Protection of Structures Against Water from the Ground.* BSI. Table 1 adapted to CIRIA Report 139 suggestions.

BS 8110: Part 1:1997. *Structural Use of Concrete. Code of Practice for Design and Construction.* BSI. Tables 2.1, 3.3, 3.4, 3.9, 3.19, 3.20 and Figure 3.2.

BS 8118: Part 1:1991. *Structural Use of Aluminium. Code of Practice for Design.* BSI. Tables 3.1, 3.2 and 3.3 and small extracts from Tables 2.1, 2.2, 4.1 and 4.2.

BS 8666:2005. *Specification for Scheduling, Dimensioning, Bending and Cutting of Steel Reinforcement for Concrete.* BSI. Table 3.

BS EN 10025. 2004 Part 2. *Hot Rolled Products of Structural Steels. Technical Delivery Conditions for Non-Alloy Structural Steels.* Table 7.

BS EN 12369. 2001. Part 1. *Wood-Based Panels. Characteristic Values for Structural Design. OSB, Particleboards and Fireboards.* Table 2.

BS EN 12369. 2004. Part 2. *Wood-Based Panels. Characteristic Values for Structural Design. Plywoods.* Table 1.

BS EN 14688. 2004. Part 2. *Geotechnical Investigation and Testing. Identification and Classification of Soil. Principles for A Classification.* Table B1.

Building Regulations, Part B 1991. HMSO. Table A2, Appendix A.

Building Regulations, Part A3 2004. MSO. Table 11.

CIBSE. 2006. *Guide A: Environmental Design.* Table 3.37 and 3.38 extracts.

Corus Construction. 2007. *Structural Sections to BS4.* TATA Steel.

CP111. 1970. *Code of Practice for the Design of Masonry in Building Structures.* BSI. Tables 3a and 4.

CPIC. 1998. *Arrangement of Work Sections,* 2nd Edition. CPIC. Selected headings.

DETR. 1997. *The Party Wall etc. Act: Explanatory Booklet.* HMSO.

Dow Corning. 2002. *Strength Values for Design of Structural Silicon Joints.*

Finnforest. 2010. *Kerto: A Wood Product for Advanced Structural Engineering.* Metsawood.

Hansen, J. B. 1961. A general formula for bearing capacity. *Danish Geotechnical Institute Bulletin.* **No. 11.** Also Hansen, J. B. 1968. A revised extended formula for bearing capacity. *Danish Geotechnical Institute Bulletin.* **No. 28.** Also Code of Practice for Foundation Engineering 1978, *Danish Geotechnical Institute Bulletin.* **No. 32.**

Highways Agency, Design Note HD25/Interim Advice Note 73/06, HMSO. Table 3.1, Chapter 3.

Howe, J. A. 1910. *The Geology of Building Stones.* Edward Arnold. Reprinted Donhead Publishing, 2000. Table XXVI.

IG Limited. 2008. *Loading Data for Steel Lintels.*

Kulhawy, F. H. 1984. Limiting tip and side resistance. *Proceedings of Symposium Analysis and Design of Piled Foundations,* edited by Meyer, J. R. California, 80–98. Tables 1 and 2.

Manual for the Design of Reinforced Concrete Building Structures to Eurocode 2 2006. ICE/IStructE. Reinforced Concrete Column Design Charts Appendix.

Manual for the Design of Steelwork Building Structures. 2002. ICE/IStructE. Section 11.3, Figures 12, 13, 14 and 15.

NA to BS EN 1990. 2005. *UK National Annex for Eurocode. Basis of Structural Design.* Cl 2.2, Table NA.A1.1, Table NA.2.1.

NA to BS EN 1991. 2002. Part 1-1. *UK National Annex to Eurocode 1. Actions on structures. General actions. Densities, Self-Weight, Imposed Loads for Buildings* Tables NA.2, NA.3, NA.4, NA.5, NA.6, NA.7, NA.8

NA to BS EN 1992. 2004. Part 1-1. *UK National Annex to Eurocode 2. Design of Concrete Structures. General Rules and Rules for Buildings* Table NA.5

NA to BS EN 1993. 2005. Part 1-1. *UK National Annex to Eurocode 3. Design of Steel Structures. General Rules and Rules for Buildings.* NA2.15, NA2.17.

NA to BS EN 1993. 2005. Part 1-8 *UK National Annex to Eurocode 3. Design of Steel Structures. Design of Joints.* Table NA.1.

NA to BS EN. 1995. 2008. Part 1-1. *UK National Annex to Eurocode 5. Design of Timber Structures. General. Common Rules and Rules for Buildings.* Table NA.3, NA.4.

NA to BS EN. 1996. 2005. Part 1-1. *UK National Annex to Eurocode 6. Design of Masonry Structures. General Rules for Reinforced and Unreinforced Masonry Structures.* Table NA.1, NA.2, NA2.7, NA.6 and Appendix E tables.

NHBC. 2011. *National House Building Council Standards.* Section 4.2.

Nickel Development Institute. 1994. *Design Manual for Structural Stainless Steel.* NIDI, 12011. Tables 3.1, 3.12, 3.5, 3.6 and A.1.

Pilkington. 2000. *Glass and Mechanical Strength Technical Bulletin.* Tables 3, 4, 5, 6, 7 and 8.

Richardson, C. 2000. The dating game. *Architect's Journal.* **23/3/00**, 56–59. **30/3/00**, 36–39. **6/4/00**, 30–31.

Rawlplug Ltd. 2007. *Specification and Design Guide.*

RMD Kwikform. 2002. *Loading Data and Charts for Super Slim Soldiers.*

Supreme Concrete. 2004. *Loading Data for Precast Prestressed Concrete Lintels.*

UKCIP02. 2002. *Climate Change Scenarios for the United Kingdom.* Tyndall Centre. Headline Indicators.

University of Bath. 2006. Inventory of Carbon and Energy (ICE). Version 1.5 Beta. Prof Geoffrey Hammond and Craig Jones. Sustainable Energy Research Team (SERT), Department of Mechanical Engineering. www.bath.ac.uk/mech-eng/sert/embodied.

See the 'Useful Addresses' section for contact details of advisory organisations and manufacturers.

Index